The Early Effects of Radiation on DNA

NATO ASI Series

Advanced Science Institutes Series

A series presenting the results of activities sponsored by the NATO Science Committee, which aims at the dissemination of advanced scientific and technological knowledge, with a view to strengthening links between scientific communities.

The Series is published by an international board of publishers in conjunction with the NATO Scientific Affairs Division

A Life Sciences B Physics	Plenum Publishing Corporation London and New York
C Mathematical and Physical Sciences D Behavioural and Social Sciences E Applied Sciences	Kluwer Academic Publishers Dordrecht, Boston and London
F Computer and Systems Sciences G Ecological Sciences H Cell Biology I Global Environmental Change	Springer-Verlag Berlin Heidelberg New York London Paris Tokyo Hong Kong Barcelona Budapest

NATO-PCO DATABASE

The electronic index to the NATO ASI Series provides full bibliographical references (with keywords and/or abstracts) to more than 30000 contributions from international scientists published in all sections of the NATO ASI Series. Access to the NATO-PCO DATABASE is possible in two ways:

– via online FILE 128 (NATO-PCO DATABASE) hosted by ESRIN, Via Galileo Galilei, I-00044 Frascati, Italy.

– via CD-ROM "NATO-PCO DATABASE" with user-friendly retrieval software in English, French and German (© WTV GmbH and DATAWARE Technologies Inc. 1989).

The CD-ROM can be ordered through any member of the Board of Publishers or through NATO-PCO, Overijse, Belgium.

Series H: Cell Biology Vol. 54

The Early Effects of Radiation on DNA

Edited by

E. M. Fielden

and

P. O'Neill

Medical Research Council
Radiobiology Unit
Chilton, Didcot
Oxon OX11 ORD
United Kingdom

Springer-Verlag
Berlin Heidelberg New York London Paris Tokyo
Hong Kong Barcelona Budapest
Published in cooperation with NATO Scientific Affairs Division

Proceedings of the NATO Advanced Research Workshop on The Early Effects of Radiation on DNA held at San Miniato, Italy, May 7–11, 1990.

ISBN-13: 978-3-642-75150-9 e-ISBN-13: 978-3-642-75148-6
DOI: 10.1007/978-3-642-75148-6

Library of Congress Cataloging-in-Publication Data
NATO Advanced Research Workshop on the Early Effects of Radiation on DNA (1990 : San Miniato, Italy)
The early effects of radiation on DNA / edited by E. M. Fielden and P. O'Neill. (NATO ASI series. Series H, Cell biology ; vol. 54)
"Proceedings of the NATO Advanced Research Workshop on the Early Effects of Radiation on DNA held at San Miniato, Italy, May 7–11, 1990"—T.p. verso. "Published in cooperation with NATO Scientific Affairs Division." Includes bibliographical references and index. ISBN-13: 978-3-642-75150-9 (alk. paper).

1. DNA—Effect of radiation on—Congresses. 2. Ionizing radiation—Physiological effect—Congresses. 3. DNA damage—Congresses. I. Fielden, E. M. II. O'Neill, P. (Peter) III. North Atlantic Treaty Organization. Scientific Affairs Division. IV. Title. V. Series. [DNLM: 1. DNA—radiation effects—congresses. QU 58 N2787e 1990] QP624.5.R33N37 1991 575.1'31—dc20 DNLM/DLC 91-4743

Softcover reprint of the hardcover 1st edition 1991

31/3140-543210 – Printed on acid-free-paper

PREFACE

Interest in the biological effects of ionising radiation closely followed the identification of such radiation. The realisation that DNA is the site of genetic information in cells subsequently focussed attention on DNA as an important target in the lethal and mutagenic effects of ionising radiation. Thus radiation effects upon DNA became an important area for fundamental scientific studies by radiation biologists, chemists and physicists. To a first approximation, the concerns of the three disciplines can be divided by time scales: the physical process of energy deposition from photon or charged particle and subsequent relaxation (~ 10^{-16} to 10^{-12} secs), followed by chemical reactions (~ 10^{-12} to 10^{2} secs), and finally, the expression of biological effect (minutes to years). Thus, the concept of 'early processes' conveys different ideas to different scientists, although they are all interrelated.

To attempt to describe in any detail all these processes is a mammoth task which is not made easier by the different conventions and experimental approaches of the three disciplines. However, the recent advances in all these scientific areas seemed, to the organisers at least, to offer the opportunity to stimulate more active interaction between physicists, chemists and biologists. With this in mind, a multi-disciplinary workshop was organised, which brought together some fifty scientists to present their own specialist interests and, through extensive discussion, explore which problems are of high priority and require input from the different disciplines to resolve them. The papers and short presentations of the recent work presented at the workshop are contained in this book, together with a distillation of the discussions. The workshop was successful and, hopefully, will lead to new approaches to the present problems and the development of interdisciplinary research in seeking those solutions.

E. Martin Fielden
Peter O'Neill

ACKNOWLEDGEMENTS

In addition to the speakers and participants who helped to make the workshop such a success the Directors would like to thank: the NATO Advanced Research Workshop Programme for its support; the Cassa di Risparmio di San Miniato for permission to use the residential and conference facilities of the Centro Studi 'Il Cappucini', San Miniato; the Scientific Committee listed below; the FRAE Laboratory, Bologna and, in particular, Drs G. Simone and M. Tamba for local organisation; the American National Institutes of Health and Department of Energy for additional financial support; Drs. W. Bernhard and A. Chatterjee for their support in applying for, and administering, the NIH and DOE grants; and last, but by no means least, Mrs. Mary Culliford for her secretarial and administrative assistance before and during the workshop, as well as in preparing the manuscripts for printing.

Scientific committee

Directors: E.M. Fielden and P. O'Neill

Members: W. Bernhard
A. Chatterjee
D. Goodhead
O. Sapora
J. Ward

CONTENTS

PART I - THE INTERDISCIPLINARY APPROACH TO RADIATION ACTION

PART II - CELLULAR DNA STRAND BREAKAGE

PART III - EARLY CHEMICAL EVENTS IN DNA DAMAGE

PART IV - DEVELOPMENT OF RADIATION MODELS

PART V - CHEMICAL MECHANISMS OF DOSE MODIFIERS

PART VI - RBE FROM THEORY TO EXPERIMENTAL

PART VII - TIMESCALE FOR DEVELOPMENT OF DNA DAMAGE

PART VIII - LET AND CHROMOSOMAL DAMAGE

PART IX - ADDITIONAL SHORT PRESENTATIONS

PART X - WORKSHOP SUMMARIES

Main presentations are listed in normal text; short presentations and discussion summaries are in italics.

MECHANISMS OF RADIATION ACTION ON DNA IN MODEL SYSTEMS - THEIR RELEVANCE TO CELLULAR DNA

J.F. Ward
Division of Radiation Biology
Department of Radiology M010
University of CA at San Diego
La Jolla
CA, 92093
U.S.A.

INTRODUCTION

Ionizing radiation is known to damage DNA by a variety of mechanisms, some or all of which may be responsible for damage induced in the DNA of mammalian cells. Strong evidence points to the significance of DNA damage as the source of the biological consequences of ionizing radiation (Makrigiorgos et al. 1990, Painter 1979, Ward 1988). There are two major reasons for understanding all the chemical reactions which lead to DNA damage in a cell (and to identify those which do not contribute): This knowledge can serve as a basis for predicting the shapes and slopes of yield dose curves for biological effects at low doses and with different radiations, and, it can permit rational approaches to improvements in radiotherapy protocols. Thus it is important that a. reactions which actually occur in vivo be delineated, and b. the damage which causes the biological effects be defined. The overall scenario which is developed should be capable of explaining variations in cellular radiosensitivity caused by radiations of different LET, different oxygen concentrations, and the effects of sensitizers and protectors.

Here the mechanisms which have been described to explain results from a variety of model systems are generally discussed. These are assessed in terms of types and yields of damage which are formed in DNA irradiated intracellularly, again in the context of the known structure(s) and environment(s) of DNA in the mammalian cell. The evidence for and against the operation of the various mechanisms in the cell is summarized.

NATO ASI Series, Vol. H 54
The Early Effects of Radiation on DNA
Edited by E. M. Fielden and P. O'Neill

MECHANISMS WHICH CAUSE DNA DAMAGE

Damage to DNA is initiated by four distinct mechanisms:

1. Direct Ionization of DNA.
2. Reactions of DNA with free radicals from water.
3. Reactions of DNA with other radicals.
4. Excitation (Cerenkov radiation).

All of these initial reactions must occur in DNA irradiated within a cell. Remaining questions are:

a. To what extent does each of these initial reactions occur?
b. What are the subsequent reactions of the primarily altered species?
c. What are the products of the reactions?

STATE OF DNA IN THE MAMMALIAN CELL

To discuss the mechanisms by which intracellular DNA can be damaged by irradiation, first the packaging of the macromolecule in the cell should be considered. Particular attention should be paid to access of water molecules, proximity and distribution of free radical scavengers, and structures which may be amenable to the migration of energy or electrons. Also, consideration should be given to any variations in structure which may cause sensitization or protection of specific regions the genome.

Within a typical mammalian cell there are 6 x 10^{-12} gms DNA, (5 x 10^{9} base pairs). This DNA is packaged in nucleosomes which consist of a chromatosome and a linker region. The chromatosome is a 165 base pair unit wrapped twice around two sets of the four histones, H2a, H2b, H3 and H4. The fifth histone, H1, acts as a lock holding the double wrap in place and enabling the whole system to pack more closely. The chromatosome with linker DNA, the length of which is species dependent 0-80 bp in length, is the nucleosome (Widom, 1989). (There are 2.5 X 10^{7} nucleosomes per mammalian cell). van der Holde has collated the lengths of linkers and finds a remarkable constancy for mammalian cells in culture of about 25 bp (a high of 30 and a low of 13).

Previously the 7Å structure of the core particle was considered from the point of view of radiation damage (Ward,

1988). In this structure a 145 bp portion of the chromatosome wraps 1.8 times around the histone core, the structure contains many layers of water molecules between the histones and the DNA. This gives added credence to the idea that radiation induced free radicals from the water surrounding the DNA play a major role in causing its damage. The structure of the actual nucleosome has not been resolved to this dimension but the impression given in the literature is that it is an extension of the core particle with no change in structure, i.e. the tightness of winding of the DNA and therefore its distance from the histones is the same as in the core particle.

It is possible that the internucleosomal linker region, about 15% of the DNA, is more accessible to diffusing free radicals and hence may be more sensitive to radiation damage.

For the majority of the cell cycle, the majority of the DNA is packaged into nucleosomes, which in turn are wrapped into the 300 Å chromatin fibres. It is only when the DNA is being replicated or transcribed that these structures are modified. There are 10^4 replication origins in a mammalian cell and the rate of bidirectional replication is 50 bp per second. After replication the original histones are shared between the new strands, and full replacement of new histones occurs after 10-20 min. It could therefore be hypothesized that newly replicated DNA, transiently having greater internucleosomal distances, may be more open and therefore more susceptible to attack by free radicals from water.

CONTRIBUTIONS OF DIFFERENT MECHANISMS TO CELLULAR DNA DAMAGE

"Direct" and "Indirect" Effects

The definitions of these terms, originally used to discuss molecules in solution, is not clear in the case of DNA, because of the tightly bound water molecules associated with the macromolecule. Two distinctions are possible.

In aqueous solution the extent to which OH free radicals from water can be scavenged from reacting with DNA (Milligan et al. 1990). The ability of the OH free radical scavenger dimethyl sulfoxide to protect against this radical damage does not increase above 1M DMSO (i.e. 8 % by weight). Clearly the DMSO can not scavenge OH radicals or their precursor H_2O^+ radical ions formed from the bound water molecules from reacting with DNA. In this instance the "direct effect" would be that damage which is unscavengeable by DMSO. Thus for DNA in solution distinction is between scavengeable and non-scavengeable damage (and the same must be true within cells).

In studies of DNA irradiated in ice (Cullis et al. 1989) a similar distinction can be made. There the majority of diffusion of radicals is precluded by the site of their formation within ice crystals, but radicals formed from bound water can still react with the DNA so that ice radiation would resemble unscavengeable solution damage.

Studies of scavengeable and non-scavengeable damage in mammalian cells found no differences in their rates of cellular repair (Evans et al. 1988) suggesting that the structures of the damages are the same in both instances--however, those neutral elution measurements do not monitor accuracy of repair.

Two other methods of studying the "direct" effect have been developed: a. Producing the radical cations directly on the DNA itself by biphotonic ionization of the bases (Opitz and Schulte-Frohlinde 1987), and b. Oxidation by $SO_4^{-\cdot}$ radical anions (Aboul-Enein A. and Schulte-Frohlinde 1988). In these cases there is no contribution to the damage from energy deposited in the bound water.

In ionizing radiation induced damage to DNA (in solution or in cells) it is not possible to scavenge the damage resulting from energy deposition in bound water. Thus the distinction in mechanisms should be between scavengeable and unscavengeable damage.

Data from model studies can be used to calculate yields of unscavengeable damage. Milligan et al. (1990) have measured the yield of DNA ssbs produced by irradiation of chromatin in the presence of 1M DMSO, i.e. maximum scavenging of OH radicals and found that it corresponds to a yield of ssb of G = 1.9.
This yield is calculated from the energy deposited directly in the DNA molecule only, since the number of water molecules from which unscavengeable damage can result (this number of waters

will be the same for both the model system and from cellular DNA and the contributions from this water will be the same in the two cases, i.e. the energy deposited in the DNA is indicative of the energy deposited in this total, DNA + bound water, in both instances.) After a dose of 1 Gray the 6 x 10^{-12} gm of DNA in a mammalian cell will have absorbed 3.6 x 10^{4} eV (by definition). With a G of ssb production of 1.9 this corresponds to 700 ssb per cell which compares favorably with the measured yield of unscavengeable ssbs of 35% of the 1,000 breaks induced in a cell by 1 Gray.

The number of water molecules whence the scavengeable OH radicals originate to cause ssb damage can be calculated. In the mammalian cell 65% of the single strand break (ssb) damage is formed by scavengeable OH radicals. The half life of these OH radicals can be calculated from the concentration of scavenger necessary to reduced the yield of scavengeable ssb by a factor of 2 and is found to be about 1 nsec (for review see Ward, 1988). The initial yield of OH radicals present at that time after irradiation should be used, i.e. G = 5 (Jonah et al 1977). In dilute solution the G of ssb from OH damage is 0.38 (Milligan et al 1990) this derives from an initial G(OH) of 2.5 (i.e the yield at the time of reaction in dilute solution). Therefore in the cell at 1 nsec this would be G = 0.38 x 5/2.5 = 0.76 from diffusing OH radicals. The yield of ssbs in the cell is 1,000 ssb per Gray per 6 x 10^{-12} gms DNA. From the OH radicals 650 ssb are produced. Using the G of ssb of 0.76, the energy necessary to cause this number of ssb in a cell is 650 x 100/.76 eV = 8.5 x 10^{4} eV. The mass of water receiving 8.5 x 10^{4} eV after 1 Gray is 8.5 x 10^{4}/6 x 10^{15} =1.4 x 10^{-11}gms. This mass is equal to about twice the mass of cellular DNA---equivalent to about 42 water molecules per deoxynucleotide. In the DNA structure it is known that, there is 0.52 gms of bound water per gm of DNA and 0.71gms of inner hydration shell water (Saenger, 1984) which may be the water from which unscavengeable damage arises. The additional water for scavengeable damage required by the above calculation corresponds to a layer less than three water molecules thick around a 2 nm diameter cylinder and the radicals would be required to diffuse less than 1 nm.

Other Radical Reactions

There have been no major studies of reactions of radicals derived from other cellular molecules with DNA. Adducts of other molecules with DNA have been described, however it is not clear whether these products are initiated by a radical on the molecule adducted or by a radical originally formed on the DNA. Yields of such adducts, between cellular protein molecules and DNA are 150 per cell per Gray (Ramakrishnan et al, 1987). There is an additional possibility that DNA protein cross-links (DPC) are formed by combination between a radical on the protein and a radical on the DNA formed within a volume of high local energy deposition (Oleinick, 1990). The types of proteins bound to DNA in DPCs are not in the main the histones but nuclear matrix proteins. This finding is additional support to the notion that the contact between DNA and histones is not close and that layers of water lie between these entities as discussed above.

Cerenkov Light

It has been shown that Cerenkov radiation can be a source of ultraviolet light induced damage in DNA irradiated by ionizing radiation (Redpath et al. 1981). The intensity of this light is dependent on the energy of the irradiation and the number of quanta reacting with cellular DNA depends on the length of the radiation path and the refractive index of the material through which the radiation passes. The contribution of Cerenkov induced damage to DNA can be calculated for high energy electrons whence the highest yield is expected. Assuming an arbitrary path length of 1 cm prior to reacting with the DNA (clearly this path length would not obtain in vivo) the amount of light reaching cellular DNA corresponds to 1.5×10^{-3} J m^{-2} of 260 nm per Gray (Redpath et al 1981). Since this is the maximum fluence to be expected and since the fluence necessary to kill mammalian cells is in the range of 10 Jm^{-2}, Cerenkov radiation can not play a part in mammalian cell killing. These studies do make an important point: The effect of Cerenkov irradiation as a cause of bacterial cell killing was discovered only by use of a UV repair deficient mutants and was not detected in repair proficient strains.

ACCESSIBILITY OF WATER

We have developed some recent evidence in support of the participation of OH radicals in damaging DNA (Webb, Ling, Ward and Milligan in preparation): It has been argued that the use of dimethyl sulphoxide (DMSO) as a means to scavenge OH radicals in cells may have other effects such as the disruption of the water structure within the cell. We have used acetonitrile to test this possibility; while it has a similar solvent effect and a similar structure to DMSO, acetonitrile reacts slowly with OH (k_{OH} = 3.5 x 10^6 $M^{-1}s^{-1}$) and with e_{aq} (k_e = 2.5 x 10^7 $M^{-1}s^{-1}$). The presence of 2M acetonitrile in Chinese hamster V79 cells has no effect on cell survival (and little effect on plating efficiency) or on the yield of dsb in the cells. The abnormally low k_{OH} of acetonitrile was checked against supercoiled DNA and found to be correct (Milligan, Abstract this Meeting). In other studies acetone (k_{OH}=7x10^7 $M^{-1}s$) at 1M also had no effect on cellular radiosensitivity (with minimal effect on PE). These results confirm the role assigned to DMSO as an OH radical scavenger in cellular systems and reemphasize that 65% of the cellular DNA damage, of chromosomal aberrations and of cell killing is caused by scavengeable OH radicals.

SUBSEQUENT REACTIONS OF INITIALLY ALTERED SPECIES

The initial site of radical or radical ion formation does not necessarily fix the type of damage or the site at which the final damaged becomes located; subsequent reactions of the primary altered moiety determine the final chemical change and the site at which the damage becomes localized could be altered by migration of energy or charge.

There are no cellular data which require the invocation of migration of energy or charge. Such propositions have been made to explain data in model systems and their extrapolation to DNA in the cell should be attempted with the complex packaging structures in mind. These mechanisms, which will be discussed by others at this meeting, include:

1. Radical transfer from base to sugar (Adinaryana et al (1988)

2. Energy migration (Al-Kazwini et al. 1989)
3. Electron or hole migration (Cullis 1989, Fielden et al. 1971, van Lith et al. 1986)

Some brief comments on these proposed mechanisms are in order: To explain yields of ssb higher than those calculated from radical attack on the deoxyribose moiety of polydeoxyadenylic acid, radical transfer from base to sugar has been invoked. If this were to occur in DNA, then one of two consequences should ensue: either damage to the sugar should lead to damaged base release (not observed Zheng et al. 1988) or a damaged base should be attached to the strand break terminal damaged sugar (not observed Henner et al. 1982).

Electron migration as shown by van Lith et al (1986) appears to occur in the hydration shell of the macromolecule. Is this mobile charge available to react with DNA or radicals on DNA? What are the consequences to this mechanism of the DNA being wrapped in nucleosomes and solenoids in a cell nucleus?

Cullis et al. (1989) invoke charge migration to form G^+ and T^- (or C^-) which are hypothesized to come together on opposite strands and thus be a precursor to a dsb. If this migration occurs in a cell then it must arise from events on a single fast electron track (since there is no effect of ultra high dose rate on mammalian cells, where radicals from separate tracks could be invoked). Chatterjee has indicated that the separation of events at low LET is 400nm; this separation in a complex cellular DNA packaging would mean that the damage consequences of two events would have a low probability of being amenable to interacting--what is the driving force to bring charged radicals together over such long distances?

While these mechanisms may operate in the model systems, their significance in mammalian DNA--in the structures present in cells, and in yields significant in the presence of other damages need to be carefully assessed. Is it necessary to deviate from the concept of direct ionization of DNA to produced cation radicals and indirect attack of OH radicals (or H_2O^+) all with the consequent lesions being formed at the site of the initial damage? Is there a need to invoke other damage mechanisms, are there inexplicable data which requires their occurrence?

SITES POSSIBLY ENCOURAGING A SPECIFICITY OF DAMAGE

In the bending of DNA around the histones some specificity has been found (Satchwell et al. (1986)): In general where the minor groove of the DNA faces inwards A+T regions are favoured inside and where the minor groove faces outward G+C rich regions are preferred outward. This could be a source of preferential damage at the base sequence level (accessibility to radical attack or for DNA-protein cross-link formation).

DNA DOUBLE STRAND BREAKS

Three mechanisms of DSB formation have been proposed (a. Siddiqi and Bothe 1987, b. Ward 1985, c. Boon et al 1984). The single radical mechanism must be invoked to explain yields of dsb in solution which have been found to occur at low and zero concentrations of radical scavengers. However the yield from the single radical source is about 1% of the ssb yield and is consequently insufficient to explain the yield (4%) at higher scavenger concentrations or in cells. It is thus necessary to invoke the locally multiply damaged sites (LMDS) mechanism (b) which involves the overlap of energy deposition events with DNA (this rationale is discussed at length in Ward, 1988).

The mechanism suggested in the ESR work (c), should be examined. Here dsb are suggested to occur by the reactions of the initially observed G^+ and T^- (in DNA irradiated at 50 mg ml^{-1} in low temperature ice). Dsbs are observed in parallel experiments in plasmid DNA (irradiated at 80 μg ml^{-1} under the same conditions). There are several points in this mechanism which are unclear: How do the ion radicals migrate to closely opposed positions? Would they then not be close enough for spin-spin interactions to perturb their ESR spectra? Why do they not intereact, neutralizing each other? The formation of deoxyribose radicals (the immediate precursors of sbs) is not confirmed in this work even when irradiation is carried out in the presence of H_2O_2 (a species X is inferred to be sugar radicals in the latter case but this is not confirmed). At least five sugar radicals (any of which could lead to a sb) are possible each having different multiline esr spectrum which would overlap seriously and reduce the detectability of this damage. Whether such radicals would be seen is a moot point, after all the total of such radicals necessary to cause the observed sb yields is only 1/10 of the base ion radicals yield. Thus, the involvement of deoxyribose radicals in these ice systems, formed either by direct ionization or OH radical attack from hydration water can not be disregarded as the source of sbs (both single and double).

COMPLEXITIES OF DAMAGE

There is a large variety of types of radiation damage in cellular DNA. This variety arises not only from the range of damaged identities of altered constituents but also from the different numbers of damages per locally multiply damaged site. Many different radiation products of DNA bases have been described in model systems and can be expected to be produced in cellular DNA. In addition, strand breaks are produced, but these appear to have only two structures: Deoxynucleoside loss leaving a 5'phosphate group and either a 3' phosphate group or a 3' phosphoglycolate (Henner et al. 1982). The distributions in size of the energy deposition events from 20 eV upwards, indicate that from 1 to 10s of radicals per site are formed, meaning that a LMDS could have 10s of damaged moieties in it. These damages will be distributed across both strands over 10s of base pairs (depending on size of the originating energy deposition event). As a consequence of this variety of damage, cells will be faced with a complexity of problems in repair.

On the basis of the following assumptions: 65/35 scavengeable vs non-scavengeable damage, base damage (bd)/sugar damage (sb) from OH radicals is 3/1 and from Direct Ionization 1/1 the ratio bd/sb in cellular DNA can be calculated to be 2.3. Thus in a doubly damaged local site the ratios of damage types are: two sbs 1; sb + base damage (bd) 4.6; and two bds 5.3, i.e. only 9% of the doubly damaged sites are dsbs. This reasoning can be extended to local sites with greater numbers of damaged moieties (multiplicities) and gives rise to the plot shown in Figure 1. In this plot the percent of sites which will give rise to a dsb is plotted as a function of the number of damages (multiplicity) of the site. The number of damages per site will depend on the amount of energy deposited in the site and the degree of overlap of the site with the DNA. Clearly if the radiation is of higher LET then the number of lesions within a LMDS will be greater and there is a high probability that in the immediate vicinity of a dsb there will be several base damaged sites.

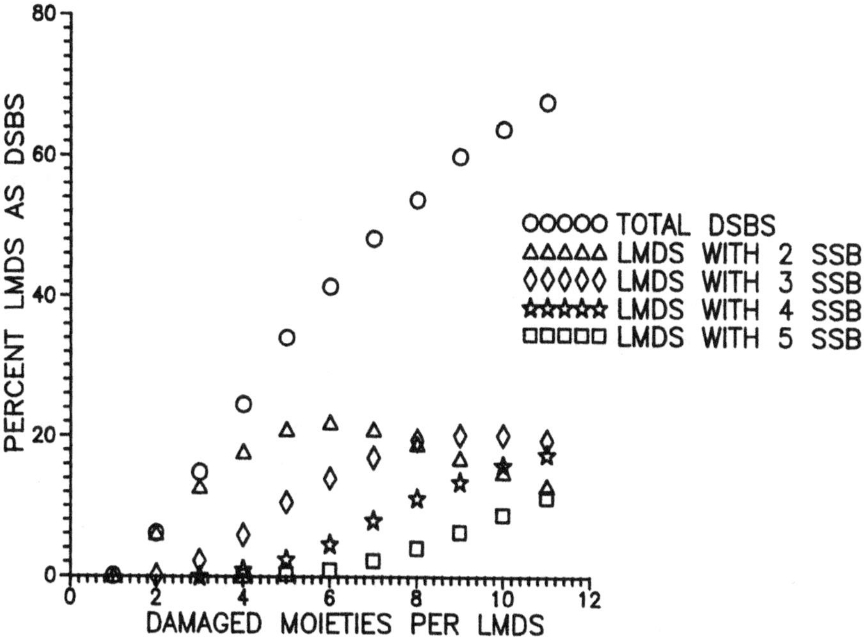

Figure 1. Percent of LMDS which result in a DSB

A large fraction of the LMDS contain only base damages particularly at low multiplicities of damage. Such multiple base damaged sites are not generally considered as having a lethal potential, however, the difficulties which a cell faces in trying to repair base damages in a LMDS have been previously pointed out (Ward 1988). If the base lesion is sufficiently close to a sb on the other strand, then the cell's attempts at excision repair of the base damage can be problematical.

There are several fates of a dsb; if the genetic information can be retained (probable if the individually damaged sites are far enough apart), then total repair could be accomplished by rejoining; if rejoining takes place without retention of base sequence, then a point mutation occurs, which may, however, be in an unimportant region of the genome; if rejoining occurs with the terminus of a different dsb, then a large deletion could occur. Sites with a low number of damages may be rejoined without problems, but as the number of lesions per damaged site increase, the "repairability" of the site will decreases. These problems arise not only when the damage is in the form of strand breaks but can also be crucial when base damage is involved.

VARIATIONS IN RADIOSENSITIVITY AMONG CELL TYPES

Following from the mechanisms of damage induction by free radicals in cells, and in particular, the varieties of damages originating from a range of sizes of energy deposition events, there is a wealth of damage types which could lead to eventual serious alterations in the cellular genome. The possibilities of repair of two damage LMDS have been discussed (Ward 1988) and the problems associated with the size of the patch of deoxynucleotides inserted at a damaged site and the balance between sb repair and bd excision discussed; if a cell attempts to remove a bd before rejoining a sb on the opposite strand then problems could ensue since a dsb will have been produced. Clearly a cells ability to deal with a more complex lesion will be even more dependent on the sequence in which the several damages are repaired. It could be hypothesized that a cells radiosensitivity could depend on the relative levels of different repair enzymes available. A gradation in damage severity can exist as shown in Table 1.

Table 1. GRADATION IN DAMAGE SEVERITY

EASIEST TO REPAIR

a. Single damaged sites BD and SB.
b. LMDS with damages on the same strand.
c. LMDS with a BD on each STRANDS. From 20bp separation to Directly opposed.
d. LMDS with 1 BD on one strand and 1SB on the other
e. LMDS with a SB on each strand.
f. LMDS with more damages spread over both strands.

MOST DIFFICULT TO REPAIR

Different cells may have differing capabilities in their abilities to repair this spectrum of damages. For instance variability is possible in: the patch size introduced, the rate at which bds generally are excised relative to sb rejoining, or the rate at which a specific bd is removed. These variabilities would be difficult to detect in cellular repair but they could readily affect cellular radiosensitivity.

OXYGEN EFFECT

The oxygen effect is generally considered to arise as a consequence of a competition between oxygen and reducing species (thiols) for radicals produced on the DNA. It has been pointed out (Holian and Garrison, 1969) that radicals produced on DNA bases would not be repaired by the chemical repair reaction; hydrogen atom donation mechanism. Radicals formed on the bases would result in an altered base whether they react with oxygen or sulfhydryl compound. Indeed a similar problem exists for H donation to sugar radicals (Raleigh et al., 1987). Hence H atom donation would only repair some of the sugar radicals. This suggests that any cell which has a deficiency in removing base damage (and is hence more radiosensitive) would show a lower oxygen enhancement ratio.

DSBS AND OXYGEN EFFECT

At the 1987 Oxygen Radical Meeting the Alper and Howard-Flanders expression developed to explain the oxygen concentration dependence of cell radiosensitivity was considered in terms of the DNA dsb being the lethal lesion in cells (Ward unpublished). The DNA dsb requires the "fixation" of two radicals (whichever mechanism of dsb formation occurs) and the "chemical repair" of either of these radicals prevents dsb formation.

From this initial premise it can be shown that the yield of dsb (of that which can be manipulated by oxygen) at any oxygen concentration is:

$$\frac{(a[O_2])^2}{(a[O_2] + b[XH])^2}$$

Where a and b are constants and [XH] is the concentration of the compounds which "chemically repair" in competition with oxygen. Clearly this expression differs from that which can be developed for single breaks:

$$\frac{(a[O_2])}{(a[O_2] + b[XH])}$$

and which is the form which gives rise to the Alper and Howard-Flanders expression. The dependence of cell radiosensitivity on oxygen concentration should be re-examined with this in mind.

ACKNOWLEDGEMENTS: Work in the author's laboratory is supported by grants from NIH; CA26279 and CA46295; and from DOE; DEFG0388ER60660.

REFERENCES

Aboul-Enein A and Schulte-Frohlinde D (1988) Biological Deactivation and Single-Strand Breakage of Plasmid DNA by Photosensitization using Tris(2,2'-Bipyridyl) Ruthenium (II) and Peroxydisulfate. Photochem. and Photobiol. 48:27-34.

Adinarayana M, Bothe E and Schulte-Frohlinde D (1988) Hydroxyl Radical-Induced Strand Break Formation in Single-Stranded Polynucleotides and Single-Stranded DNA in Aqueous Solution as Measured by Light Scattering and by Conductivity. Int. J. Radiat. Biol. 54:723-738.

Al-Kazwini AT, O'Neill P, Adams GE, and Fielden EM (1990) Radiation Induced Energy Migration within Solid DNA: The Role of Misonidazole as an Energy Trap. Radiat. Res. 121:149-153.

Cullis PM, Jones GDD, Sweeney MC, Symons MCR and Wren BW (1989) DNA Damage in Frozen Aqueous Solution: Sequence Dependence and End Groups. 6:149-152. And other papers cited therein by the same authors.

Evans JW, Limoli CL, and Ward JF (1987) Production and Repair of Locally Multiply Damaged Sites in Gamma Irradiated DNA of Cells and Nuclei. Radiation Research (8th ICRR). 1:149.

Fielden EM, Lillicrap SC and Robins AB (1971) The Effect of 5-Bromouracil on Energy Transfer in DNA and Related Model Systems: DNA with Incorporated 5-BUdR. Radiat. Res. 48: 421-431.

Henner WD, Grunbert SM, and Hazeltine WA (1982) Sites and Structure of γ Radiation-Induced DNA Strand Breaks. J. Biol. Chem. 257:11750-11754.

Holian J and Garrison WM (1969). Reconstitution Mechanisms in the Radiolysis of Aqueous Biochemical Systems: Inhibitive Effects of Thiols. Nature (London) 221:57.

Jonah CD and Miller JR (1977). Yield and Decay of the OH Radical from 200 ps to 3 ns. J. Phys. Chem. 81:1974-1976.

Makrigiorgos G, Adelstein, SJ, and Kassis, AI (1990) Auger Electron Emitters: Insights gained from in vitro experiments. Radiat. and Environ. Biophys. 29:75-91.

Milligan JR, Calabro-Jones, PM, Lai DT, Nelson TR and Ward, J.F. (1990) G Value for SSB in Gamma Irradiated DNA. 38th Annual Radiat. Res. Soc. Mtg, New Orleans. Abstract Ev5.

Oleinick N (1990) DNA-Protein Cross-Links, . In: Wallace SS and Painter RB (eds) Ionizing Radiation Damage to DNA: Molecular Aspects. Alan Liss Publisher (in Press).

Opitz J and Schulte-Frohlinde D (1987). Laser-Induced Photoionization and Single-Strand Break Formation for Polynucleotides and Single Stranded DNA in Aqueous Solution: Model Studies for the Direct Effect of High Energy Radiation of DNA. J. Photochem. 39:145-163.

Painter RB (1979) The Role of DNA Damage and Repair in Cell Killing Induced by Ionizing Radiation. In: Meyn RE and Withers HR (eds) Radiation Biology in Cancer Research. Raven Press, New York. pp. 59-68.

Raleigh JA, Fuciarelli AF and Kulatung CR (1987) Potential Limitation to Hydrogen Atom Donation as a Mechanism of Repair in Chemical Models of Radiation Damage. In: Cerutti PA, Nygaard OF and Simic MG (eds) Anticarcinogenesis and Radiation Protectors. Plenum, New York. pp.33-39.

Ramakrishnan N, Chiu S-M, and Oleinick NL (1987) Yield of DNA-Protein Cross-Links in γ-Irradiated Chinese Hamster Cells. Cancer Res. 47:2032-2035.

Redpath JL, Zabilansky E, Morgan T, and Ward JF (1981) Cerenkov Light and the Production of Photoreactivatable Damage in X-irradiated E. Coli. Int. J. Radiat. Biol. 39:569-575.

Saenger W (1984) In: Principles of Nucleic Acid Structure. Springer Verlag, Berlin and New York.

Satchwell SC, Drew HR and Travers A (1986) Sequence Periodicities in Chicken Nucleosome Core DNA. J. Molec. Biol. 191:659-675.

Siddiqi MA and Bothe E (1987), Single- and Double Strand Break Formation in DNA Irradiated in Aqueous Solution: Dependence on Dose and OH Radical Scavenger Concentration. Radiat. Res. 112:449-463.

van der Holde, KE (1989) "Chromatin", Springer Verlag.

van Lith D, Eden J, Warman JM and Hummel A. (1986) Electron Migration in Hydrated DNA and Collagen at Low Temperatures. J. Chem Soc., Faraday Trans.I 82:2945-2950. and references by the same authors quoted therein.

Ward, JF (1984) Biochemistry of DNA Lesions. Radiat. Res. 104:S103-S111.

Ward, JF (1985) Ionizing Radiation Induced DNA Damage: Identities and DNA Repair. In: Simic MG, Grossman L and Upton AC Mechanisms of DNA Damage and Repair. Plenum Press, New York, pp. 135-138.

Ward JF (1988) DNA Damage Produced by Ionizing Radiation in Mammalian Cells: Identities, Mechanisms of Formation and Repairability. Prog. Nucleic Acids Res. and Molec. Biol. 35: 95-125.

Widom J. (1989) Toward a Unified Model of Chromatin Folding, Ann. Rev. Biophys. Chem. 18:365-95.

Zheng S, Newton GL, Gonick G, Fahey RC, and Ward JF (1988) Radio-protection of DNA by Thiols: Relationship Between the Net Charge on a Thiol and Its Ability to Protect DNA. Radiat. Res. 114:11-27.

BIOPHYSICAL MODELS OF RADIATION ACTION - DEVELOPMENT OF SIMULATION CODES

Herwig G. Paretzke
GSF - Institute for Radiation Protection
D-8042 Neuherberg, FRG

ABSTRACT

Radiation carcinogenesis from different radiation fields at low doses and low dose rates remains to be an important research topic in radiation protection. Most likely particular modification of DNA play a decisive role in the initiation steps of carcinogenesis, and, thus, detailed analyses must be performed of the interaction events of different radiation fields with DNA and of their quantitative correlation with the yields of important cellular responses (e.g. transformations). Such analysis must comprise detailed simulations in a computer of all relevant physical interaction processes, the simulation of the subsequent direct chemical modifications of the DNA and the surrounding cellular structures and of those occurring through subsequent diffusion controlled chemical reactions, and a correlation study of the frequency of important events deemed relevant in this context with frequencies of important cellular responses (e.g. DNA strand breaks, chromosome changes, transformations) using mechanistic working hypotheses for the initial carcinogenetic actions of different radiation fields.

This paper describes the state of the art in the simulation of charged particle tracks from various radiation fields (photons, neutrons, internal emitters) in biological cells for the calculation of direct and indirect modifications of the DNA. A significant progress could be achieved through the recent availability of interaction cross sections for condensed molecules and of very powerful parallel computers. Presently the main shortcomings in the track structure approach to understand radiation action are the lack of solid information on the migration and decay of excited states in DNA, and of good working hypotheses for relevant sensitive target structures in a cell and the chemical changes needed in them to trigger initiating steps in radiation carcinogenesis.

INTRODUCTION

In radiation research one is usually interested in the consequences of a radiation interaction either for the radiation field (Class I prob-

NATO ASI Series, Vol. H 54
The Early Effects of Radiation on DNA
Edited by E. M. Fielden and P. O'Neill

lems) or for the irradiated matter (Class II problems) (Inokuti, 1983). The study of class I questions is comparatively simple theoretically (e.g. because of the existence of confining sum-rules (Bethe, 1930; Inokuti, 1971)) and experimentally (e.g. by measurement of the angle and energy of particles with state of the art detectors. Therefore much knowledge could be obtained already in this field and many major problems concerning the effects of single and multiple collisions on a radiation field itself may even be considered as generally solved. As a consequence, the modification of a radiation field can be interpreted reasonably well and has found widespread applications in industry, medicine, etc (e.g. in X-ray diagnostics). In addition, there exist quantities (e.g. activity, fluence, stopping power, attenuation coefficients) and units accepted almost universally in science, technique, and even in daily life (after the Chernobyl accident) describing radiation sources and fields properly.

Unfortunately, the scientific area concerned with the fate of the irradiated matter, i.e. with class II problems (fig. 1), is much less developed and even in a rather rudimentary state. Work on improvement of this situation represents still a major intellectual challenge; this is true both for theory and experiment as well as for concepts, quantities and units. The usefulness for radiation research of the quantities dose, RBE, LET, Y, etc. is highly questionably (Paretzke, 1989); quantities describing charged particle track structures in a general way are not existent. In the attempts to learn more about the changes induced in matter by tracks left behind from an incident radiation field, the analysis of computed simulations of such tracks in their different stages plays a promenent role. The reason for this importance of tracks in matter and the state-of-the-art in development of computer codes simulating charged particle tracks for the testing of action hypotheses and optimization of instrumentation is discussed below.

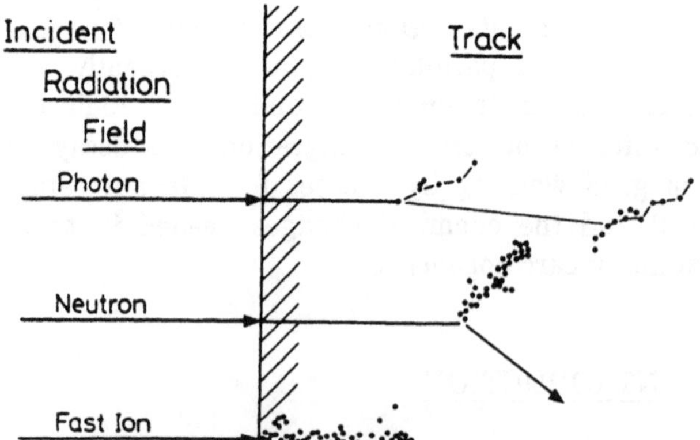

Fig. 1. Class I problems deal with effects on the incident radiation field, class II problems with the consequences of tracks left behind in matter

ROLE OF BIOPHYSICAL MODELS OF RADIATION ACTION

Ionizing radiation can induce and cure cancer and produce many other biophysical effects of theoretical and practical importance (e.g. acute, teratogenic, genetic effects). Unfortunately, because of the unsatisfying data situation and the theoretical many-body nature, even the mechanisms of physical and early biochemical action are not yet fully understood. Therefore, they cannot be quantified with the accuracy required for integral models for carcinogenesis, etc., at low doses on which finally important administrative or medical decisions are based on (e.g. the linear hypotheses for radiation carcinogenesis). Unfortunately, different fields of radiation (e.g. x-rays, γ-rays, internal alphaemitters, neutrons) can have quite different biological risks per unit of absorbed dose, and for some types of radiations and endpoints a human data base even at higher doses is virtually non-existing. In the most important low dose, low dose rate regime from experimental and/or epidemiological data for principal, statistical reasons a direct input into a better knowledge of biophysical radiation effects can never be expected. In this regime from biophysical, etc. modelling a better quantification of low dose effects can finally be expected. Therefore, the production and testing of mechanistic or empirical hypotheses or models for the action of radiation is an important task in radiation research. In this regime from biophysical, etc. modelling a better quantification of low dose effects can finally be expected.

It might be necessary to stress that the physical aspect of such models must be treated in adequate detail. That is because different effectivenesses of different types ("qualities") of radiation, of different doses, dose rates and fractionation patterns can have their origin only in different initial structures of the tracks left behind in matter by the irradiation. There cannot be any other reason. The only and decisive information to a target of its irradiation occurs through physical processes of energy transfer leading to electronic, vibrational rotational, dissociative, etc. excitation of target molecules (fig. 2). This defines the inital conditions for all subsequent chemical, biological, and medical re-actions to this sudden disturbance. There can be many other factors (e.g. presence of oxygen, tumor promoters, etc.) influencing these reactions later, but the quantitative result of all these modifications, again, is subject to the actual boundary condition set in about a femto to pico second at the physical stage of radiation action.

Because of this, it is also necessary to consider always the whole, correlated ensemble of events and processes initiated by a radiation interaction and describe it with joint probability functions. A separate treatment of particular aspects of track structure is not possible in general (but may be an acceptable approximation in certain problems). Such an analysis including correlations is only possible in track simulation calculations. With experimental measurements, e.g. microdosimetric counters, cloud chamber measurements, this cannot be achieved because only the sum of ionizations and in one single volume is given or because of the lacking capability to give adequate information on the different types of new species in biological matter.

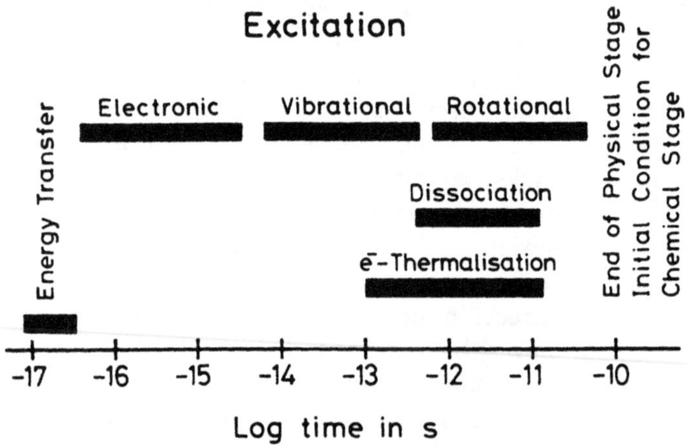

Fig. 2. Time sequence of processes during the physical stage of radiation action on matter

ROLE OF TRACK SIMULATIONS IN BIOPHYSICAL MODELS

Upon radiation interaction with matter an immensely complex, dynamic track consisting of chemical, physical, and, finally, biological changes is produced. It is an important, fundamental task in biological radiation research to identify in this time dependent pattern those changes which are the relevant and essential ones for a particular effect under consideration. This process leads to the necessary information reduction and concentration on the important properties of charged particle tracks and is an indispensable prerequisite for the iterative generation of action hypotheses, for the derivation of general principles, and for the classification, characterization and measurement of radiation fields adequate for their effects in matter. Such identification and information reduction is the motivation for, and, thus implicitely underlies each biophysical model. Biophysical models are usually either dosimetric or track related (e.g Goodhead, 1984, 1988, 1989).

There are several classes of dosimetric models. All of them essentially consider the energy deposited in one small sensitive volume or mass element as the relevant quantity (fig. 3) (e.g. Kellerer and Rossi, 1974;

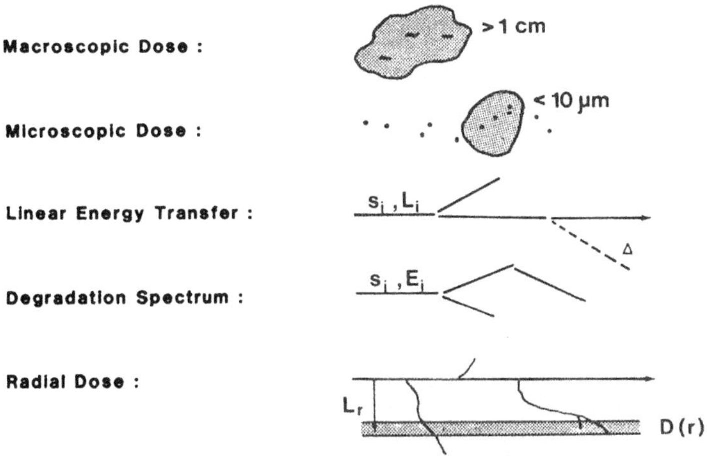

Fig. 3. Examples of essentially dosimetric concepts

Feinendegen, 1990). Unfortunately, this micro- or macroscopic quantity "energy absorbed per mass (=dose)" has increasingly been found to be an inappropriate, bad descriptor of the radiation interaction in biological radiation research. This is indicated e.g. by the need for the introduction of a correction factor (called "Quality factor Q") - which can vary by two orders of magnitude - when absorbed dose D is correlated with biological radiation effects (ICRU, 1986) through the corrected "equivalent" dose $H = D \cdot Q$. Another reason is the fact that at low doses and for small targets, (e.g. the volume of the DNA in a mammalian cell) the specific energy actually absorbed in a particular target is no longer proportional or even dependent at all on the macroscopic absorbed dose D; this is quantitatively shown in fig. 4 (after ICRU, 1983). Biophysical models based on the quantity (restricted) linear energy transfer (e.g. Harder et al., 1988) appear to agree better with experimental data. Good agreement is also reported by Watt (1989) using his "primary ionisation" model of biological damage which is a special form of models based on the charged particle degradation spectrum. The radial dose model of Katz et al. (1972) has successfully been applied to the empirical interpretation of track effects of heavy charged particles in a large variety of targets; this lead to a significant progress in track structure research and was main stimulus for the development of simulation codes (Benton and Starace, 1989). All available track structure simulation codes can compute energy deposition spectra for any specified small sensitive element to a reasonable accuracy (e.g. Nikjoo et al., 1989; Olko et al., 1989).

The track related biophysical models retain more detail of the interaction pattern, i.e. they do not reduce the pattern to just the energy spent to produce it in one single volume. Again here we find a variety of expressions of such models (fig. 5) (e.g. Chadwick and Leenhouts, 1981;

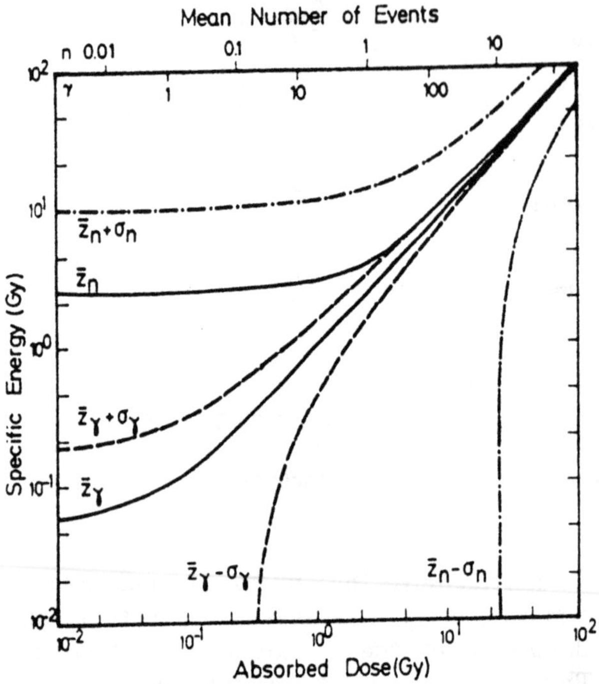

Fig. 4. Specific energy as a function of absorded dose in a 1 μm sphere (mean value and ± 1σ) for Co-60- and n-irradiation. The mean number of hits per sphere is indicated on the upper abscissa

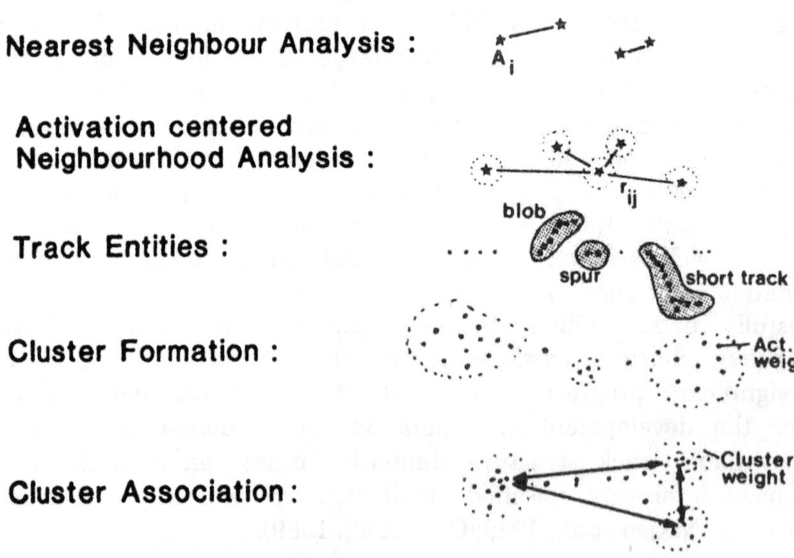

Fig. 5. Examples of track related biophysical concepts

Chatterjee and Holley, 1989; Günther and Schulz, 1983; Mozumder and Magee, 1966; Hamm et al., 1978). Here the use of advanced cluster algorithms (for a review see Ryzin, 1977) could provide promising new approaches. For use in this type of biophysical models, track structure simulation codes must try to provide the full information on the initial locations of relevant new species and their dynamic reactions afterwards.

CONCEPTS OF SIMULATION CODES

Track structure simulation codes basically are designed to provide in a computer the detailed information on the time dependent locations and types of relevant molecular changes in the tracks left behind in matter by ionzing radiation for further analysis biophysical models. Following Fano (1970) the structure of a charged particle track should be described in terms of the spatial positions of identifiable, localized events (e.g. a sufficiently long living excitation, ionization, radical, dissociation product) within an atom or molecule. At the end of the physical stage these primary events, i.e. the new physico-chemical species, will start to react with each other and with the bulk material (Zaider and Brenner, 1984). This will lead to the disappearance of some primary species and the occurrence in space and time of newly formed chemical or biological species. These later generations of track species can differ in types, locations, and number. Thus the structure of tracks is time dependent and can only be described by joint probability distributions correlating all species present in a certain space-time element. The objective of any radiation track structure theory is to describe and predict the spatial distribution of localized event with a minimum of detailed analysis and of assumptions concerning preliminary processes (Fano, 1970). This correlation function then represents the adequate input for the study of further actions of radiation in biophysical models. Simulation codes essentially try to compute these time dependent event structures.

The presently most important problem in track simulations for the physical stage is the lack of theoretical and experimental knowledge on the migration of energy transferred from radiation to condensed matter along molecules or in electron energy bands (either mono-electronic or in collective states) (Ritchie et al., 1989; Baverstock et al., 1988, 1989; Bednar, 1989) and the decay of excited molecular states in the condensed phase into new chemical species; progress is badly needed since class II problems in radiation research are crucially determined by the types and locations of new chemical species in matter and not by the amount and location of energy lost from the incident radiation field. Other very important problems are the behaviour of low energy electrons in condensed matter (e.g. Schou, 1988) and their final thermalization (e.g. Goulet and Jay-Gerin, 1988, 1989; Inokuti, 1983).

Presently, there are essentially only eight groups concerned with the development of simulation codes (there are several more applying and modifying the codes):

M. Berger, previously National Institute of Standards and Technology, Gaithersburg, USA, (e.g. Berger, 1963)
B. Großwendt, Phys. Techn. Bundesanstalt, Braunschweig, FRG (e.g. Grosswendt, 1983)
A. Ito, University-Institute of Medical Science, Tokyo, Japan (e.g. Ito, 1988)
I. Kaplan, Karpov Institute of Physical Chemistry, Moscow, USSR (e.g. Kaplan, Miterev, 1989)
H.G. Paretzke, GSF-Institute for Radiation Protection, Neuherberg, FRG (e.g. Paretzke, 1987)
M. Terrisol, Universite-Centre de Physique Atomique, Toulouse, France (e.g. Terrisol, 1978)
J.E. Turner, R.N. Hamm, R.H. Ritchie, Oak Ridge National Laboratory (ORNL), USA (e.g. Ritchie et al, 1989)
M. Zaider, D. Brenner, Columbia University, New York, USA (e.g. Zaider et al, 1983)

A good overview on these codes can be found in IAEA, 1989. Most simulation codes have been developed to compute track structures in the model substance water (liquid or vapour) and for homogeneous, unstructured target. Only the new GSF-code PARTRAC can produce simulated tracks in structured targets (e.g. DNA in a cell which is in a tissue, by applying complex geometry algorithms) which can be made up from different materials (e.g. the gas and the wall materials of a low pressure proportional counter). Chemical reactions following the initial physical disturbance can be considered in PARTRAC and the ORNL-, CPA-, and Columbia-code. A short outline of the present status of and samples of track simulation results are given below for incident photon, neutron, electron, proton, alphaparticle and heavy ion fields.

TRACK SIMULATION FOR PHOTONS AND NEUTRONS

In the energy range used in most radiobiological experiments photons can interact with target materials by photoelectric- or Compton-effect or by pair production; in all cases electrons (and in the later case also positrons) are produced. The tracks of these secondary particles usually determine the action of the irradiation and not the single interaction event creating them. Nevertheless it might be worthwhile to point out that it is presently still a problem to compute with great accuracy partial Compton scattering cross sections for electrons bound in a molecule (Hubbell, 1979); this is necessary because the decay consequences of the removal of an electron from a molecule depend on its previous position. For photoeffect electrons such positions in the molecular shell structure can be estimated e.g. from ESCA work of Siegbahn et al. (1969), mass spectroscopy work of Tan et al. (1978), and theoretical models (e.g. Gelius, 1972) (see also Terrisol et al., 1989). Detailed track structure calculations for photons can be performed by the codes of most groups mentioned above (e.g. Nikjoo et al., 1989).

When neutrons penetrate biological matter they can produce photons and fast ions by several nuclear processes; usually recoiled protons are the most prominent ions. The present status of the pertinent interaction cross sections has recently been reviewed by Pihet and Menzel (1989). A simulation code computing the initial and slowing down spectra of the secondary heavy particles produced in neutron fields has been developed e.g. by Caswell and Coyne (1966). The simulation of the track structures of these charged particles is feasable for photon produced electrons and for stripped fast ions. It becomes increasingly difficult with decreasing ion energy and with increasing number of electrons bound to the projectile (see below). Most research groups mentioned above have simulation codes available at least for the proton component (e.g. Zaider et al, 1983).

ELECTRON TRACK SIMULATION

Electrons are the most important particles in class II problems in radiation research; their action determines the outcome of almost any exposure to ionizing radiation (Inokuti, 1983). It is in particular the low energy range (below, say, 1 keV) which plays a decisive role (fig. 6) in biological radiation action.

Unfortunately, for generalization, there are not yet sufficient experimental data available for electron scattering in condensed matter in this energy range (Schou, 1988); most useful data on low Z materials come from the laboratory of Sanche (1987). They clearly show that low energy electron scattering processes are modified in the condensed phase as compared to the gas phase. The few attempts of theoretical calculations of such cross sections needed in track structure research (see e.g. Ritchie et al., (1989). Zaider et al., (1989), Long et al., (1989)) are still far from providing in satisfying results.

Even less is known about the new species resulting from the decay of excited states in condensed matter of produced by these scattering electrons (again more information is available for the class I problem of radiation field modification). Most simulation codes assume the same mole-

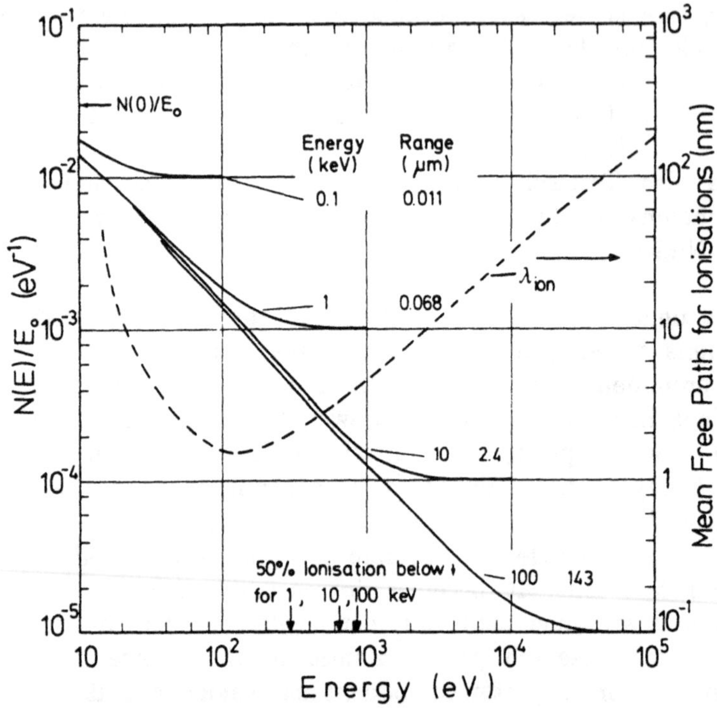

Fig. 6. Number of electrons N(E) created at least with start-energy E by primary electrons of initial energy E_0=0.1, 1, 10, 100 keV. Also indicated is their range, the mean free path between subsequent ionizations and the 50%-median of the contribution of all electrons to the total ionization (after Paretzke and Berger, 1978)

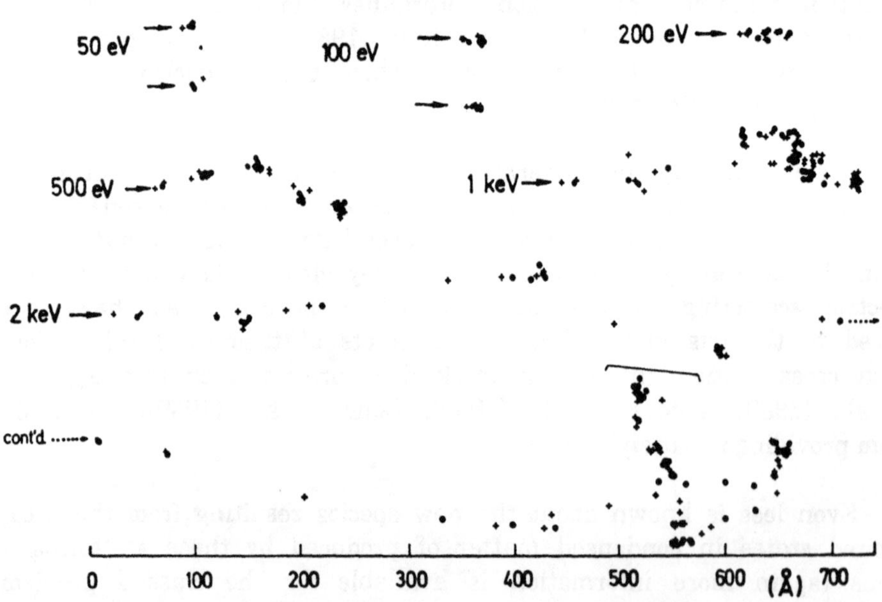

Fig. 7. Simulated electron tracks in water (Paretzke, 1988)

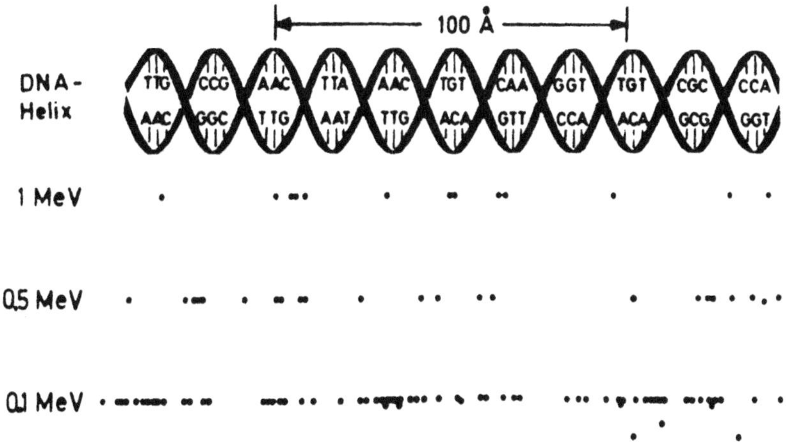

Fig. 8. Simulated proton tracks in water (Paretzke, 1988)

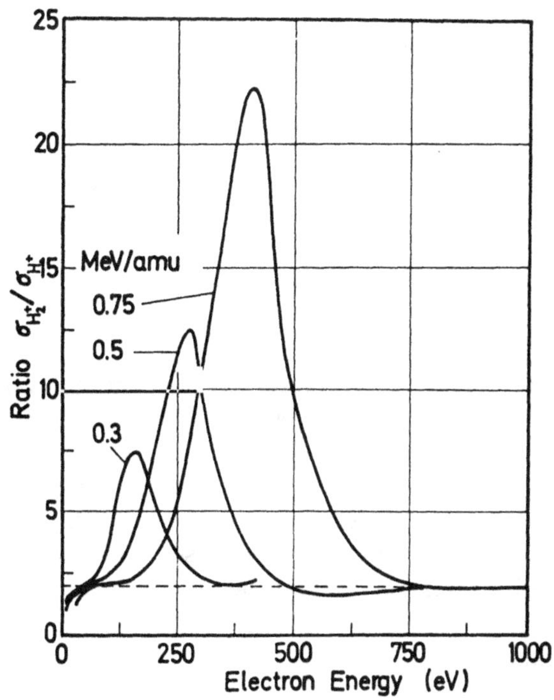

Fig. 9. Ratio of secondary electron spectra from H_2^+ impact on H_2 to that from H^+ impact for 0.75, 0.5 and 0.3 MeV/amu ions as a function of electron energy (Wilson and Toburen, 1973). The peaks result from electron emission from the projectile

cular decay pattern to hold in condensed targets as it has been derived for the gas phase (see e.g. Paretzke, 1988). Under this hypothesis electron tracks in water might resemble those given in fig. 7 which were derived for water vapour and scaled to the liquid only with respect to the density. It is evident that very slow electrons produce only a few events which, however, are densily clustered. With increasing initial energy an increasing number of events is created but some events might be "wasted" with respect to biological effects since they have to act alone (that is why electron irradiation is often called "sparsely ionizing"). Again, all research groups given above have developed simulation codes for electrons.

ION TRACK SIMULATION

The problem of ion track simulation may be divided into two parts: the interactions of a fast or slow ion itself with the target material, and the interaction of the hereby produced electrons and photons (also Rutherford scattered recoil ions can be observed rarely). In principle these two contributions to the track structure are strongly correlated and ought to be treated together (e.g. a secondary electron slows down under the influence of the Couloub fields of the positve ions produced earlier by a fast incident heavy particle). In practice, however, presently all simulation codes deal with the primary and the secondary contributions to track structure in a strictly independent, additive way. The main reason for this implifying approach is the strong increase in computer power required for a more accurate simulation. The increasing availability of highly cost-effective parallel computers, which are ideally suited for this type of Monte Carlo calculations, will significantly and soon contribute to advances in track structure theory.

For fast, stripped ions the differential ionization cross sections can reasonably be estimated even for condensed matter. Therefore the proton tracks for 0.5 and 1 MeV in the model substance water, as indicated in fig. 8, might not be too far from reality. With decreasing energy, however, the incident ion captures electrons into own shells. This leads to the additional possibility of electron emission also from the fast moving ion (see e.g. fig. 9) and to higher frequencies of multiple ionizations in one collision event. Both processes cannot be scaled from cross sections for events in proton tracks.

Within condensed target the ejection cross section for the numberous (ca. 95%) slow electrons cannot be measured directly, and the theory becomes complicated also because the moving ion is likely to be in a highly excited state (because of many rapid interactions). Presently, none of the ion track simulation codes can take these effects into consideration properly. If this important effect is neglected and the secondary electron emission cross sections are scaled by z_{eff} (resulting e.g. from stopping power measurements) one can use proton simulation cross sections to pro duce a first rough approximation to simulated tracks of heavier charged particles (fig. 10).

It will represent a major step forward in track structure theory when such heavy charged particle tracks can be simulated with sufficient reliability. Only then they can be used in biophysical models for the interpretation of radiobiological experiments performed with these ions (for a review see Kraft, 1989).

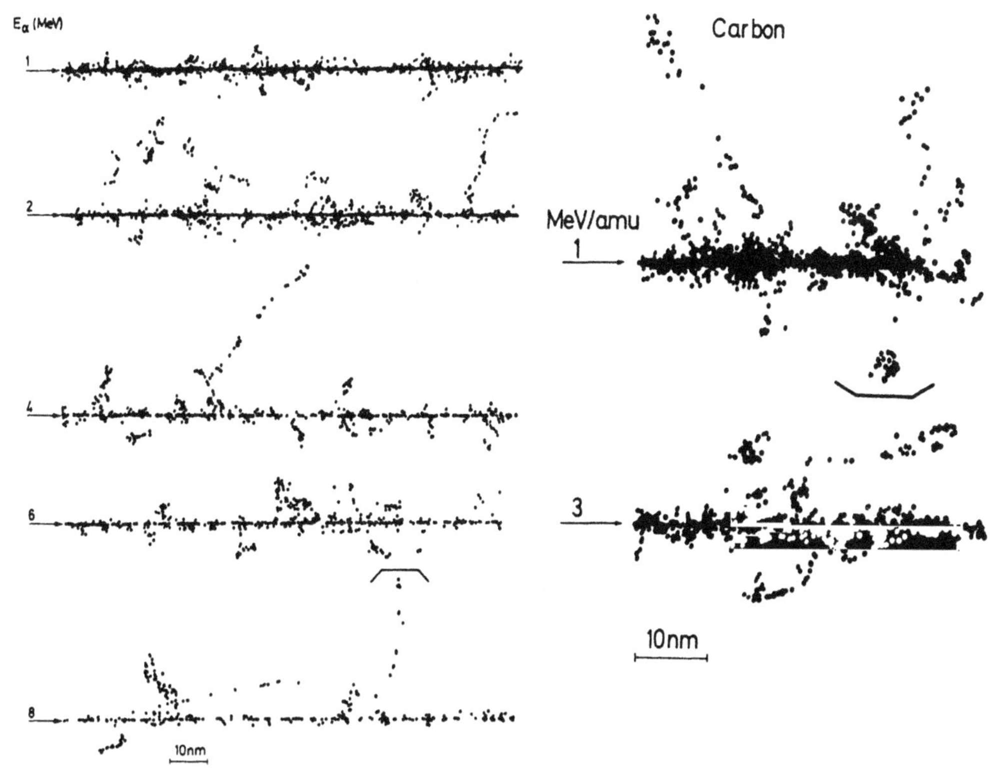

Fig. 10. Approximation of simulated tracks of 1-8 MeV alphaparticles and of 12 and 36 MeV C-ions in water (the emission cross section were scaled from proton cross sections by z_{eff}, multiple ionizations were neglected; Paretzke, 1988).

REFERENCES

Baverstock KF, Cundall RB (1988) Long-range energy transfer in DNA. Radiation Phys Chem 32:553-556

Baverstock KF, Will S (1989) Evidence for the dominance of direct excitation of DNA in the formation of strand breks in cells following irradiation. Int J Radiat Biol 55: 563-568

Bednar J (1989) Solitons in radiation chemistry and biology. J Radioanalyt Nucl Chem 133:185-197

Benton EV, Starace AF (eds) (1989) Special issue: Recent Advances in track physics. Nucl. Tracks and Radiat. Meas 16:81-224

Berger M (1963) Monte Carlo calculation of the penetration and diffusion of fast charged particles. In: Methods in computational physics, Alder B, Fernback S, Rotenberg M (Eds), Vol. 1: 135-215, Academic Press, N.Y.

Bethe H (1930) Zur Theorie des Durchgangs schneller Korpuskularstrahlen durch Materie. Ann Phys 5: 325-400

Chadwick KH, Leenhouts HP (1981) The molecular theory of radiation biology, Springer Verlag, Berlin Heidelberg New York

Chatterjee A, Holley WR (1989) Energetic electron tracks and DNA strand breaks. Nucl Tracks and Radiat Meas 16:127-133

Feinendegen LE (1990) The cell dose concept; potential application in radiation protection. Phys Med Biol 35:597-612

Gelius U (1972) Electron Spectroscopy. Shirley DA (ed) p 311

Goodhead DT (1984) Deductions from cellular studies on inactivation, mutagenesis and transformation. In: Radiation Carcinogenesis: Epidemiology and biological significance, Boice JD, Fraumeni jr JF (eds) p. 369-385, Raven, New York

Goodhead DT (1988) Spatial and temporal distribution of energy. Health Phys 55: 231-240

Goodhead DT (1989) Initial physical damage produced by ionizing radiation. Int J Radiat Biol 56: 623-634

Goulet T, Jay-Gerin J-P (1988) Thermalization distances and times for subexcitation electrons in solid water. Radiat Res 118: 46-62

Großwendt B (1983) Track structure studies of low-energy electrons in TE gas. In: 8th Symp on Microdosimetry, Booz J, Ebert HG (eds) CEC-Publication EUR 8395, p. 79-88 Luxemburg

Günther K, Schulz W (1983) Biophysical theory of radiation action. Academie-Verlag, Berlin

Hamm RN, Wright HA, Katz R, Turner JE, Ritchie RH (1978) Calculated yields and slowing down spectra for electrons in liquid water: implications for electron and photon RBE. Phys Med Biol 23:1149-1161

Harder D, Blohm R, Kessler M (1988) Restricted LET remains a good parameter of radiation quality. Radiat Prot Dosim 23:79-82

Hubbell JH, Veigele WJ, Briggs EA, Brown RT, Cromer DT, Howerton RJ (1975) Atomic form factors, incoherent scattering functions and photon scattering cross sections. J Phys Chem Ref Data 4:471

IAEA (1989) Atomic and Molecular Data for Radiotherapy,Proceedings of an Advisory Group Meeting, IAEA-TECDOC-506 Vienna

ICRU (1986) Report of a Joint Task Group of the ICRP and ICRU to the ICRP and ICRU: The Quality Factor in Radiation Protection. Report 40, Bethesda, USA

Inokuti M (1971) Inelastic collision of fast charged particles with atoms and molecules - the Bethe theory revisited. Rev Mod Phys 43:297-347

Inokuti M (1983) Radiation physics as a basis for radiation chemistry and biology. Appl Atomic Collision Physics, Vol. 4, 179-236

Ito A (1988) Electron track simulation for microdosimetry. In: Monte Carlo transport of electrons and photons, TM Jenkins, WR Nelson, A Rindi (eds) Plenum Press, N.Y., London: 361-384

Kaplan IG, Miterev AM (1989) Characteristics of the interaction of ionizing radiation with a molecular medium and the role of tracks in radiation chemistry. Usp Khim 55:713-742

Katz R, Sharma SC, Homayoonfar M (1972) The structure of particle tracks. In: Topics in Radiation Dosimetry, Supp 1: 317-383, FH Attix (ed), Academic Press, New York, London

Kellerer AM, Rossi HH (1974) The theory of dual radiation action. In: Current Topics in Radiation Res, Vol VIII: 85-158, North-Holland Publ, Amsterdam

Kraft G (1989) Interpretation of radiobiological experiments performed with heavy charged particles. Loc cit IAEA (1989): 117-127

Long KA, Paretzke HG, Müller-Plathe F, Diercksen GHF (1989) Calculation of double differential cross sections for the interaction of electrons with a water molecule, clusters of water molecules and liquid water. J Chem Phys 91: 1569-1577

Mozumder A, Magee J (1966) Model of tracks of ionizing radiations for radical reaction mechanisms. Radiat Res 28:203-214

Nikjoo H, Goodhead DT, Charlton DE, Paretzke HG (1989) Energy deposited in small cylindric targets by ultrasoft x-rays. Phys Med Biol 34:691-705

Olko P, Booz J, Paretzke HG, Wilson WE (1989) Energy deposition in the nanometer sites based on the track structure calculations. loc cit IAEA (1989): 105-116

Paretzke HG (1987) Radiation track structure theory. In: Kinetics of nonhomogeneous processes. GR Freeman (ed), Wiley & Sons, N.Y.: 89-170

Paretzke HG (1988) Simulation von Elektronenspuren im Energiebereich 0,01-10 keV in Wasserdampf, GSF-Bericht 24/88, Neuherberg

Paretzke HG (1989) Parameters characterizing charged particle track structures. Loc cit IAEA (1989): 72-79

Paretzke HG, Berger MJ (1978) Stopping power and energy degradation for electrons in water vapour. In: Proc 6th Symp on Microdosimetry, Booz J, Ebert HG (eds) Harwood Academic Publ, London,p. 749-758

Pihet P, Menzel HG (1989) Atomic data required in accurate measurements of kerma for neutrons with low pressure proportional counters. Loc cit IAEA(1989): 91-105

Ritchie RH, Hamm RN, Turner JE, Wright HA, Ashley JC, Basbas GJ (1989) Physical aspects of charged particle track structure. Nucl Tracks and Radiat Meas 16:141-155

Sanche L (1987) Experimental studies on the interactions of slow electrons with molecules in solid films. In: Radiation Research, Fielden EM, Fowler JF,Hendry JH, Scott D (eds) Vol 2: 260-265

Schou J (1988) Secondary electron emission from solids by electron and proton bombardment. Scanning Microscopy 2: 607-632

Siegbahn K, Nordling C, Johansson, G, Hedman J, et al (1969) ESCA applied to free molecules. North Holland Publ Comp, Amsterdam

Tan KH, Brion CE, Van der Leeuw PE, Van der Wiel MJ (1978) Absolute oscillator strengths (10-60 V) for the photoabsorption, photoionisation, and fragmentation of H_2O. Chem Phys 29: 299-309

Terrisol M (1978) Méthode de simulation du transport d'electrons d'énergies comprises entre 10 eV and 30 keV. Thèse de Doctorat ès-Sciences, No. 839, Toulouse.

Terrisol M, Bordage MC, Caudrelier V, Segur P (1989) Cross sections for 0.025 eV-1 keV electrons and 10 eV-1 keV photons. Loc cit IAEA (1989): 219-233

Van Rijzin J (ed)(1977) Classification and Clustering. Academic Press, New York

Watt DE (1989). An approach towards a unified theory of damage to mammalian cells by ionising radiation for absolute dosimetry. Radiat Prot Dosim 27: 73-84

Wilson WE, Toburen LH (1973) Electron emission in H_2^+ - H_2 collisions from 0.6 to 1.5 MeV. Phys Rev A7: 1535-1544

Zaider M, Brenner DJ, Wilson WE (1983) The application of track calculations to radiobiology. I. Monte Carlo simulation of proton tracks. Radiat Res 95: 231-247

Zaider M, Brenner DJ (1984) On the stochastic treatment of fast chemical reactions. Radiat Res 100: 245-256

Zaider M, Fry JL, Orr DE (1989) A semiempirical tight-binding calculation of the dielectric response function of water. Nucl Tracks and Radiat Meas 16: 159-167

LINKS BETWEEN RADIATION TRACK STRUCTURE, RADIOCHEMICAL SPECIES, AND CELL SURVIVAL

D. J. Brenner and C. R. Geard
Center for Radiological Research,
Columbia University,
630 West 168th Street,
New York, NY 10032, U.S.A.

ABSTRACT

A stochastic track-structure dependent model is presented based on DNA double-strand breaks (DSB) interacting in a time and distance dependent manner, and in competition with DSB repair, to form exchange-type chromosomal aberrations. Cellular survival is assayed on a cell-by-cell basis. A single model is capable of quantitatively explaining all the main features of low and high LET mammalian survival curves, at low and high doses.

INTRODUCTION

It has been clear since the work of Puck (1958) that chromosomal damage is the direct cause of radiation-induced cellular lethality. In addition, chromosomal damage is central in both mutagenesis (e.g. Schwartz et al, 1985) and oncogenesis (e.g. Evans, 1981). In terms of cell killing, we can clearly say that asymmetrical chromosomal aberrations are lethal, and that such aberrations are by far the commonest cause of cellular lethality (Wlodek and Hittelman, 1988, Carrano, 1973). Of course asymmetric aberrations include both exchange-type aberrations and acentric fragments, but there is considerable evidence, especially in terms of the shape of

NATO ASI Series, Vol. H 54
The Early Effects of Radiation on DNA
Edited by E. M. Fielden and P. O'Neill

dose-response curves for the latter (e.g. Revell, 1966) that acentric fragments also result from exchange processes.

The nature of the entities that are interacting to form these chromosomal aberrations is also clear. For example, the close correlation between the post-irradiation time course of chromosomal break removal, and the time-course of DNA double strand break (DSB) removal (e.g. Cornforth and Bedford, 1983, Wlodek and Hittelamn, 1988), makes it clear that the interacting entities are indeed DSB.

A final piece of information is that the yield of initial chromosomal breaks as a function of x-ray dose is approximately linear (Cornforth and Bedford, 1983, Wlodek and Hittelman, 1988). This same result can be derived from DNA elution experiments (Okayasu and Iliakis, 1989, Wlodek and Hittelman, 1988), though such experiments are hard to interpret. Thus DSB are probably not themselves the result of interaction type processes, except perhaps in the sense of a damaged strand causing induced damage in the other strand (e.g. Meselson and Radding, 1975).

Putting these 'pieces of the jigsaw' together, the general picture of radiation induced cellular damage must be:

1. DSB are formed by a subset of single tracks traversing cells.
2. DSB interact with one another in a time- and distance-dependent manner yielding exchange-type chromosomal aberrations (ETCA), thus leading to a reduction with time of DSB, and an increase in time of ETCA; these time- and distance-dependent DSB interactions reflect the fact that chromatin in the interphase nucleus is a dynamic rather than a static entity.
3. Step 2 occurs in competition with DSB repair, which also leads to a reduction with time of DSB.
4. Each ETCA has a certain probability (from symmetry arguments, probably 50%) of causing cell lethality. Some of the remaining aberrations and / or rearrangements will then be responsible for mutagenesis or oncogenesis.

In this paper, a time-dependent simulation is presented in which

* the positions of ionizations produced by various LET particles are calculated by track-structure simulation;
* ionizations are assumed to induce DSB, with a given low probability;
* DSB are either repaired or undergo binary interaction in a time and distance-dependent manner, producing exchange-type chromosomal aberrations (ETCA);
* Each ETCA has a 50% chance of producing lethality.

Our basic approach, integrating track structure and mechanistic information, will be a stochastic one, in which the passage of radiation through individual cells is simulated, followed by the production and interaction of DSB, all on a cell by cell basis. For each cell, the time development of the DSB and ETCA will be followed, followed by a determination of the viability of each individual cell, based on the number of ETCA in that cell at an appropriate time.

The use of a stochastic, cell-by-cell approach allows for a realistic treatment of the relation between cell survival and aberration formation. In general, biophysical models make a link between the average number of lesions, $\bar{\varepsilon}(D)$, produced in cells at a dose D, and cellular survival. This is based on the idea that each increment in the average number of lesions eliminates a proportionate fraction of clonogenic cells, and thus survival, S, at a dose, D, can be written

$$S(D) = \exp[-K\bar{\varepsilon}(D)], \tag{1}$$

where K is a constant.

However, a non-linear relation such as Eq. (1) implies that the response of a particular cell is determined by the average number of lesions in all the exposed cells, rather than the number of lesions in the particular cell itself.

(Note that the number of lesions at a given dose is not Poisson distributed). Formally,

$$S(\bar{\varepsilon}) = \overline{S(\varepsilon)} \qquad (2)$$

holds only if the relation between S and $\bar{\varepsilon}$ is linear.

Such arguments would imply that Eq. (1) is appropriate only for very small numbers of lesions, i.e. at very low doses and low LET's, where

$$\exp[-K\bar{\varepsilon}(D)] \simeq 1 - K\bar{\varepsilon}(D). \qquad (3)$$

Thus it is not possible to extend an approach based on average yields of lesions to high doses or to high LET's (as, respectively, attempted by Zaider and Rossi, 1980, and Kellerer and Rossi, 1972) where, for example, there is saturation - an inefficiently high concentration of radiation damage. In such situations the mean number of lesions or chromosomal aberrations cannot be an adequate predictor of cell survival. Thus, for example, the radon problem, which is basically the result of alpha particles at LET's above 150 keV/μm, cannot be reasonably addressed with approaches based on Eq. (1), but could be addressed with the stochastic approach outlined here.

SIMULATION TECHNIQUES

Here we describe the techniques for 1) the simulation of the passage of various radiations through cells and 2) the simulation of the time and distance-dependent interactions of double-strand breaks to form exchange-type chromosomal aberrations.

We have chosen to model the data of Bird et al (1980) on the inactivation of V-79 cells in late-S phase by 250-kVp x rays and monoenergetic track-segment charged particles (i.e. of a single, constant, LET). The LET's varied from 20 to 170 keV/μm, though higher LET's have also been modelled. This data set was chosen primarily because track-segment experiments offer a well-defined high-LET physical situation, whilst the use of synchronized cells obviously reduces biological variability.

Our technique involves several steps. For a given dose of a given radiation, for each cell analyzed, the following steps are adopted, each being described in more detail below.

1) For each LET, a very long simulated track is generated using a Monte-Carlo track-structure computer code. The simulated track consists of the spatial coordinates of all the ionizations produced by the primary track and the secondary delta rays.
2) For a given dose, the number, ν, of events (i.e. independent track segments traversing the cell) is chosen by random-number techniques from the Poisson distribution.
3) ν independent segments of track are extracted from the previously generated track simulation and superimposed on the cell nucleus, each starting in the same direction, and from a randomly-chosen position on the upper surface of the cell. The nucleus is assumed to be spherical with a diameter of 7 μm, chosen to match the nuclear volumes of V-79 cells derived from the work of Bird et al (1980) and Raju et al (1987).
4) It is assumed that 1 in L of the ionizations produces a DSB at the position of that ionization. In this approach, no mechanisms are postulated for the production of DSBs, which could be as a result of direct and/or indirect action. The value of L was fixed at 1500 (see below), on the basis of some preliminary simulations with 250-kVp x rays, and was not subsequently changed.
5) Simulate stochastically the time-dependent diffusion and distance-dependent interaction of the DSBs to form ETCAs, taking into account the natural repair of DSBs ("sub-lethal damage" repair).
6) Assay each cell for survival, assuming that each ETCA present has a 1 in 2 chance of inactivating that cell - on the general grounds that half of all chromosomal exchange-type aberrations would be expected to be lethal.

7) Repeat the above procedure for an appropriately large number of cells (typically 200 to 1000, depending on the survival level) and thus estimate the overall survival for a given dose of a given type of radiation.

Track Simulation

The computer code used to produce the simulated track segments has been described in detail elsewhere (Brenner, 1988, Zaider et al, 1983). The code uses Monte-Carlo techniques to simulate, on an event-by-event basis, the locations and types of all the energy deposition events produced by the primary particle and by the secondary and higher-order delta rays, whilst travelling through water. Options are available for transport through water vapor or condensed water, the latter being used here. For this calculation, only the position of the ionizing events were recorded. The ions that were simulated here are of sufficient energy that their differential cross sections can be assumed to be those of an equivalent-velocity proton (with an appropriate charge scaling factor).

The 250-kVp x rays were simulated assuming an incident photon spectrum taken from Johns (1961), and calculating a secondary electron distribution using the computer code PHOEL2 (Turner et al, 1980). Electrons with energies chosen from this distribution were then transported using our Monte-Carlo computer code.

Choice of Numbers of Events in Each Cell

The probability that exactly ν independent tracks pass through the nucleus is chosen from the Poisson distribution, where the mean number of tracks passing through the nucleus for a given dose of a given radiation is given by

$$\theta = 5d\, r^2 / 4y_F \qquad (4)$$

where d is the dose in Gy, r is the nuclear radius in micrometers and y_F is the frequency-averaged lineal energy in keV/μm. For the heavy ions of interest here, passing through a nucleus of diameter 7μm, y_F is essentially the same as the LET, whilst for 250-kVp x rays, a measured value of 1 keV/μm was used.

Superposition of Random Track Lengths on Cell Nuclei

Once the number of events, ν, was determined for a given cell, ν random segments of track were superimposed on the cell nucleus. This was accomplished by randomly choosing a primary ionization from within the central region of the previously-generated very long length of track of the appropriate ion. A point was then randomly chosen within the nucleus, and a linear coordinate transformation of the track coordinates made to make the chosen ionization coincident with this point. A similar procedure was adopted for the 250-kVp x rays, except that a secondary-electron energy was first chosen from the photon-produced electron energy spectrum (see above). In addition, if the track was for x rays, it was subjected to a random rotation. All the points on the track were then checked to see whether they were inside the nucleus. This procedure was repeated ν times to generate ν tracks.

Estimation of fraction of ionizations producing DSB

The parameter L is the only one in the model that can truly be said to be adjustable. This is understandable in the sense that the actual mechanisms of DSB production are not addressed here. Thus L will be taken to fit experimental data. The order of magnitude of L can be estimated from results reported by Goodhead et al (1979) for inactivation of asynchronous V-79 cells by very short-ranged 270-eV x rays. For example, a survival level from single-photon events of about 0.3 for a dose of 2 Gy implies a mean number of inactivating events of about 1.2. At this dose, about 10^4 x rays will be deposited in each cell nucleus producing, on average, about 10 ionizations each. A value of **L** of ~ 600 would then imply that about 1.2 in 10^4 of these x rays would produce two DSB in a cell which, being initially formed within a few nanometers of each other, would be likely to produce an exchange-type aberration.

In these calculations, we varied **L** until we obtained agreement with the x-ray survival data of Bird et al (1980), and obtained a value of 1500, which is certainly consistent with the estimate obtained from the soft x ray data - which were for asynchronous V-79 cells, rather than the late-S cells of interest here.

Stochastic simulation of the diffusion and interaction of DSBs

Techniques for simulating stochastically the diffusion, repair, and interaction of species have been described elsewhere (Brenner, 1988, Brenner and Zaider, 1985). The technique is based on the result (derived from Smoluchowski's equation) that the probability, p(r,t), that two species, initially a distance r apart, "interact" by a later time t is

$$p(r,t) = \frac{a}{r} \, \mathrm{erfc} \left(\frac{r - a}{\sqrt{4Dt}} \right) \qquad (5)$$

where **D** is the relative diffusion coefficient of the species and erfc(x) is the complementary error function. The parameter "**a**" is related to the reaction rate between DSBs and is a computationally-efficient way of introducing the reaction rate into the simulation. In the simulation, it is assumed that the two DSBs interact to form a ETCA whenever they reach a separation smaller than "**a**", often termed the "encounter radius".

We will discuss the estimation of the parameters **a** and **D** in the next section; here we describe the procedure, modified from that initially suggested by Clifford et al (1982), for simulating the diffusion and interaction of DSBs to form ETCAs, based on Eq. (5). A schematic flow chart is shown in Fig. 1.

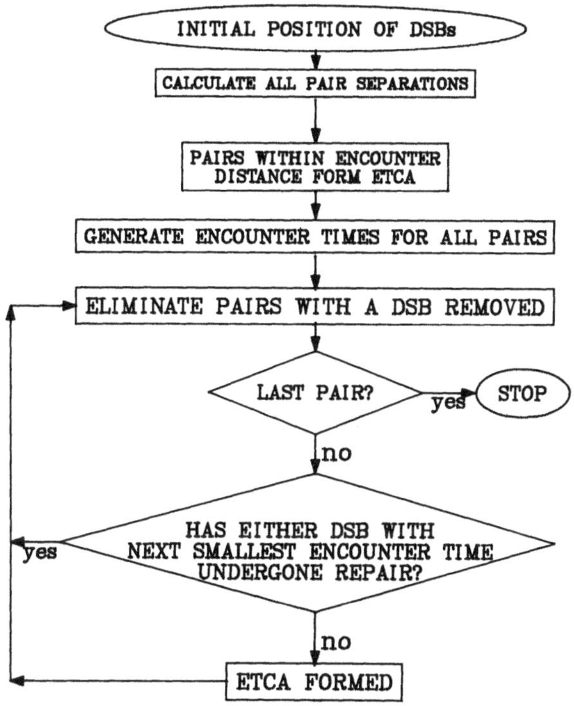

Fig. 1. Flow Diagram for Simulation of DSB Diffusion, Repair, and Interaction to Form ETCA.

The first step is to calculate the initial distances between all pairs of DSBs in the nucleus. All pairs initially closer than the encounter distance are assumed to interact immediately to form ETCAs. For every pair of DSBs that survive these initial interactions, a random time, based on their separation, when this pair <u>might</u> interact is generated from the distribution described by Eq. (5). These times are then sorted, and the pair corresponding to the shortest time becomes the first candidate for ETCA formation.

Before they are assumed to have interacted, however, each of the two chosen DSBs is checked to see whether it has undergone repair. This repair check is performed by assuming that the probability that a DSB is not repaired at time t is described by the exponentially decreasing function $\exp(-t/t_0)$, where t_0 is a characteristic repair time for the cell line. Thus for a potential interaction of two DSBs, two random numbers are uniformly generated on the unit interval. If either is greater than $\exp(-t/t_0)$, where t is the time since that DSB was last checked, then no ETCA is formed, as one or both of the DSBs has undergone repair. The time t_0 was taken as 2000 sec, which is a typical characteristic recovery time for sublethal damage to V-79 cells (Zaider and Brenner, 1984, Freeman et al, 1983, Bedford et al, 1980, Weibezahn and Coquerelle, 1981). It should perhaps be noted here that the estimates of the characteristic recovery time from these references varies from 700 to 4700 sec.

Having analyzed the shortest reaction time in the list of potential reaction times, the list is searched for all reaction times involving either of the DSBs just considered; these are removed from the list. The next shortest potential reaction time is then chosen, its DSBs checked for repair, and so on. For each cell, at each interaction time, the number of DSBs and ETCAs present is tallied, enabling distributions and averages over many cells to be estimated.

Estimation of DSB diffusion constant and encounter radius

Eq. (5) gives the probability that two species, initially a distance r apart, will interact (i.e. come within the encounter radius, a) by time t. A related quantity in the Dual Radiation Action Model (Kellerer and Rossi, 1978) is the so-called gamma function, $\gamma(r)$, which gives the probability that two energy transfers, a distance r apart, will ultimately result in a lesion. This function, $\gamma(r)$, has been evaluated for V-79 cells in late-S phase (Brenner et al, 1987), and is shown in Fig. 2. Thus this evaluated function may be used to estimate the parameters **D**xt and **a** in Eq. (5); here **D** is the relative diffusion coefficient for the DSB, **t** is some appropriately long time when all time development has ceased, and **a** is the "encounter radius" for DSB interaction.

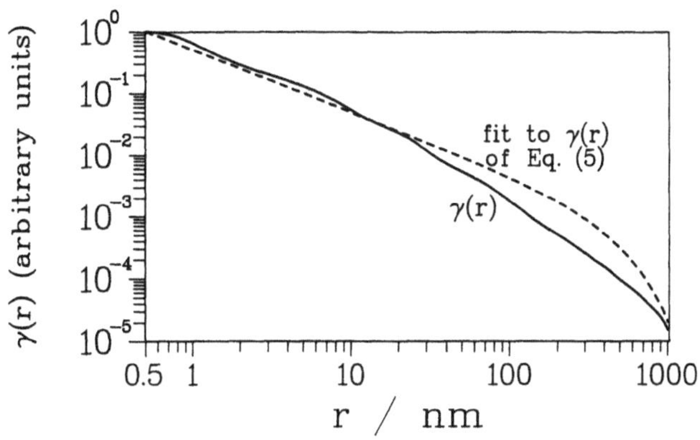

Fig. 2. Interaction Function $\gamma(r)$ for V-79 Cells in Late S, and Fit to Eq. (5). Curves are Normalized at 0.5 nm.

A reasonable fit of $\gamma(r)$ to Eq. (5) was obtained with a=0.5 nm and Dxt=12 x 10^4 nm^2. We have taken t=10^4 sec (i.e. 5 times the characteristic repair time) as a time when all ETCA formation has ended, and thus estimated D as 12 nm^2/sec. The relatively large value of the diffusion constant is, of course, a reflection of the dynamic state of chromatin in the interphase nucleus.

Survival Assay

A time of 10^4 seconds (5 times the characteristic repair time t_0) was chosen as an appropriate time to assay for cell viability, on the grounds that all ETCA formation will have ceased. For each cell, each ETCA is assumed to have a 1 in 2 chance of causing cell lethality (see above). Each ETCA is considered in turn and a random number, R, generated uniformly on the unit interval. If R is less than ½, then the cell is considered inactivated, and analysis of the next cell initiated. If R is greater than ½, this ETCA is assumed not to have inactivated the cell, and the next ETCA in the cell (if any) is considered. If all the ETCAs in that cell do not inactivate it (or there are no ETCAs in the cell), then it is tallied as a survivor.

To save computer time, each cell, containing its given number of ETCAs, is independently analyzed for survival ten times, in the above manner.

RESULTS and CONCLUSIONS

Simulations were performed for various doses of 250-kVp x rays, and LET's of 20, 40, 90, 127 and 170 keV/μm. In addition, though there are no data for direct comparison, calculations were performed at 500 keV/μm. Fig. 3 shows the time development of the mean numbers of DSB and ETCA for 5-Gy doses of 250-kVp x rays, 20 keV/μm and 127 keV/μm radiations. Fig. 4 shows results for a given LET, 20 keV/μm, for various doses.

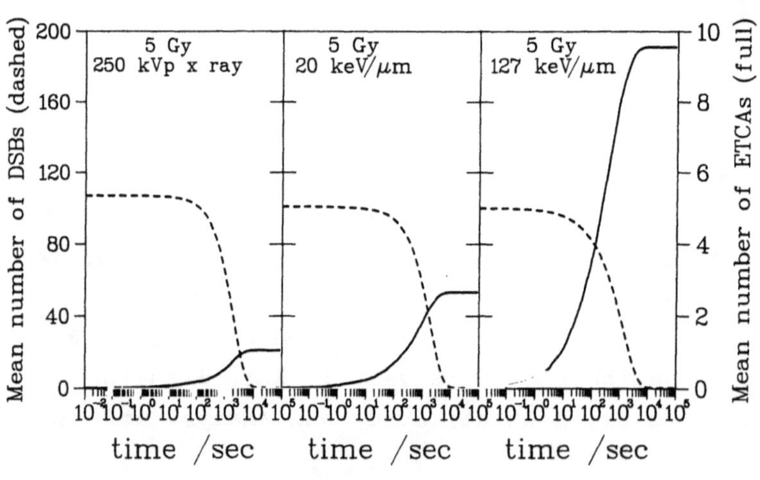

Fig. 3. Time-dependence of DSB Decay and ETCA Formation for 5 Gy of Radiations of Various LET's.

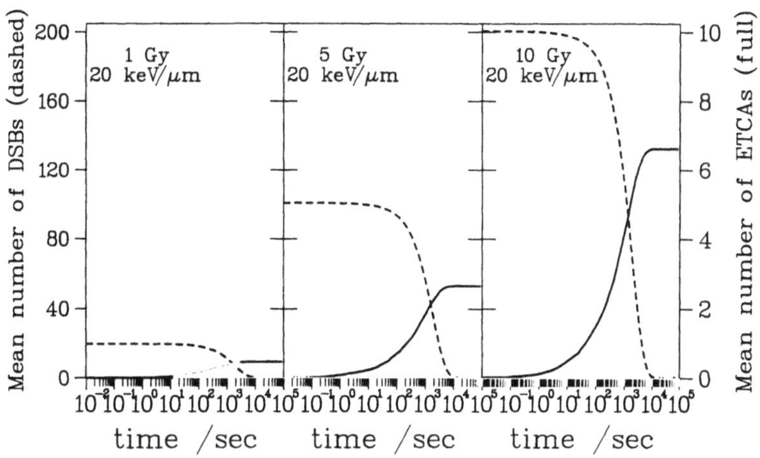

Fig. 4. Time-dependence of DSB Decay and ETCA Formation for Various Doses of 20 keV/μm Radiation.

As might be expected, the time development of ETCA formation depends quite strongly on both LET and dose. As a function of LET, repair of DSB has less time to occur at high LET, because the DSB are formed closer, and so interact with each other more rapidly. This results in a higher final yield of ETCA at high LET even when, for the same dose, there will be about the same initial yield of DSB. Thus 5 Gy of x rays produce an average of about 1 ETCA whereas the same dose of 127 keV/μm radiation produces an average of nearly 10 ETCAs - despite the fact that both radiations initially produce almost the same number of DSB.

It may also be noted that 10 Gy of 20 keV/μm radiation produces (see below) about the same level of survival as 5 Gy of 127 keV/μm radiation, despite the mean number of ETCAs (6.8 vs 9.5) being rather different. This is due to the phenomenon of saturation, and illustrates the point made in the Introduction about the pitfalls of relating mean numbers of ETCAs to survival.

Survival results are shown in Fig. 5 and compared with the measured results of Bird et al (1980). The techniques used here do not yield an analytic functional form for the shape of the survival curve. Each survival point is calculated independently (in an analogous manner to the experiment protocol) and is subject to a variance inherent in a stochastic approach. Thus the calculated results in Fig. 5 consist only of points, which are joined by straight lines, solely to guide the eye. In all cases, sufficient cells (in the range 200 to 1000) were analyzed such that the standard deviation of the estimated survival was less than 10%.

Fig. 5 also shows calculated data at 500 keV/μm. The effectiveness of 500 keV/μm radiation is comparable to that of 40 keV/μm radiation, in agreement with the data of Thacker et al (1979) for inactivation of unsynchronized V-79 cells.

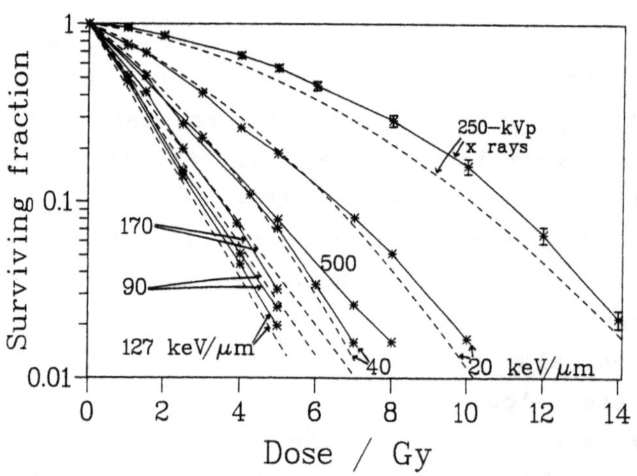

Fig. 5. Calculated (Points Connected by Straight Lines) and Measured (Dashed - Smooth Curves Through Data of Bird et al, 1980) Survival as a Function of Dose and LET for V-79 Cells in Late S Phase.

Thus, it appears that the phenomena of low-LET curvilinear survival, medium-LET exponential survival curves, and saturation at high LET, can all be modelled within one consistent framework. The model reflects the basic notions of dynamic DSB interactions in the interphase nucleus competing with DSB repair to generate chromosomal aberrations.

These results, therefore, support the notion that the mechanisms of radiation action are the <u>same</u> for low, high and very high LET, the differences in response being due only to differing energy deposition patterns. Thus, a knowledge of the response of a system to low LET, together with a knowledge of the track structure of a high-LET radiation, should suffice to predict the response of the system to the high-LET radiation.

REFERENCES

Bedford JS, JB Mitchell, MH Fox (1980) Variations in responses of several cell lines to low dose-rate irradiation In Radiation Biology in Cancer Research (RE Meyn, HR Withers, Eds) pp 251-262 Raven Press, New York, 1980

Bird RP, N Rohrig, RD Colvett, CR Geard, SA Marino (1980) Inactivation of Synchronized Chinese Hamster V79 Cells with Charged-Particle Track Segments Radiat Res **82**:277-289

Brenner DJ (1988) Stochastic Calculations of the Fast Decay of the Hydrated Electron in the Presence of Scavengers -- Tests of Model Consistency Rad Phys Chem **32**:157-162

Brenner DJ, RP Bird, M Zaider, P Goldhagen, PJ Kliauga, HH Rossi (1987) Inactivation of Synchronized Mammalian Cells with Low-Energy X Rays - Results, Significance Radiat Res **110**:413-427

Brenner DJ, M Zaider (1985) Stochastic and Deterministic Treatments of the Time Decay of Species Created by Heavy-Charged Particle Interactions Radiat Prot Dosimetry **13**:127-131

Carrano AV (1973) Chromosome Aberrations and Radiation-Induced Cell Death II Predicted and Observed Cell Survival Mut Res **17**:355-366

Clifford P, NJB Green, MJ Pilling (1982) Monte Carlo Simulation of Diffusion, Reaction in Radiation-induced Spurs Comparison with Analytic Models J Phys Chem **86**:1322-1327

Cornforth MN, JS Bedford (1983) X-Ray Induced Breakage and Rejoining of Human Interphase Chromosomes Science **222**:1141-1143

Evans HJ (1981) Genes, Chromosomes and Neoplasia: An Overview In Genes Chromosomes and Neoplasia (F Arrighi, PN Rao, E Stubblefield, Eds) p 511 Raven Press, New York

Freeman ML, E Sierra, EJ Hall (1983) The repair of sublethal damage in diploid human fibroblasts: A comparison between human and rodent cell lines Radiat Res **95**:382-391

Goodhead DT, J Thacker, R Cox (1979) Effectiveness of 03 keV carbon ultrasoft X-rays for the inactivation and mutation of cultured mammalian cells Int J Radiat Biol **36**:101-114

Johns HE (1961) The Physics of Radiology: 2nd ed, p263, Charles C Thomas, Springfield, IL

Kellerer AM, HH Rossi (1978) A generalized Formulation of Dual Radiation Action Radiat Res **75**:471-488

Meselson MS, CM Radding (1975) A general model for genetic recombination Proc Nat Acad Sci USA **72**:358-361

Okayasu R, G Iliakis (1989) Linear DNA elution Dose Response Obtained in CHO Cells with Non-Unwinding Filter Elution After Appropriate Selection of the Lysis Conditions Int J Radiat Biol 55:569-582

Puck TT (1958) Action of Radiation on Mammalian Cells III Relationship Between Reproductive Death and Induction of Chromosome Anomalies by X-irradiation of Euploid Human Cells in Vitro Proc Nat Acad Sci **44**:772-780

Raju MR, SG Carpenter, JJ Chmielewski, ME Schillaci, ME Wilder, JP Freyer, NF Johnson, PL Schor, RJ Sebring, DT Goodhead (1987) Radiobiology of Ultrasoft X-rays I Cultured Hamster Cells V79, Radiat Res **110**:396-412

Revell SH (1974) Evidence for a Dose-squared Term in the Dose-Response Curve for Real Chromatid Discontinuities Induced by X-Rays and Some Theoretical Consequences Thereof Mut Res **3**:34-53

Schwartz JL, WF Morgan, P Brown-Lindquist, V Afzal, RR Weichselbaum, S Wolff (1985) Comutagenic Effects of 3 Aminobenzamide in Chinese Hamster Ovary Cells Cancer Res **45**:1556-1559

Thacker J, A Stretch, MA Stephens (1979) Mutation and inactivation of cultured mammalian cells exposed to beams of accelerated heavy ions II Chinese hamster V79 cells Int J Radiat Biol **36**:137-148

Turner JE, RN Hamm, HA Wright, JT Modolo, GMAA Sardi (1980) Monte Carlo calculations of initial energies of Compton electrons and photoelectrons in water irradiated by photons with energies up to 2 MeV Health Phys **39**:49-55

Weibezahn KF, T Coquerelle (1981) Radiation Induced DNA Double Strand Breaks are Rejoined by Ligation and Recombination Processes Nucleic Acid Res **9**:3139-3150

Wlodek D, WN Hittelman (1988) The Relationship of DNA and Chromosome Damage to Survival of Synchronized X-Irradiated L5178Y Cell (Parts I and II) Radiat Res **115**:550-575

Zaider M, DJ Brenner (1984) Comments on "V79 Survival following simultaneous or sequential irradiation by 15-MeV neutrons and Co photons" by Higgins et al [Radiat Res 95, 45-56 (1983)] Radiat Res **99**:438-441

Zaider M, DJ Brenner, WE Wilson (1983) The application of track calculations to radiobiology--I Monte Carlo simulation of proton tracks Radiat Res **95**:231-247

Zaider M, HH Rossi (1980) Saturation Effects for Sparsely Ionizing Particles In: Radiation Physics, Biophysics, Radiation Biology, Radiological Research Laboratory DOE Annual Report, Columbia University, pp 126-134

A KINETIC INVESTIGATION OF THE MECHANISMS OF RADIATION-INDUCED STRAND BREAKAGE IN DNA MODEL SYSTEMS

George D.D. Jones* & Peter O'Neill
Division of Molecular Processes,
MRC Radiobiology Unit,
Chilton,
DIDCOT,
Oxon
OX11 0RD
United Kingdom

Time-resolved reductions in the light scattering intensity (LSI) of N_2O or oxygenated aqueous solutions of single stranded homologous polynucleotides, following pulse irradiation, have been studied as indicies of strand breakage (s.b.) (for a review of the LSI method see Schnabel 1979). With doses of 3-24Gy/pulse, a number of kinetically distinct modes of breakage are apparent. For the polypyrimidines, poly C and poly U, these include i) a small proportion of fast strand breaks ($\tau^1/_2$ = 50μs) arising from the direct action of OH-radicals at the sugar phosphate backbone, and ii) major, slower first order routes to breakage that proceed through base radical mediated pathways (Figure 1) (Lemaire et al.

Figure 1

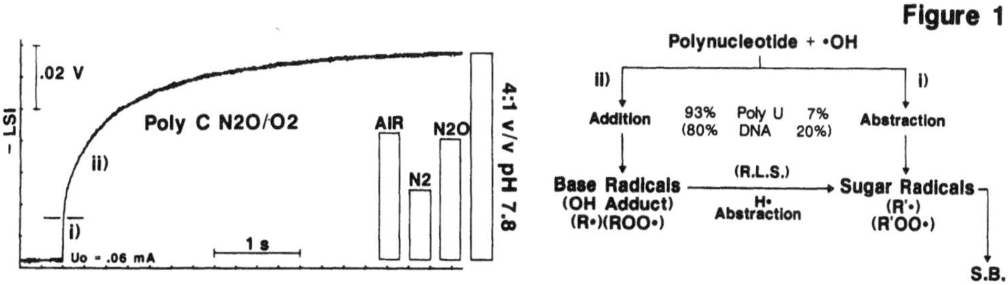

1984). The two routes to s.b. yield non-equivalent OER ssb-values with the initial, fast breaks exhibiting values of 3->4, whilst the major route to s.b. reflects an OER of 1.5-2. In the absence of O_2, the rate of major s.b. generally increases (~5-fold for poly C and ~2-fold with poly U) and mirrors the decay of the principle transient optical absorptions assigned to the major pyrimidine 6-yl radical species (Figure 2). Irradiation of poly U under anoxia at pH4 results in a further 2-fold increase in the rate of major s.b., which is again reflected in the optical studies, plus an additional minor, fast strand breaking component. Thus, the rate-limiting step of major s.b. for the polypyrimidines is the native base radical's abstraction of a hydrogen atom from an adjacent sugar moiety (cf Hildenbrand & Schulte-Frohlinde 1989). In the presence of oxygen it has been shown that a base peroxyl rad-

NATO ASI Series, Vol. H 54
The Early Effects of Radiation on DNA
Edited by E. M. Fielden and P. O'Neill

ical mediates the abstraction and that this step is also rate limiting (Jones & O'Neill 1990; Bothe et al. 1986). For the purine polynucleotide, poly A, in addition to the immediate, fast losses of LSI, only minor slower losses of scattering intensity are subsequently observed. From this it is concluded that the yield of base radicals contributing to s.b. is much less than for poly C or poly U (cf Önal et al. 1988). Also an estimate of the OER for total breakage of ≥4 confirms that the principle route to s.b. is by direct attack of OH-radicals at the backbone.

The reactions involved in the base mediated route to s.b. generates a modified base in the vicinity of the break (Figure 3). Such

Figure 2

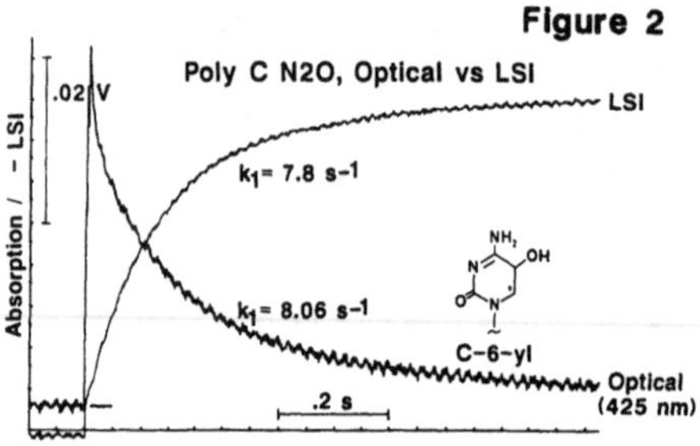

Figure 3

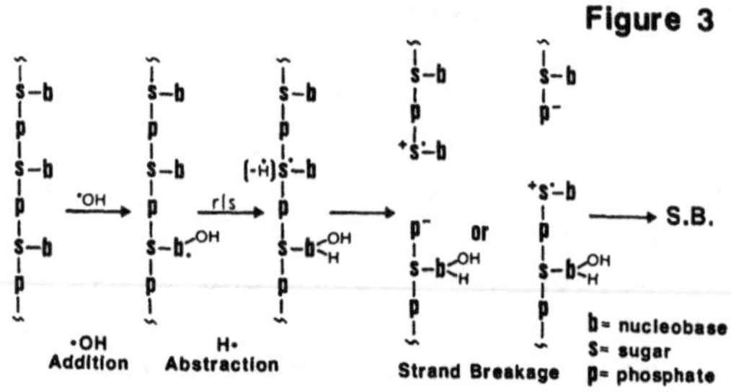

amplification of damage, following a single "hit", should possibly be taken into account when considering the fate of radical clusters produced within small volumes as a result of primary energy deposition events (e.g. spurs, blobs and short tracks) within the cellular DNA macromolecule (ie. LMDS entities as proposed by Ward 1988). Preliminary kinetic and steady state studies with simple double stranded systems (poly A: poly U pH7.8; poly C pH4) suggest that base radicals are still able to transfer some of the damaging spin to the backbone in the duplex environment.

Bothe E, Behrens G, Böhm E, Sethuram B, Schulte-Frohlinde D (1986) Hydroxyl radical-induced strand break formation of poly(U) in the presence of oxygen: a comparison of the rates as determined by conductivity, e.s.r. and rapid-mix experiments with a thiol. Int J Radiat Biol 49:57-66
Hildenbrand K, Schulte-Frohlinde D (1989) E.s.r. studies on the mechanism of the hydroxyl radical-induced strand breakage of polyuridylic acid. Int J Radiat Biol 55:725-738
Jones GDD, O'Neill P (1990) Kinetics of radiation-induced strand breakage in polynucleotides in the presence of oxygen: a time-resolved light-scattering study. Int J Radiat Biol 57:1123-1139
Lemaire DGE, Bothe E, Schulte-Frohlinde D (1984) Yields of radiation-induced main chain scission of poly U in aqueous solution: strand break formation via base radicals. Int J Radiat Biol 45:351-358
Önal AM, Lemaire DGE, Bothe E, Schulte-Frohlinde D (1988) γ-Radiolysis of poly(A) in aqueous solution: efficiency of strand break formation by the primary water radicals. Int J Radiat Biol 53:87-796
Schnabel W (1979) Application of the light scattering detection method to problems of polymer degradation. In: Grassie N (ed) Developments in polymer degradation. Applied Science Publishers, London, p 35
Ward JF (1988) DNA damage produced by ionizing radiation in mammalian cells: identities, mechanisms of formation, and repairability. Prog Nucl Acid Res Mol Biol 35:95-125

DISCUSSION ON INTERDISCIPLINARY APPROACH TO RADIATION ACTION

P. O´Neill
MRC Radiobiology Unit
Chilton
Didcot OX11 ORD
U.K.

This group of lectures, presented in the opening session chaired by Dr E.M. Fielden, set the tone of the meeting through the introduction of an interdisciplinary approach to the field covering the three major disciplines, namely physics, chemistry and biology. The content of the lectures covered a broad area of interests ranging from mechanisms of radiation action on DNA and their relevance to cellular DNA (Ward), development of computer codes for biophysical models of radiation action (Paretzke) through to links between track structure, chemistry and cell survival (Bremner). Overall, the speakers addressed the questions relating to the initial energy deposition events in terms of identifying the important reactive chemical species which contribute to the biologically relevant DNA damage.

The timescale covered by the content of these lectures extends over at least 20 orders of magnitude. The initial effects of radiation action occur with the deposition of energy whereby it was suggested (Paretzke) that these events may be considered as a fingerprint of the ultimate damage. A major point of discussion focussed on the extrapolation of the track codes for water vapour to the condensed phase (i.e. unit density). How significant is such an extrapolation in the development of future models?

Through the next generation of particle track codes, now in their infancy, it was hoped to be able to simulate the subsequent interactions of hydroxyl radicals with heterogeneous targets in order to bridge the gap from the physics to the chemical stages of radiation action. In fact, a water layer of three molecules thick around the DNA was identified as the critical aqueous region associated with DNA

NATO ASI Series, Vol. H 54
The Early Effects of Radiation on DNA
Edited by E. M. Fielden and P. O'Neill

when considering diffusible water radicals (Ward). Several speakers during the meeting also addressed the possibility of radiation-induced energy migration within DNA whereby the ultimate damage is localised at sites distant from that where the initial energy is deposited. Furthermore, many of the present biophysical models are developed to simulate differences in radiation quality but do not as yet take into account differences in radiosensitivity due to dose modifying agents.

The formation of locally multiply damaged sites (Ward) is often discussed as critical lesion. A short presentation by Dr Jones highlighted the possibility of amplification of DNA damage through a base radical mediated formation of a strand breakage in DNA. Further, the question was raised (Paretzke) as to whether the oxygen enhancement ratio is due to the physics or the ensuing chemistry. The importance of these areas of overlap between physics and chemistry was noted as important for future developments of more sophisticated models of radiation action.

A major discussion theme which was addressed by the speakers and was continued throughout the meeting, concerns the confusion that often arises from the use of the terms of "indirect" and "direct" effect when discussing the mechanistic aspects of radiation action. In the case of DNA as the target, the physicists generally relate "direct" effects to energy deposition within the DNA envelopewhich include at least some of the water of hydration, whereas chemists generally consider "direct" effect only in terms of energy deposition with the DNA molecule itself. The confusion arises from defining the effect for the role of water molecules associated with or in the vicinity of the DNA. From the novel model presented by Paretzke the ratio of direct to free water radical effects for the formation of double strand breaks (dsb) was calculated to be 65%:35%. However, an assumption used to arrive at this ratio highlighted the problem since a dsb arising from one ssb due to energy deposition within the DNA (direct) and one ssb arising from a water radical within the same track would have been scored as a dsb arising through the direct effect. It was proposed from the role of scavengers (e.g. DMSO) which protect against radiation-induced strand breakage that the terms "scavengeable" and "non-scavengeable" damage (see Ward) could be used although even these terms were recognised to have limitations. On this basis, the experimentally-determined ratio of non-scavengeable to scavengeable damage in aerated cellular systems is of the order 35%:65%. It is considered that double strand breaks arising from different mechanistic pathways (e.g. two ssb arising from energy deposition in DNA together with associated ·OH-induced damage) may fall into the category of scavengeable damage in contrast to the "direct effect" classification as discussed above.

A further confusion with the use of "direct" effect arises from the fact that, in some cases, identical DNA damages could arise from either energy deposition within the DNA (e.g. ionisation) or interaction of water radicals with DNA (see Fielden and Schulte-Frohlinde). However, the use of the terms "scavengeable" and "non-scavengeable" damage, based upon phenomenology, were considered to be, at least, more instructive than the use of "direct" and "indirect" damage.

A stimulating discussion centred around the presentation of Bremner on the basis for double strand break interaction in a time and distance dependent manner leading to exchange type chromosomal aberrations. The importance of the distance over which the dsb interact was questioned as well as the chemistry/biochemistry of such interactions. It was stressed that the formation of exchange type chromosomal aberrations from dsb is in competition with their repair. Since dsb repair kinetics are complicated (e.g. bi- or triphasic), the assumption on repair kinetics was questioned as to whether the appropriate function had been used. Consideration of the difference of repair kinetics of dsb produced by radiations of different LET may lead to further enhancement of the model.

A short presentation by Milligan highlighted the constancy in the ratio of the yields of dsb:ssb upon irradiation of plasmid DNA in the presence of increasing concentration of radical scavengers. This constancy of dsb:ssb yield measured in a variety of systems remains intriguing.

In summary, this session opened the meeting in fine style, paving the way for many of the topics raised to be discussed during other sessions. From the discussion, it was apparent that the approaches used by the different disciplines was appreciated by the meeting and initiated dialogue and exchanges of ideas which could only be addressed through interdisciplinary approaches. Progress was made in trying to establish a common dialogue between the various disciplines as was exemplified in the discussions of definitions of radiation effect.

MEASUREMENT OF DNA DOUBLE STRAND BREAKS IN MAMMALIAN CELLS: COMPARISON BETWEEN PULSED FIELD GEL ELECTROPHORESIS AND NON-UNWINDING FILTER ELUTION

George Iliakis, Lorenz Metzger, Ryuichi Okayasu, Gabriel Pantelias and Orlando Cicilioni

Thomas Jefferson University
Department of Radiation Oncology and Nuclear Medicine
Laboratory of Experimental Radiation Oncology
Philadelphia, PA 19107
USA

Abstract

DNA damage induction dose response curves were measured in synchronous and plateau-phase CHO cells following exposure to sparsely ionizing radiation using pulsed-field gel electrophoresis and non-unwinding filter elution; assays that have been shown to respond specifically to the presence of double strand (dsb) breaks in the DNA. The fraction of DNA-associated activity released (FAR) from the agarose plug, where cells were embedded, and the fraction of activity eluted (FAE) from the filter were measured and used as parameters for the quantitation of the DNA damage present after pulsed-field gel electrophoresis and non-unwinding filter elution, respectively. Dose response curves for FAR and FAE were, as expected in general sigmoid, but under certain lysis conditions, the data obtained,in the low dose region could be fitted by straight lines. FAR and FAE per Gy fluctuated throughout the cell cycle in a quantitatively similar manner suggesting that both techniques respond to similar cell cycle associated effects. High values were obtained for these parameters in cells irradiated in G_1 or G_2+M; the values decreased as cells entered S, and reached a minimum in the middle of this phase. Calibration of pulsed field gel electrophoresis by means of ^{125}I-decay demonstrated that the fluctuations in the FAR per unit dose do not reflect variations in the production of DNA dsb. It is, therefore, proposed that the observed phenomena reflect alterations in the physicochemical properties of the DNA induced by the inception of replication, and caused by the presence of forks and eyes in the molecules. In both assays, the resulting DNA damage induction dose response curves could be altered by modifying the lysis conditions employed. In general, the sensitivity of the assay increased when NLS (sodium-N-laurylsarcosine) was substituted for SDS (sodium dodecyl sulphate) and when high temperatures (50-60°C) were used during lysis. These findings indicate that correlations established between the magnitude of the generated signal, by either technique, and DNA dsb, hold only for the particular cell population used and the specific lysis conditions employed. Extrapolation of such correlations for the interpretation of results obtained with different cell populations treated under different conditions may lead to erroneous conclusions.

NATO ASI Series, Vol. H 54
The Early Effects of Radiation on DNA
Edited by E. M. Fielden and P. O'Neill

ntroduction

There is compelling evidence suggesting that the target to radiation-induced cell killing is the DNA (Painter 1980) and that the lethal lesions are DNA double strand breaks. For example, a direct correlation between unrepaired DNA dsb and cell killing was established in DNA dsb repair deficient mutants of yeast (Resnick and Martin 1976, Frankenberg and Frankenberg-Schwager 1981, Frankenberg et al 1981), and radiosensitive mutants of mammalian cells have been isolated that are partially defective in their ability to repair this lesion (Jeggo and Kemp 1983, Kemp et al 1984, Giaccia et al 1985, Wlodek and Hittelman 1987, Iliakis and Okayasu 1990, Iliakis and Pantelias 1990). The importance of DNA dsb in cell killing was also indicated by the observation that production of DNA dsb by restriction endonucleases leads to the induction of chromosome aberrations (Bryant 1984, 1985, Natarajan and Obe 1984), that are known to directly correlate with cell death (Sasaki and Norman 1967, Dewey et al 1970, 1971, Grote and Revell 1972, Carrano 1973, Lloyd et al 1975, Joshi et al 1982).

Despite the potential importance of DNA dsb in radiation induced cell killing, the standard centrifugation methods used in the past for their detection in mammalian cells have either been insensitive (Lehmann and Stevens 1977, Matsudaira et al 1977), or very laborious (Blöcher 1982). More recently, a DNA filter elution assay, originally introduced to measure induction by radiation and other cytotoxic agents of total DNA breaks (Kohn and Ewig 1973), has been adapted, by pH adjustment, to detect DNA dsb at low doses of radiation (Bradley and Kohn 1979). In this assay, the increase in the fraction of DNA eluted from a filter in a certain time interval, in irradiated versus non-irradiated cells (e.g. Bradley and Kohn 1979, Woods 1981, Van der Shans et al 1983, Radford 1985, Radford and Hodgson 1985, Okayasu et al 1988), or the logarithm of the retention ratio of control to irradiated DNA measured at a specific fraction (Peak et al 1988), is assumed to reflect the amount of dsb present in the DNA.

However, the hypothesis of a general direct correlation between the signal measured in the elution assay and the induction of DNA dsb (Radford 1985, Radford and Broadhurst 1988, Sweigert et al 1988) has been challenged (see Hutchinson 1989 for a recent outline), and evidence was presented that this correlation may change with cell cycle phase and lysis conditions (Okayasu et al 1988). Thus, in a series of radiation experiments carried out with synchronized CHO cells (Okayasu et al 1988), a large decrease was observed in relative elution during S, that did not correlate with cell killing, and which was paralleled by alterations in the sedimentation properties of the DNA in neutral sucrose gradients. It was proposed that these phenomena derive from alterations in the physicochemical properties of the DNA, presumably caused by DNA replication-associated alterations in DNA structure and chromatin conformation. Furthermore, a set of experiments designed to test the effect of various lysis conditions on DNA filter elution (Okayasu and Iliakis 1989) indicated that simple substitution of the detergent used can cause large alterations in the fraction of DNA eluted from irradiated cells and thus affect the shape of the resulting dose response curve.

The above outlined phenomena and the associated uncertainties thus introduced in the measurement of DNA dsb (Okayasu et al 1988, Okayasu and Iliakis 1989) prompted a search for alternative assays in order to test whether the phenomena established using non-unwinding filter elution derive from fundamental alterations in the properties of the DNA molecule, and can, therefore, be observed using different techniques, or are inherent to the elution assay. Pulsed field gel electrophoresis (PFGE), originally devised for the separation of DNA molecules up to several million base pairs (Schwarz and Cantor 1984, Carle and Olson 1984, Carle et al 1986, Chu et al 1986, Lalande et al 1987), has recently been shown to be a powerful assay for the measurement of induction and repair of DNA double strand breaks (dsb) in mammalian cells. Among the variations of pulsed field gel electrophoresis developed (for reviews see Lai et al 1989, Anand 1986) the clamped homogeneous electric field (CHEF) gel electrophoresis (Blöcher 1990), the transverse alternating field electrophoresis (TAFE) (Stewart et al 1988, Agar and Dewey 1990) and the asymmetric field inversion gel electrophoresis (AFIGE) (Denko et al 1989, Stamato and Denko 1990, Iliakis et al 1990a, 1990b) have been successfully employed for the measurement of dsb in the DNA of mammalian cells.

Here, we present experiments comparing DNA damage induction dose response curves obtained under various lysis conditions with cells in different phases of the cell cycle using AFIGE and non-unwinding filter elution, and show that the degree of these alterations has a similar magnitude for both techniques. Calibration using ^{125}I decay (Radford and Hodgson 1985) of AFIGE further showed that induction of DNA dsb is similar in the various phases of the cell cycle, suggesting that the variations in response observed throughout the cycle are unlikely to derive from alterations in the induction of DNA dsb.

Materials and Methods

Cell culture conditions and irradiation

Chinese hamster ovary (CHO) cells, strain 10B, grown as monolayers on plastic in McCoy's 5A medium supplemented with 5% fetal bovine serum and antibiotics were used. All incubations were at 37°C in a humidified incubator, in an atmosphere of 5% CO_2 and 95% air. Plateau-phase cells were obtained by growing 3×10^5 cells in 100 mm dishes (9 ml medium) for 4 days, without medium change. Filtered supernatant (conditioned medium, C-med), acquired from plateau-phase cultures was used in some experiments to maintain cells in G_1-phase. Cells used in experiments were labeled with 0.74 kBq/ml ^{14}C-Tdr plus 5 μM Tdr, added either at the time of culture preparation (exponentially growing cells) or two days later (plateau-phase cells).

Cells were irradiated on ice using a Siemens therapeutic X-ray machine operated at 250 kV, 15 mA with a 2 mm Al filter, effective photon energy about 70 keV. Irradiation was carried out at 25 and 50 cm, and at dose rates of 8.4 or 2.0 Gy/min, respectively. Dosimetry was performed with a Victoreen dosimeter.

The methods used to prepare synchronized cell populations in various phases of the cell cycle have been described previously (Iliakis et al 1987, Okayasu et al 1988). G_1 cells were produced by growing cultures to plateau-phase and were used after 3h incubation in fresh growth medium. Cultures in other phases of the cell cycle were prepared by trypsinizing plateau-phase cultures and incubating the cells for 10 h in fresh growth medium supplemented with 1 μg/ml aphidicolin. Under these conditions, cells progress through the cycle and accumulate at the G_1/S border. Cells in early S, mid S, late S and G_2+M were harvested at various times after release of the aphidicolin block. The appropriate times were determined by parallel flow cytometry. Cells were stained for flow cytometry by direct suspension (about $5x10^5$ cells) in a solution (1 ml) containing 0.1 M Tris, 0.1M NaCl, 5 mM $MgCl_2$, 0.05% Triton X-100 and 2 μg/ml DAPI. For measurement, an arc lamp flow cytometer was used (Partec, PAS II) interfaced to a computer and equipped with the necessary software to calculate the fraction of cells in G_1, S, and G_2+M-phase. Histograms showing the distribution of cells at the stages of the cell cycle used in this work have been published (Okayasu et al 1988).

A protocol previously developed for the incorporation of ^{125}IdU using synchronized cell populations (Iliakis et al 1987) was modified as needed and was employed to generate cells that had incorporated ^{125}I in the form of ^{125}IdU in various phases of the cell cycle (Iliakis et al 1990b). ^{125}IdU was administered at a concentration of 69 kBq/ml (0.187 μCi/ml) in the presence of 2.5 μM thymidine. ^{125}I incorporation in cells was measured immediately after collection in a scintillation counter calibrated using known activity standards as described previously (Iliakis et al 1987). Cells collected in various phases of the cell cycle were either frozen to accumulate dsb in the DNA, or were embedded in agarose plugs, lysed as described in the next section; and allowed to accumulate DNA damage at 4°C. The average number of decays per cells accumulated in a certain period of time was calculated from the initial number of decays per minute and the half life of ^{125}I (60.14 days).

Non-unwinding filter elution

The non-unwinding DNA filter elution technique developed by Bradley and Kohn (1979) was used. Modifications were made in the type and the concentration of the detergents used, as described in the corresponding sections in the next chapter. Proteinase K at 0.2 mg/ml and glycine was always included in the lysis solution. In some experiments cells were incubated during lysis in an oven at 60°C for 1h. During this time, samples were covered with aluminium foil to prevent evaporation. The fraction of DNA-associated activity eluted from the filter in 17 h (flow rate 1.5 ml/h) was used as a parameter to quantify the amount of DNA damage induced, and this value was plotted against the radiation dose throughout this paper (Radford and Hodgson 1985). The filter was loaded with 5 x 10^5 cells, and this number was kept constant in all experiments. The results obtained were plotted as fraction of activity eluted from the filter (after subtraction of the fraction of DNA eluted in non-irradiated controls; termed background) versus absorbed radiation dose.

Pulsed Field Gel Electrophoresis

The field inversion gel electrophoresis technique developed by Stamato and Denko (1989) was used, with minor modifications. The methodology can be divided into a) the preparation of the DNA in the plugs and b) the actual electrophoresis. These steps are described here briefly. Cells were trypsinized, washed and resuspended in serum-free medium. Subsequently, they were mixed at 37°C with an agarose solution prepared in serum-free McCoy's 5A medium (1% final concentration) (InCert agarose, FMC), pipetted into 3 mm glass tubes, cooled down in ice until solidification and cut into 5 mm pieces (10^5 cells per plug) in a cold room. In the experiments described here, irradiation was carried out at this stage on ice. Subsequently, plugs were incubated overnight at 50°C for lysis (unless stated otherwise) in a solution containing 0.5 M EDTA, 0.01 M TRIS, 2% NLS (unless stated otherwise) and 0.1 mg/ml proteinase K at pH 8.0. After lysis, plugs were washed for 1 h in 0.1 M EDTA, 0.01 M TRIS, pH 8.0 and treated for 1 h at 37°C with 0.1 mg/ml ribonuclease A (boiled for 10 min to destroy DNase contamination). Electrophoresis was carried out at 10°C in half strength TBE buffer (45 mM TRIS, 45 mM boric acid, 1.5 mM EDTA, pH 8.2) in BRL (H4) horizontal boxes, in 0.5% agarose (SeaKem GTG, FMC) gels (25 cm length) cast in the presence of 0.5 μg/ml ethidium bromide. Asymmetric field inversion gel electrophoresis (AFIGE) was performed by applying cycles of 1.25 V/cm for 900s in the direction of net DNA migration, and of 5 V/cm for 75s in reverse direction. The electrical conditions were adjusted with the help of an IBI (Minipulse) switching unit and a custom made resistor box connected as previously described (Denko et al 1989). Electrophoresis was carried out for 40 h. The fraction of activity released from the plug (FAR) was taken as a measure for the amount of DNA damage induced and was determined by counting (in a scintillation counter, Packard) separately the activity in the plug and in the lane.

Results

Effect of lysis temperature and detergent on DNA damage induction dose-response-curves

The effects of lysis temperature and detergent on the shape of the DNA elution dose response curves were examined in plateau phase CHO cells. In Figure 1 elution results are shown obtained with cells lysed in 0.002% SDS or 2% NLS, either at room temperature (RT) or at 60°C. NLS was found to result in steeper DNA elution dose response curves at either lysis temperature employed. Both detergents were more effective when lysis was carried out at 60°C. It is interesting that the data obtained with cells lysed at 60°C could be fitted by straight lines, whereas the results obtained at room temperature required linear quadratic regression for fitting. These findings suggested that lysis conditions can significantly affect the shape of the DNA elution dose response curves.

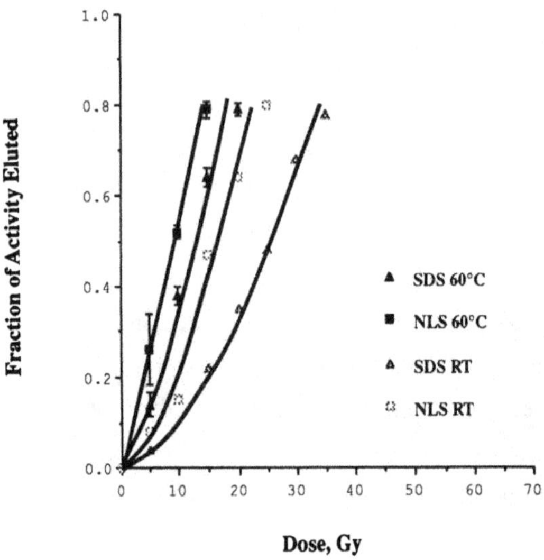

Figure 1. DNA elution dose response curves of plateau phase cells lysed with the detergents and at the temperatures indicated.

Since variations in the shape of the dose response curves may have profound consequences in the interpretation of the results obtained with the non-unwinding filter elution, experiments were performed to examine whether similar phenomena can also be observed using AFIGE. The outcome of a series of such experiments carried out also with plateau phase cells is summarized in Figure 2. Lysis at room temperature for 1 h with 1% SDS resulted in a very flat dose response curve; a significantly higher dose response curve was obtained when lysis was carried out for 5 h. When lysis was carried out at 50°C the results obtained after one and five hour lysis were very similar, but the values for the FAR were higher than those obtained after 5 h lysis at 20°C. All dose

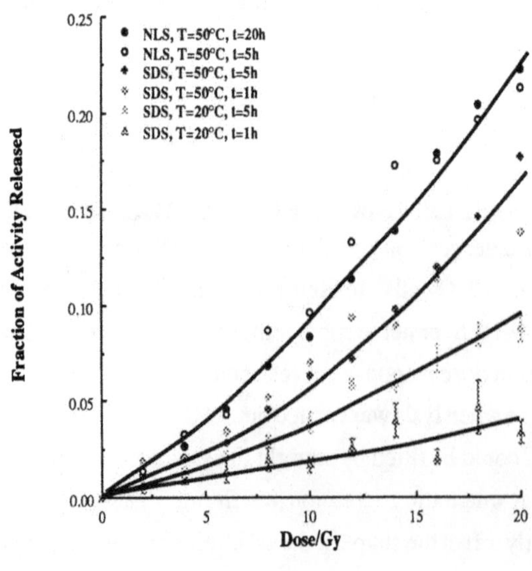

Figure 2. FAR as a function of radiation dose in plateau phase cells lysed with 1% SDS for 1h or 5h, either at 20°C or 50°C as indicated. Shown also are data obtained under similar conditions with cells lysed for 5h and 20h with 1% NLS at 50°C.

response curves obtained with SDS under the lysis times and temperatures described here were flatter than the dose response curves obtained after lysis with NLS for 5 h or 20 h at 50°C. These results are qualitatively similar to those shown in Figure 1 and suggest, on the one hand, that complete lysis may not be possible at room temperature even after five hours incubation with SDS, and on the other hand, that lysis with SDS may systematically lead to flatter dose response curves in certain cell systems.

Variations in DNA damage induction dose-response-curves throughout the cycle

In figure 3, DNA elution dose response curves are shown obtained with cells irradiated at various times after release from aphidicolin block in various stages of the cell cycle. Lysis was carried out in 0.002% SDS. Cells at G_1/S had a DNA content and a DNA elution dose response curve practically identical to that of cells irradiated in G_1-phase (results not shown). Irradiation at stages of the cycle where the majority of cells had synthesized 10%, 30%, or 50% of the DNA resulted in progressively flatter dose response curves. These results indicate that the fraction of DNA eluted after exposure to a given dose of X-rays decreased as cells progressed through the first half of the S-phase.

The results obtained with cells irradiated during the second half of S-phase demonstrate that as cells completed DNA replication an increase in DNA elution was observed. Although, due to the gradual loss of cell synchrony, the values for the average amount of DNA synthesized can only be approximate in the later part of S, the results clearly indicate that as cells reach the end of S-phase the DNA elution dose response increases. Cells that had been accumulated in G_2-phase

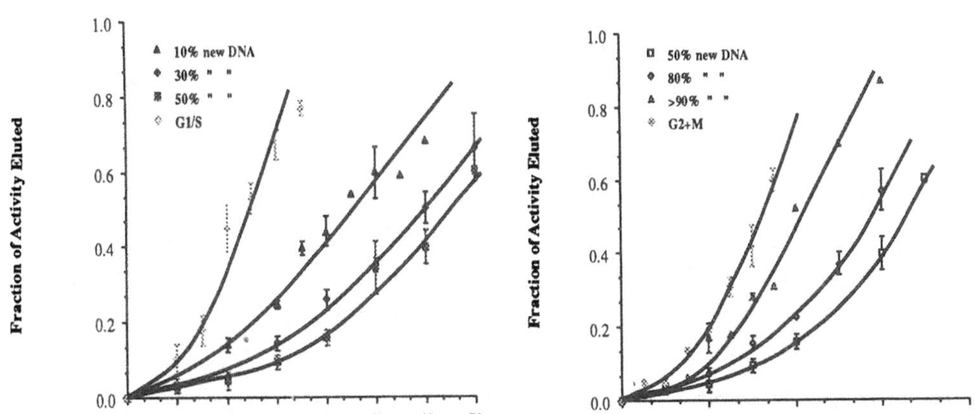

Figure 3. DNA elution dose response curves of cells irradiated at various stages of the cell cycle. The fraction of DNA synthesized is given as parameter to identify the position of the majority of cells in S.

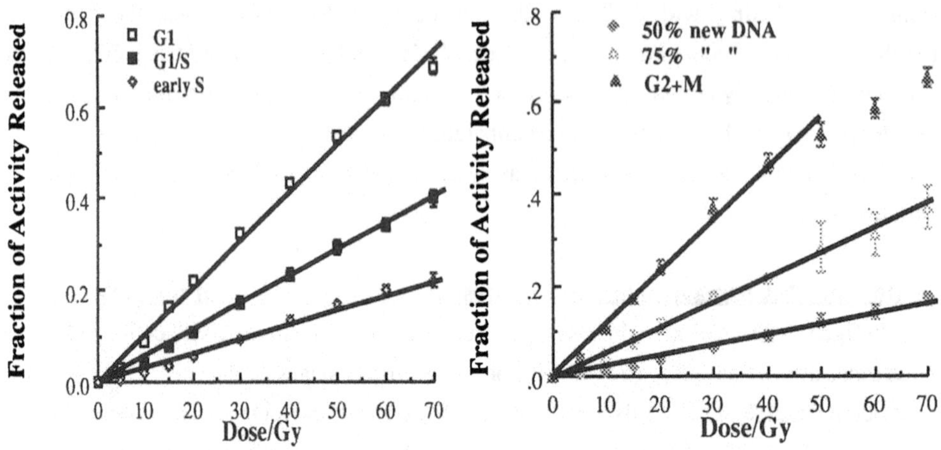

Figure 4. FAR as a function of radiation dose in cells irradiated in various stages of the cell cycle as indicated. The lines drawn show linear fits forced through the origin.

(nocadazole at 0.04 μg/ml prevented cell division and shake-off before trypsinization removed mitotic cells) gave a dose response curve (closed diamonds) close to that observed with G_1-cells. Thus, a cyclic response in the elution properties of the DNA was indicated, starting with a maximum at the beginning of S-phase reaching a minimum in the middle of S-phase and rising again to a maximum after completion of DNA synthesis, to levels comparable to those observed with early S-cells. Similar results were obtained with synchronous cells lysed with NLS at 60°C, but the DNA elution dose response curves measured under these conditions could be fitted by straight lines.

Experiments were performed to examine whether phenomena similar to those shown in Figure 3 can also be observed with AFIGE. The results obtained with cells irradiated in G_1, G_1/S and early S (25% new DNA synthesized) are shown in the left panel of Figure 4, whereas the results obtained with cells irradiated in mid S (50% new DNA), late S (75% new DNA) and G_2+M are given in right panel of Figure 4. The fraction of activity released shows a dramatic decrease as cells enter S, with a minimum reached in mid S. FAR increased again as cells progressed towards the end of S phase and entered G_2. The response obtained with cells irradiated in G_2+M was similar to that obtained with cells irradiated in G_1.

The observed variation in response for the FAR throughout the cell cycle is reminiscent to the above discussed fluctuations in relative elution of DNA from the filter in the non-unwinding filter elution assay. A comparison between the variation observed throughout the cell cycle in relative elution, as measured by non-unwinding filter elution, and the FAR, as measured by AFIGE, in irradiated cells is shown in Figure 5. Variation in FAR as a function of cell cycle phase is expressed as variation in the slope of the resulting dose response curve, and variation in relative elution as described previously (Okayasu et al 1988). The relative fluctuations observed are similar

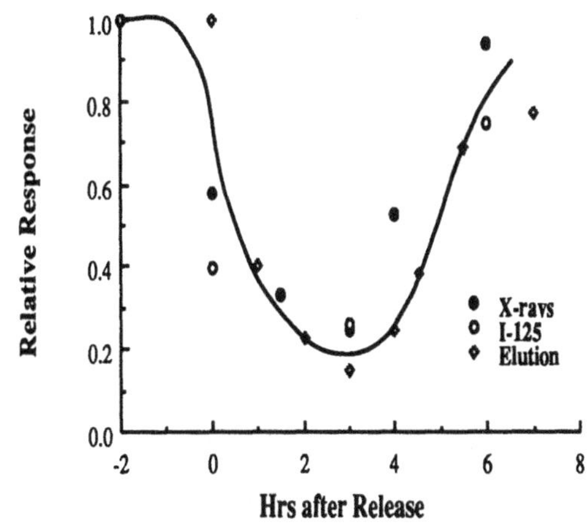

Figure 5. Variation in the slope of the dose response curves for FAR in various stages of the cell cycle for cells exposed to x-rays or ^{125}I decay. Results shown have been normalized to the slope of the dose response curve obtained with G_1 cells. The times shown indicate the time of release of cells from the aphidicolin block. Arbitrarily, G1 cells are shown at -2h. Shown in the figure is also the variation in relative elution throughout the cell cycle.

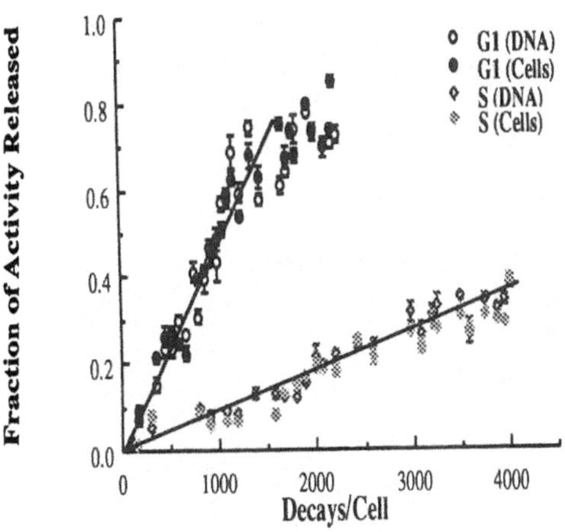

Figure 6. FAR as a function of the number of ^{125}I decays accumulated either in frozen cells, or in the DNA of cells lysed immediately after collection at various stages of the cell cycle.

for both techniques suggesting that they may be caused by similar phenomena at the molecular level.

The fluctuations in the FAR or in the fraction of DNA eluted per Gy can be interpreted as alterations throughout the cell cycle in the induction of DNA dsb per Gy per dalton, or as alterations in the physicochemical properties of the DNA that modify elution and electrophoretic mobility. To distinguish between these two possibilities calibration of either assay throughout the cell cycle was required. It has been previously suggested that decay of ^{125}I incorporated into DNA in the form of ^{125}IdU, can be used to calibrate assays measuring DNA dsb (Radford and Hodgson 1985), based on the observation that approximately one such lesion is induced per decay (Krisch and Sauri 1975, Krisch et al 1977). In Figure 6, results are summarized showing FAR during AFIGE as a function of the number of accumulated ^{125}I decays in cells assayed either in G_1 or in mid-S. Two sets of data are shown for each phase of the cell cycle: The closed symbols depict the results obtained when ^{125}I decays accumulated in frozen cells, whereas the open symbols depict the results obtained when ^{125}I-decays accumulated in the DNA of lysed cells. Similar results were obtained with both protocols suggesting a similar effectiveness of ^{125}I-decay in DNA dsb induction in chromatin in the cellular milieu and in the partly purified DNA. The FAR increased with increasing number of ^{125}I-decays, and reached a plateau at high doses similar to that observed with cells exposed to high doses of X-rays (Iliakis et al 1990a). However, the results obtained for FAR values up to 0.5 could be adequately fitted with a straight line, similar to observations that were made with cells exposed to X-rays. The results shown in Figure 6, as well as the slope values derived from them, suggest that the FAR per ^{125}I decay fluctuated significantly throughout the cell cycle. It was high for cells allowed to accumulate damage in G_1, and low in mid-S cells. In fact complete data throughout the cycle indicated that the fluctuations observed were similar to those found after exposure to X-rays (see Figure 5). This finding suggests that the same number of dsb in the DNA causes different

Table 1. Number of DNA dsb per Gy per diploid DNA complement (equal to the DNA complement of a G_1 CHO cell) and number of dsb per Gy per dalton in cells exposed to X-rays in various stages of the cell cycle, as calculated from the results shown in Figures 4 and 6 as well as from results presented elsewhere (Iliakis et al 1990a, 1990b).

cell cycle position	dsb/Gy/ diploid DNA complement	dsb/Gy/dalton
G_1	21 ± 3	$(7.0\pm1.0)\times10^{-12}$
G_1/S	31 ± 3	$(10.3\pm1.2)\times10^{-12}$
S	21 ± 3	$(7.0\pm1.0)\times10^{-12}$
G_2+M	27 ± 8	$(9.0\pm2.9)\times10^{-12}$

release of DNA into the lane depending upon the cell cycle stage. Through combination of the results shown in Figures 5 (and other results that are not shown here) with the results shown in Figure 3 the number of DNA dsb induced per Gy in various stages of the cell cycle can be calculated. The results obtained are summarized in Table 1. Induction of DNA dsb /Gy was similar in G_1 and S cells and not statistically different in G_2+M cells; it ranged around 21 dsb/cell/Gy. However, in G_1/S cells an induction of about 30 dsb/cell/Gy was measured.

Discussion

The results presented in the previous section indicate that the fluctuations in the FAR throughout the cell cycle are very similar to the fluctuations in relative elution for a given dose of radiation. This observation suggests that both assays may respond to the same cell cycle associated alterations of the biophysical and/or biochemical properties of the DNA molecule. This periodic modulation in the FAR and the FAE throughout the cell cycle, can be attributed either to modulations in the induction of DNA dsb in cells irradiated in various phases of the cell cycle, or, to alterations in the electrophoretic properties of the DNA, presumably caused by the inception of DNA replication. Since variations in the FAR, similar to those observed after exposure to x-rays, were also found after exposure to ^{125}I decay throughout the cell cycle, and the induction of DNA dsb per Gy and dalton was similar in G_1 and S cells, it seems reasonable to assume that the observed fluctuations reflect alterations in the elution properties and the electrophoretic mobility of the DNA molecules during S. These fluctuations may be caused by structural and conformational changes in the molecules resulting from the inception of DNA replication. It is interesting that similar dose response curves for FAR and FAE were measured in cells irradiated either before inception or after completion of DNA replication. The results obtained with cells exposed to ^{125}I-decay in various phases of the cell cycle indicate that the function that correlates the generated signal (FAR, or FAE) with the amount of dsb in the DNA varies throughout the cycle. Specifically, the sensitivity of the assay in measuring DNA dsb appears to be at maximum in G_1 and at minimum in mid-S. We propose that the variation in the sensitivity of AFIGE in measuring DNA dsb throughout the cell cycle also reflects alterations in the electrophoretic mobility of the DNA molecules.

It is intriguing that effects qualitatively similar to those observed with the non-unwinding filter elution after modification of the lysis conditions, could also be simulated with AFIGE (see Figures 1 and 2). This finding suggests that different detergents acting at various temperatures affect irradiated DNA structures in ways that cannot be predicted or characterized at the present time (see also Radford 1990). However, such uncertainties need to be acknowledged and accounted for in any quantitative analysis of the results and before introduction of the resulting fitting formalisms in the development of mechanistic models of the radiation action.

In summary: The similarities observed between alterations in the FAR, in cells exposed either to x-rays or ^{125}I decay, and of relative elution in cells exposed to x-rays throughout the cell cycle, suggest that both techniques respond to the same cell cycle associated alterations of the biophysical and/or biochemical properties of the DNA molecule. These alterations are assumed to derive from the inception of replication upon entry of cells into S, and do not directly reflect alterations in the production by radiation of DNA dsb in various stages of the cell cycle. Although both techniques appear to measure DNA dsb the correlation between the generated signal and the amount of DNA dsb present is variable and depends upon cell cycle position and lysis conditions employed. Therefore, quantitative interpretation of the results obtained with either technique requires calibration carried out under similar lysis conditions and with identical cell populations.

Acknowledgements

This work was supported by NCI Grants CA 45557, CA 42026 awarded from NIH, DHHS. Special thanks go to Mr. Giridhar Vedala and Ms. Mary Jackson for excellent technical assistance, and to Ms. Nancy Mott for secretarial help.

References

Agar, D.D., Dewey, W.C., Gardiner, K., Harvey, W., Johnson, R.T., and Waldren, C.A., 1990, Measurement of radiation-induced DNA double-strand breaks by pulsed field gel electrophoresis. *Radiation Research* (In press).

Agar, D.D. and Dewey, W.C., 1990, Calibration of pulsed field gel electrophoresis for measure ment of DNA double strand breaks. *International Journal of Radiation Biology*. (In press).

Anand, A., 1986, Pulsed field electrophoresis: a technique for fractionating large DNA molecules. *Trends in Genetics,* **11**, 278-283.

Blöcher, D., 1982, DNA double strand breaks in Ehrlich tumor cells at low doses of x-rays: I. Determination of induced breaks by centrifugation at reduced speed. *International Journal of Radiation Biology*, **42**, 317-328.

Blöcher, D., 1990, In CHEF electrophoresis a linear induction of dsb corresponds to a nonlinear fraction of extracted DNA with dose. *International Journal of Radiation Biology*, **57**, 7-12.

Bradley, M.D., and Kohn, K.W., 1979, X-ray induced DNA double strand break production and repair in mammalian cells as measured by neutral filter elution. *Nucleic Acids Research,* **7**, 793-804.

Bryant, P.E., 1984, Enzymatic restriction of mammalian cell DNA using Pvu II and Bam H1: Evidence for the double strand break origin of chromosomal aberrations. *International Journal of Radiation Biology*, **46**, 57-65.

Bryant, P.E., 1985, Enzymatic restriction of mammalian cell DNA: Evidence for the double strand breaks as potentially lethal lesions. *International Journal of Radiation Biology*, **48**, 55-60.

Carle, G.F. and Olson, M.V., 1984, Separation of chromosomal DNA molecules from yeast by orthogonal-field alternation gel electrophoresis. *Nucleic Acids Research*, **12**, 5647-5664.

Carle, G.F., Frank, M., and Olson, M.V., 1986, Electrophoretic separations of large DNA molecules by periodic inversion of the electric field. *Science*, **232**, 65-68.

Carrano, A.V., 1973, Chromosome aberration and radiation-induced cell death: II. Predicted and observed survival. *Mutation Research*, **17**, 355-366.

Chu, G., Vollrath, D., and Davis, R.W., 1986, Separation of large DNA molecules by contour-clamped homogeneous electric field. *Science*, **234**, 1582-1585.

Denko, N., Peters, B., Giaccia, A., and Stamato, T., 1989, An asymmetric field inversion gel electrophoresis system for the separation of large DNAs. *Analytical Biochemistry*, **178**, 172-176.

Dewey, W.C., Furman, S.C., and Miller, H.H., 1970. Comparison of lethality and chromosome damage induced x-rays in synchronized Chinese hamster cells in vitro. *Radiation Research*, **43**, 561-581.

Dewey, W.C., Miller, H.H., and Leeper, D.B., 1971, Chromosomal aberrations and mortality of x-irradiated mammalian cells: Emphasis on repair. *Proceedings of the National Academy of Science*, USA., **68**, 667-671.

Frankenberg, D., and Frankenberg-Schwager, M., 1981, Interpretation of the shoulder of dose-response curves with immediate plating in terms of repair of potentially lethal lesions during a restricted time period. *International Journal of Radiation Biology*, **39**, 617-631.

Frankenberg, D., Frankenberg-Schwager, M., Blöcher, D., and Harbich, R., 1981, Evidence for DNA double strand breaks as the critical lesions in yeast cells irradiated with sparsely or densely ionizing radiation under oxic or anoxic conditions. *Radiation Research*, **88**, 524-532.

Giaccia, A., Weinstein, R., Hu, J., and Stamato, T.D., 1985, Cell cycle dependent repair of double strand DNA breaks in a gamma ray sensitive Chinese hamster cell line. *Somatic Cell and Molecular Genetics*, **11**, 485-491.

Grote, S.J. and Revell, S.H., 1972, Correlation of chromosome damage and colony forming ability in syrian hamster cells in culture, irradiated in G1. *Current Topics in Radiation Research*, **7**, 303-308.

Hutchinson, F., 1989, On the measurement of DNA double-strand breaks by neutral elution. *Radiation Research*, **120**, 182-186.

Iliakis, G., Pantelias, G.E., Okayasu, R., and Seaner, R., 1987, ^{125}IdUrd induced chromosome fragments, assayed by premature chromosome condensation, and DNA double strand breaks have similar repair kinetics in G_1-phase CHO cells. *International Journal of Radiation Biology*, **52**, 705-722.

Iliakis, G.E., Metzger, L., Denko, N., and Stamato, T.D., 1990, Detection of DNA double strand breaks in synchronous cultures of CHO cells by means of asymmetric field inversion gel electrophoresis. *International Journal of Radiation Biology*. Submitted.

Iliakis, G.E. and Pantelias, G.E., 1990, Production and repair of chromosome damage in an X-ray sensitive CHO mutant visualized and analyzed in interphase using the technique of premature chromosome condensation. *International Journal of Radiation Biology*. In press.

Iliakis, G.E. and Okayasu, R., 1990, Radiosensitivity throughout the cell cycle and repair of potentially lethal damage and DNA double-strand breaks in an X-ray-sensitive CHO mutant. *International Journal of Radiation Biology*. In press.

Iliakis, G.E., Cicilioni, O., and Metzger, L., 1990, Measurement of DNA double strand breaks in CHO cells at various stages of the cell cycle using pulsed field gel electrophoresis: Calibration by means of ^{125}I decay. *International Journal of Radiation Biology*. Submitted.

Jeggo, P.A., and Kemp, L.M., 1983, X-ray-sensitive mutants of Chinese hamster ovary cell lines: Isolation and cross-sensitivity to other DNA-damaging agents. *Mutation Research*, **112**, 313-327.

Joshi, G.P., Nelson, W.J., Revell, S.H. and Shaw, C.A., 1982, X-ray induced chromosome damage in live mammalian cells, and improved measurement of the effect on their colony forming ability. *International Journal of Radiation Biology,* **41**, 161-181.

Kemp, L.M., Sedgwick, S.G., Jeggo, P.A., 1984, X-ray sensitive mutants of CHO cells defective in double strand break rejoining. *Mutation Research,* **132**, 189-196.

Kohn, K.W., and Ewig, R. A. G., 1973, Elution analysis, a new approach to the study of DNA single-strand interruptions in cells. *Cancer Research,* **33**, 1849-1853.

Krisch, R.E. and Sauri, C.J., 1975, Further studies of DNA damage and lethality from the decay of iodine-125 in bacteriophages. *International Journal of Radiation Biology,* **27**, 553-560.

Krisch, R.E., Krasin, F., and Sauri, C.J., 1977, DNA breakage, repair, and lethality accompanying ^{125}I decay in micro-organisms. *Current Topics in Radiation Research Quarterly,* **12**, 355-368.

Lai, E., Birren, B.W., Clark, S.M., Simon, M.I. and Hood, L., 1989, Pulsed field gel electrophoresis. *Biotechniques,* **7**, 34-42.

LaLande, M., Noolandi, J., Turmel, C., Rousseaux, J. and Slater, G.W., 1987, Pulsed-field electrophoresis: application of a computer model to the separation of large DNA molecules. *Proceedings in the National Academy of Science, USA,* **84**, 8011-8015.

Lehman, A. and Stevens, S., 1977, The production and repair of double strand breaks in cells from normal humans and from patients with ataxia telangiectasia. *Biochimica et Biophisica Acta* , **474**, 49-60.

Lloyd, D.C., Purrot, R.J., Dolphin, C.W., Bolton, D., Edwards, A.A., and Corp, M.J., 1975, The relationship between chromosome aberration and low LET radiation dose to human lymphocytes. *International Journal of Radiation Biology,* **28**, 78-90.

Matsudaira, H., Furuno, I., Ueno, A.M., Shinohara, K., and Yoshizawa, K., 1977, Induction and repair of strand breaks and 3'hydroxy-terminals in the DNA of mammalian cells in culture following irradiation. *Biochimica et Biophysica Acta,* **476**, 97-107.

Natarajan, A. T., and Obe, 1984, Molecular mechanisms involved in the production of chromosomal aberrations: III. Restriction endonucleases. *Chromosoma (Berlin),* **90**, 120-127.

Okayasu, R., Blöcher, D., and Iliakis, G., 1988. Variation through the cell cycle of DNA neutral filter elution dose response in x-irradiated synchronous CHO-cells. *International Journal of Radiation Biology,* **53**, 729-747.

Okayasu, R., and Iliakis, G., 1989, Linear DNA elution dose response curves obtained in CHO cells with non-unwinding filter elution after appropriate selection of the lysis conditions. *International Journal of Radiation Biology,* **55**, 569-581.

Painter, R.B., 1980, The role of DNA damage and repair in cell killing induced by ionizing radiation. *Radiation Biology in Cancer Research,* Eds. R. E. Meyn and H. R. Withers. Raven Press. pp. 59-68.

Peak, J.G., Blazek, R.E. Hill, C.K. and Peak, M.J., 1988, Measurement of double-strand breaks in Chinese hamster cell DNA by neutral filter elution: Calibration by ^{125}I Decay. *Radiation Research,* **115**, 624-629.

Radford, I. R., 1985, The level of induced DNA double-strand breakage correlates with cell killing after x-irradiation. *International Journal of Radiation Biology,* **48**, 45-54.

Radford, I. R. and Hodgson, G. S., 1985, ^{125}I-induced DNA double strand breaks: use in calibration of the neutral filter elution technique and comparison with x-ray induced breaks. *International Journal of Radiation Biology,* **48**, 555-566.

Radford, I.R. and Broadhurst, S., 1988, Aphidicolin synchronization of mouse L cells perturbs the relationship between cell killing and DNA double-strand breakage after X-irradiation. *International Journal of Radiation Biology*, **53**, 205-215.

Radford, I.R., 1990, Lysis solution composition and non-linear dose-response to ionizing radiation in the non-denaturing DNA filter elution technique. *International Journal of Radiation Biology*, **57**, 479-483.

Resnick, M.A., and Martin, P., 1976, The repair of double strand breaks in the nuclear DNA of *saccharomyces cerevisae* and its genetic control. *Molecular Gene Genetics*, **143**, 119-129.

Sasaki, M.S., and Norman, H., 1967, Selection against chromosome aberrations in human lymphocytes. *Nature*, **214**, 502-503.

Schwartz, D. and Cantor, C., 1984, Separation of yeast chromosomes by pulsed field gradient electrophoresis. *Cell*, **37**, 67-75.

Stamato, T.D., and Denko, N., 1990, Asymmetric field inversion gel electrophoresis: a new method for detecting DNA double strand breaks in mammalian cells. *Radiation Research*, **121**, 196-205.

Stewart, G., Furst, A., and Avdalovic, N., 1988, Transverse alternating field electrophoresis (TAFE). *BioTechniques*, **6**, 68-73.

Sweigert, A., Rowley, A., Warters, R., and Dethlefsen, L., 1988, Cell cycle effect on the induction of DNA double strand breaks by x-rays. *Radiation Research*, **116**, 228-244.

Van der Schans, G.P., Paterson, M.C., and Cross, W.G., 1983, DNA strand breaks and rejoining in cultured human fibroblasts exposed to fast neutrons or gamma rays. *International Journal of Radiation Biology*, **44**, 75-85.

Wlodek, D., and Hittelman, W.N., 1987, Repair of double-strand DNA breaks correlates with radiosensitivity of L5178Y and L5178Y-R cells. *Radiation Research*, **112**, 1-10.

Woods, W.G., 1981. Quantitation of the repair of gamma-radiation-induced-double-strand DNA breaks in tumor fibroblasts. *Biochimica et Biophisica Acta*, **655**, 342-348.

A MOUSE LYMPHOMA LINE THAT UNDERGOES RAPID INTERPHASE DEATH (APOPTOSIS) SHOWS EXTREME SENSITIVITY TO RADIATION-INDUCED DNA DOUBLE-STRAND BREAKAGE

Ian R. Radford
Molecular Science Group
Peter MacCallum Cancer Institute
481 Lt Lonsdale Street
Melbourne Victoria 3000
Australia

The relationship between radiation-induced DNA double-strand breakage (dsb) and cell killing for two mouse lymphoma lines was compared with that for hamster V79 cells. One of the lymphoma lines (STRij-4-2.2), which undergoes rapid interphase death (apoptosis) following cytotoxic insult, showed extreme sensitivity to gamma ray or DNA-associated iodine-125 decay-induced DNA dsb (7 $\pm$ 1 iodine-125 decays per lethal event). Surprisingly, the other lymphoma line (WEHI-22.1), which does not undergo rapid interphase death, was also much more sensitive to DNA dsb than were V79 cells (17 $\pm$ 1 versus 61 $\pm$ 2 decays per lethal event). Flow cytometric data suggested that both lymphoma lines may ultimately undergo rapid interphase death but the induction of this process in WEHI-22.1 may depend upon blockage in G2-phase. It is concluded that there are marked differences between the radiation responses of lymphoma and fibroblast lines.

1. Introduction

Loss of metabolic function in lethally irradiated mammalian cells is generally a slow process that is dependent upon mitotic activity (reviewed in Hopwood and Tolmach 1979). Even after moderately large radiation doses, cells will usually complete at least one mitosis prior to their disintegration. Although very little is known about the molecular events occurring in dying cells, the close correlation between the presence of a chromosome aberration and cell death (Joshi et al. 1982) is suggestive of a causal relationship. In marked contrast, cell types such as normal resting lymphocytes, spermatogonia, oocytes, and some stem cells (see Kondo 1988 for references) die rapidly in interphase, without undertaking

NATO ASI Series, Vol. H 54
The Early Effects of Radiation on DNA
Edited by E. M. Fielden and P. O'Neill

mitosis, following irradiation. A form of programmed cell destruction, which has been well-characterized at the morphological level and termed "apoptosis" (reviewed in Wyllie 1987), appears to be initiated by irradiation of these cell types. None of these cells are particularly suitable for biochemical studies of radiation-induced rapid interphase death, for reasons such as difficulty and expense of obtaining large numbers of purified cells and problems in labelling their DNA. However, Harris and Lowenthal (1982) have shown that although normal cycling lymphocytes appear to undergo mitotic killing some mouse lymphoma lines undergo rapid cell death following irradiation and may provide a useful model of the process occurring in normal cells.

Several lines of evidence suggest that DNA dsb is the critical radiation-induced lesion that ultimately leads to production of chromosome aberrations and mitosis-related cell death (Bryant 1984, Radford 1985). Data from normal and transformed fibroblasts of rodent and human origin indicated that the number of DNA dsb required to produce a lethal event was similar for all cell types examined (Radford 1986a). Accordingly, it was of interest to test whether such conclusions about the lethality of DNA dsb were applicable to lymphoma cells that undergo rapid interphase death.

In this study, two mouse lymphoma lines, one showing a rapid loss of viability after irradiation and the other exhibiting a much slower decline, are compared with the Chinese hamster V79 cell line.

2. Materials and Methods

2.1. Cells, culture conditions, viability assay, and flow cytometry

The mouse T lymphoma lines STRij-4-2.2 (ST4) and WEHI-22.1 (W22) were kindly provided by Dr. A. Harris (Walter and Eliza Hall Institute, Melbourne) and were grown in the alpha modification of Eagle's medium containing 20mM Hepes, pH 7.4, and supplemented with 10 per cent fetal calf serum. These lines had population doubling times of 9.5 and 10.5 h respectively and both had a pseudo-diploid karyotype. Cell survival was determined by cloning in growth medium containing 0.3% noble agar (Difco). As shown by Lowenthal (1982), ST4 cell cloning was dependent upon the addition of sheep red blood cells (0.4% final concentration). After growth for 7 (ST4) or

8 days (W22), colonies were stained by addition of 0.01% crystal violet in 1.5% acetic acid and then counted. Control cloning efficiencies for both lines were around 60-70%. The growth and cloning of Chinese hamster V79-743 cells has been described elsewhere (Radford and Hodgson 1985). Cell viability was measured by mixing equal volumes of cell suspension and growth medium containing 3 μg/ml acridine orange and 10 μg/ml ethidium bromide. Under a fluorescence microscope set up for fluorescein excitation, live cells stain green (acridine orange) whilst cells that have lost membrane function stain red (ethidium bromide) (Lee et al. 1975). At least 500 cells were scored for each determination. The cycle distribution of cell populations was measured using the BrdUrd-antibody technique (Dean et al. 1984).

2.2. Radioactive labelling of cellular DNA

ST4 and W22 cells were uniformly labelled, 2 days after subculturing, by adding [methyl-^{14}C] thymidine or [5-^{125}I] iododeoxyuridine to the growth medium for around 14 h. The respective radioisotope concentrations for ST4 cells were 1.39 kBq/ml (5 μM thymidine) and around 0.11 kBq/ml (2.5 μM thymidine). For W22 cells the respective concentrations were 1.85 kBq/ml (5 μM thymidine) and around 0.12 kBq/ml (2.5 μM thymidine). All other details regarding the processing and freezing and thawing of ^{125}I-labelled cells can be found in Radford and Hodgson (1985). One change was that cells were frozen in growth medium containing 10% DMSO. After rapid thawing, fresh growth medium was slowly added. Slow dilution of the DMSO-containing solution was particularly critical in the case of ST4 cells as they appear to be very susceptible to osmotic shock.

2.3. Irradiation and H_2O_2 incubation

Cell cultures were chilled on ice for approx. 30 min before irradiation in a ^{137}Cs source. The dose rate was approx. 0.9 Gy/min. After irradiation, the cells were either held on ice until lysis for filter elution or cloned immediately. H_2O_2 treatment was performed by mixing 0.25 ml of cells (5 x 10^6/ml) with an equal volume of growth medium containing different concentrations of H_2O_2 (BDH Aristar grade). After 10 min incubation on ice, the cells were diluted into medium, centrifuged, and then rinsed again prior to analysis by elution assay or cloning. The concentration of stock H_2O_2 solutions was determined using a molar extinction coefficient at 240 nm of 39.58 M-1 cm-1 (del Rio et al. 1977).

2.4. Filter elution

Non-denaturing filter elution and the low-sensitivity version of the alkaline filter elution technique, using [^{3}H] dThd-labelled mouse L cells given 45 or 10 Gy respectively of ^{137}Cs irradiation as an internal reference, were performed as described previously (Radford and Hodgson 1985, Radford 1986b).

3. Results

3.1 Cell viability after irradiation

The three cell lines were given equitoxic (as measured by clonogenic assay) doses of gamma irradiation and assayed for viability by vital dye staining after various times of

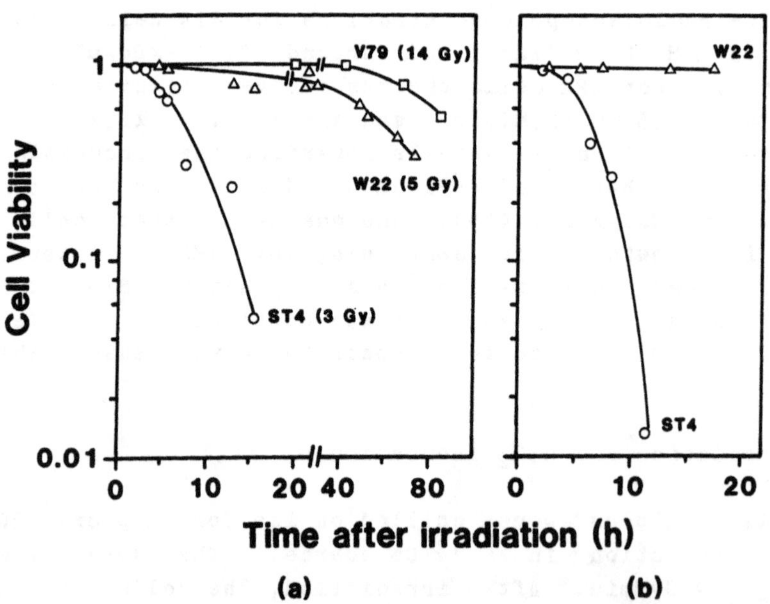

Figure 1. Decrease in viability (exclusion of ethidium bromide) of ST4 (○), W22 (△), and V79(□) cells exposed to equitoxic doses of gamma irradiation (a) or DNA-associated ^{125}I-decays (b) during subsequent incubation at 37°C. Each data set was obtained from at least 2 experiments.

incubation under normal growth conditions (figure 1a). In confirmation of the data of Harris and Lowenthal (1982), ST4 cells showed a rapid decline in viability. Within 4 h of incubation there was a significant decrease in viability and by 16 h virtually all cells were dead. By contrast, there was no significant decrease in the viability of W22 and V79 cells until around 30 and 60 h respectively after irradiation. An initial assumption was that death in W22 cells was probably mitosis-related.

The rate of loss of cell viability was also examined for ST4 and W22 cells that had accumulated approx. 30 and 60 DNA-associated ^{125}I-decays respectively (figure 1b). The results again indicated a clear difference in response between the two cell lines.

3.2. Cell killing and DNA dsb by gamma irradiation

Gamma ray survival data for the three lines are shown in figure 2a. The ST4 data are fitted by a simple exponential function of dose with a D_0 of 0.54 Gy. W22 cells are considerably more resistant than ST4 but are still extremely radiosensitive when compared with V79 cells. The W22 dose-response is intially linear but shows a pronounced curvature at higher doses. However, neither lymphoma line shows the large low-dose shoulder that typefies the response of many fibroblast-like lines such as V79.

Replicate cultures, to those used for the survival studies, were assayed for DNA dsb induction using the non-denaturing filter elution technique (figure 2b). In view of the marked differences in sensitivity to killing between these cell lines (figure 2a) and the previous finding for fibroblast lines that DNA dsb induction dose-response curves reflected relative radiosensitivity (Radford 1986a), it was a surprise to find little difference between the responses of these lines. In each case, the responses could be fitted to a linear-quadratic function of dose. It should be noted that although the ST4 DNA dsb dose-response is non-linear and comparable in shape to that of a cell line such as V79, over the survival curve dose range (0-4.5 Gy) the response may approximate linearity.

The data of figure 2 suggested either that DNA dsb may not be the critical radiation-induced lesion in lymphoma cells or that these cells are much more sensitive than fibroblasts to such damage. Subsequent work examined these points.

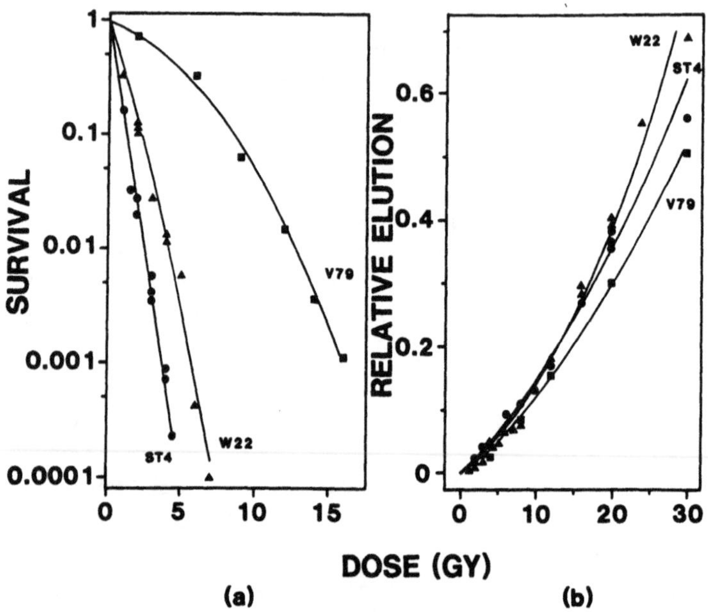

Figure 2. (a) Clonogenic survival of ST4 (●), W22 (▲), and V79 (■) cells as a function of gamma radiation dose. The ST4 data were fitted to a simple exponential function of dose, whilst the W22 and V79 data were fitted to a linear quadratic exponential function of dose. (b) DNA double-strand breakage induction by gamma irradiation as measured by non-denaturing filter elution. The relative elution data shown have had the background (no irradiation) value subtracted.

3.3. Cell killing by DNA-associated ^{125}I-decays

Labelling cells to a known specific activity with [^{125}I] dUrd, followed by decay accumulation during liquid nitrogen freezer storage, has proved a useful means of producing cell populations containing a known number of DNA dsb (see Radford and Hodgson 1985). This method was used to compare the number of DNA dsb required to produce a lethal event in lymphoma and V79 cells.

Batches of lymphoma and V79 cells were labelled to initial specific activities of around 4 and 12 decays/cell/day respectively and assayed for clonogenic survival at intervals during subsequent storage at liquid nitrogen temperature (figure 3). Consistent with previous data (Radford and Hodgson 1987), V79 cells required 61 ± 2 ^{125}I-decays (which can be equated with DNA dsb) to produce a lethal event. However, ST4 and W22 cells required only 7 ± 1 or 17 ± 1 ^{125}I-decays respectively to produce a lethal event. This data is thus consistent with the gamma ray results in suggesting that the lymphoma lines are much more sensitive than fibroblasts to radiation-induced DNA dsb.

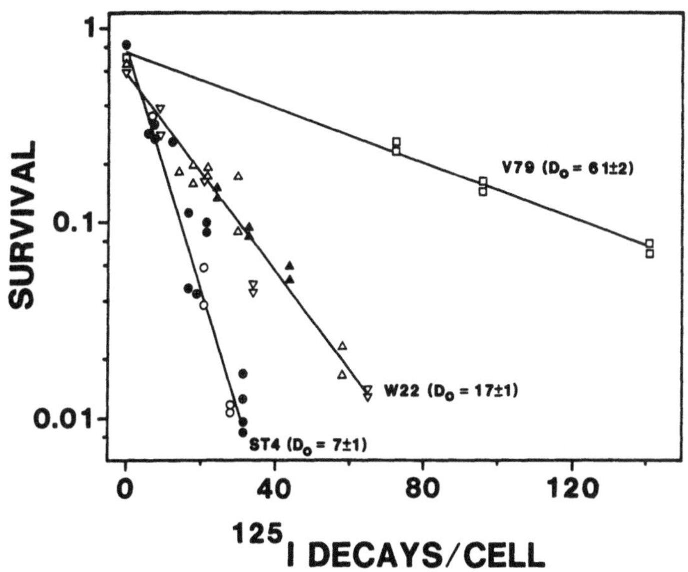

Figure 3. Clonogenic survival of ST4 (circles), W22 (triangles), and V79 (□) cells that have accumulated DNA-associated ^{125}I-decays during liquid nitrogen storage. Three batches of ST4 and W22 cells (indicated by the different symbols) and one batch of V79 cells were analyzed. Representative curves and D_0 values for a fitting to a simple exponential function of number of accumulated decays are shown.

The extreme sensitivity to DNA dsb of ST4 cells, which undergo rapid interphase death, was perhaps predictable. However, it was surprising that W22 cells, which show a much slower loss of viability after irradiation (figure 1), displayed a sensitivity to DNA dsb that was closer to ST4 than to V79 cells. This result prompted a comparison of cell death in ST4 and W22 cells.

3.4. Electron microscope and flow cytometer studies

The rapidity and dramatic nature of the morphological changes that occur in ST4 cells following irradiation can be visualized under the electron microscope (figure 4). Most ST4 cells that have been incubated for 2 h after 3 Gy of gamma irradiation show changes characteristic of apoptosis such as prominent membrane blebbing, nuclear fragmentation, and cell shrinkage. As noted by others (e.g. Afanas' ev et al. 1986), these overt morphological changes precede loss of membrane function (see figure 1a). Examination of irradiated W22 cell populations suggested that similar changes occurred after longer incubation times. In figure 4 a W22 cell that had been incubated for 30 h after 5 Gy of gamma irradiation shows membrane blebbing and margination of the chromatin into masses that abut the nuclear membrane.

A possible explanation for the difference in the time-course for induction of apoptosis-like cell destruction was sought by examining the effect of gamma irradiation on cell cycle progression in W22 cells (figure 5a). Following irradiation, there is a rapid exit of most cells from G1-phase, along with a temporary blockage in S-phase, and an accumulation of cells in G2-phase. Subsequently, cells exit from S-phase and progressively accumulate in G2-phase. A DNA content profile in figure 5b shows that at 13 h after irradiation most cells are in late S-phase or G2-phase. By 20 h after irradiation there was no detectable DNA synthesis occurring and around 90% of cells were blocked in G2-phase. At least up until 20 h after irradiation there was no significant increase in cell number. At time points beyond about 20 h incubation, accurate estimation of cycle phase distribution was prevented by the presence of large amounts of cell debris. This debris is generated during processing of samples for flow cytometry and presumably reflects the fragility of dying cells. Similarly, it was not possible to prepare samples for flow cytometry from irradiated and incubated ST4 cell populations.

3.5. Cell killing and DNA strand breakage by H_2O_2

The extreme sensitivity of both ST4 and W22 cells to radiation-induced killing raises the possibility that lesions

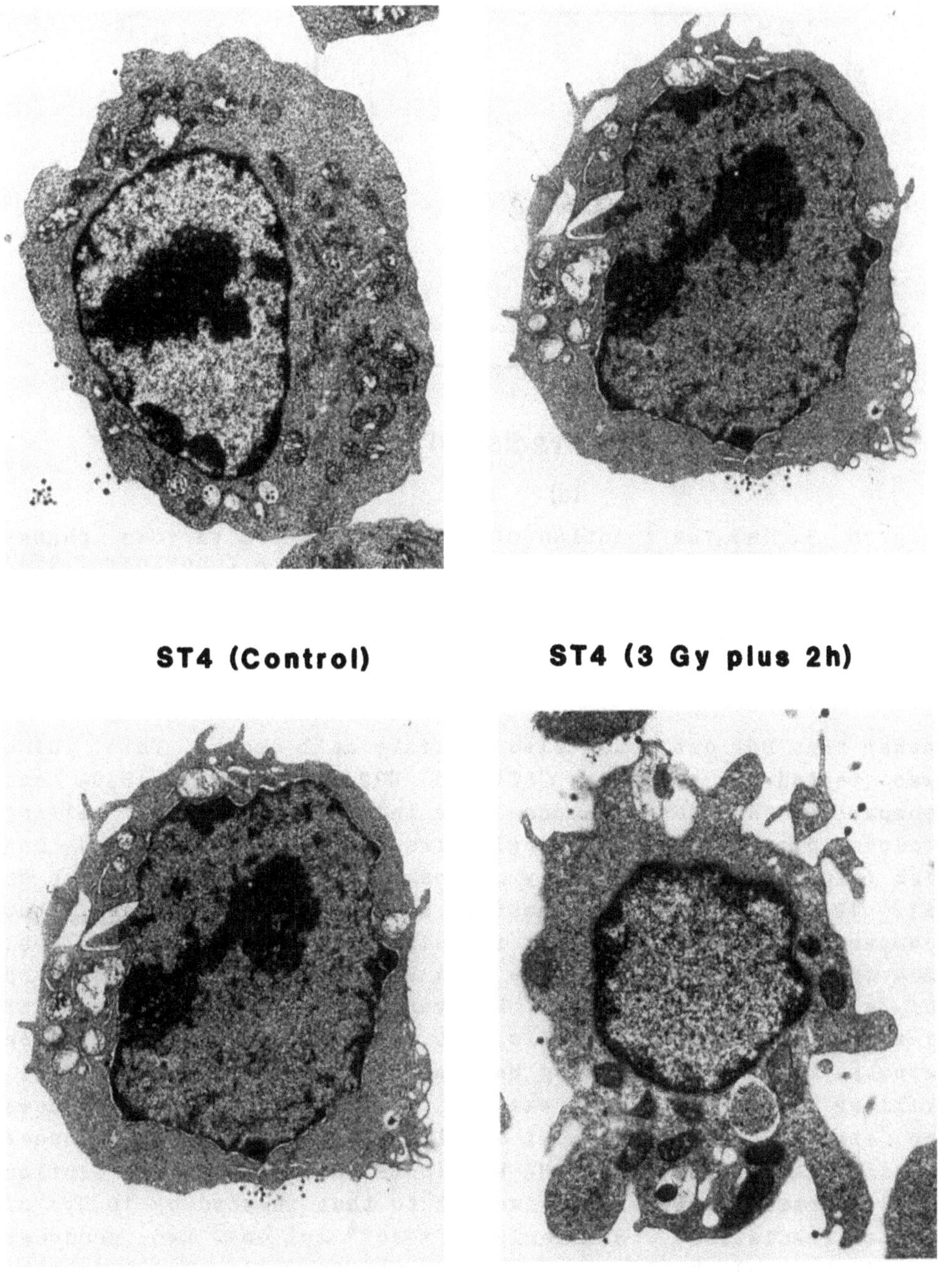

Figure 4. Electron micrographs of control (unirradiated) ST4 (x 8,250) and (x 7,550) W22 cells and cells given 3 (x 9,950) or 5 Gy (x 15,000) of gamma radiation and then incubated at 37°C for 2 or 30 h respectively.

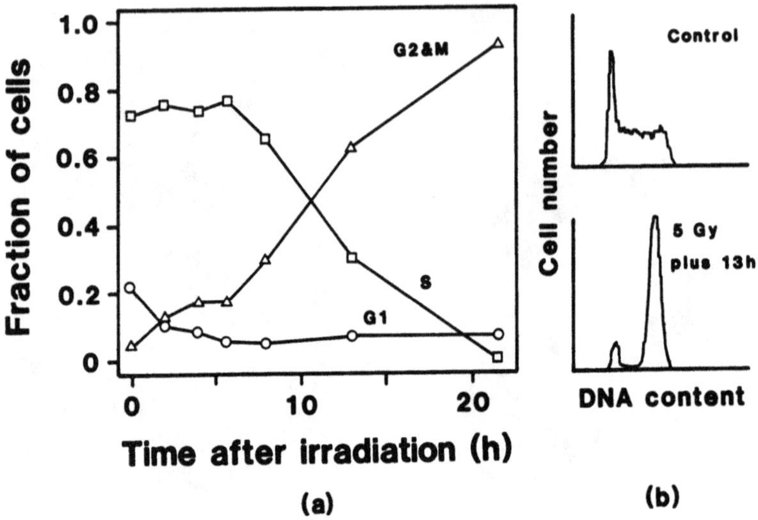

Figure 5. (a) The fraction of W22 cells in the various phases of the cell cycle (○, G1; □, S; △, G2 + M) as a function of time at 37°C after 5 Gy of gamma irradiation. Data were obtained from two experiments. (b) Representative flow cytometer DNA content profiles of control and irradiated and incubated W22 cell populations.

other than DNA dsb could also initiate cell death. This point was tested by treating ST4 and W22 cells with H_2O_2 and comparing the dose-responses for induction of cell killing (clonogenic assay), DNA single-strand breakage (ssb), and DNA dsb (figure 6). As shown by others (Ward et al. 1985, Prise et al. 1989), H_2O_2 is effective at producing DNA ssb, but comparatively ineffective at inducing cell killing and DNA dsb. ST4 cells were more sensitive than W22 cells for the induction of all three end points. However, the important piece of information in this figure is that, for both cell lines, there appears to be a correlation between the dose-responses for cell killing and DNA dsb. For example, 10^{-4} mol dm $^{-3}$ H_2O_2 produces no detectable killing or DNA dsb induction in ST4, but induces a considerable level of DNA ssb (note that a relative elution of 1 represents DNA ssb equivalent to that induced by 10 Gy of gamma irradiation). Similarly, 5×10^{-4} mol dm^{-3} H_2O_2 produces negligible cell killing and DNA dsb in W22 cells, but induces a high level of DNA ssb.

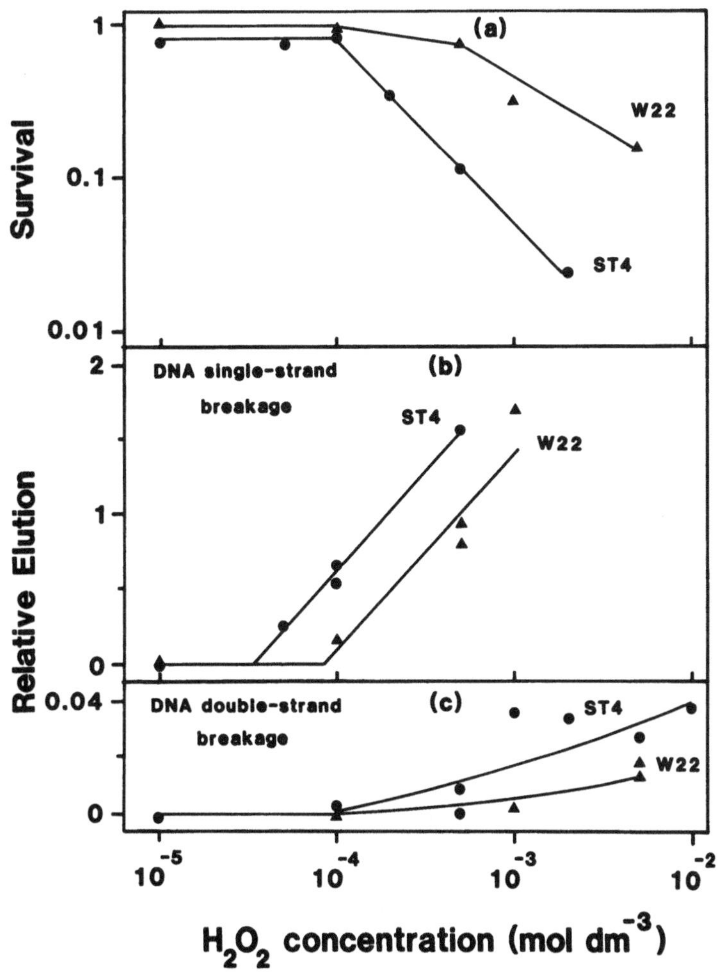

Figure 6. Dose responses for induction of cell killing (clonogenic assay) (a), DNA single-strand breakage (b), and DNA double-strand breakage (c) after incubation at ice-bath temperature of ST4 (●) or W22 (▲) cells with H_2O_2 for 10 minutes.

4. Discussion

The mouse lymphoma lines ST4 and W22, used in this study, show markedly different rates of loss of viability following irradiation. However, an initial assumption that W22 and V79 cells may have comparable radiobiological properties, based on their similar slow decline in viability after irradiation, proved incorrect. Both lymphoma lines were extremely sensitive to radiation-induced DNA dsb. ST4 cells required only 7 ± 1 ^{125}I-decays to produce a lethal event. This represents approx.

a 10-fold increase in sensitivity compared to V79 cells. Elucidation of the mechanism that is able to respond to such a small number of lesions is clearly of great interest. These results may explain the high radiosensitivity of organs such as ovary, testis, and thymus which contain cell populations that undergo rapid interphase death (see Hendry 1988 for references).

It has been implicitly assumed so far in this paper that rapid interphase death is initiated by DNA damage. Consistent with this assumption, it has been shown that treatment with various DNA-damaging drugs, such as bleomycin, induces a response comparable to radiation (Harris and Lowenthal 1982, Lowenthal 1982). Further support for this view comes from the finding herein that DNA-associated ^{125}I-decays (which produce highly localized damage) are most effective at initiating the process. In addition, studies using H_2O_2 treatment suggest that DNA ssb is not a lethal lesion to ST4 or W22 cells and that DNA dsb is probably responsible for cell killing. The former result is consistent with the knowledge that resting lymphocytes contain sizeable levels of DNA ssb (Carson et al. 1986) and that these lesions are produced during processes such as RNA transcription and DNA synthesis. Hence, although mitosis-related cell death and rapid interphase death are probably different processes DNA dsb appears to be the initiating lesion for both.

Despite showing different rates of loss of viability, the final process of cell destruction (apoptosis) in irradiated ST4 and W22 cell populations appeared to be morphologically similar. Flow cytometric data showed a progressive accumulation of irradiated W22 cells in G2-phase. There was an approximate temporal correlation between the blocking of most cells in G2-phase and the start of the decline in cell viability. This suggests that apoptosis might be initiated following prolonged blockage of W22 cells in G2-phase. Although flow cytometric analysis of irradiated ST4 cells could not be performed, the rapidity of the induction of apoptosis in these cells suggests that it is not dependent on blockage in a particular phase of the cell cycle. The greater number of ^{125}I-decays required to produce a lethal event in a W22 cell as compared with an ST4 cell (17 ± 1 versus 7 ± 1), might reflect an increased time for DNA repair prior to surveillance by the apoptosis-initiating mechanism.

Much of the above discussion is highly speculative, reflecting our minimal knowledge of the molecular events that occur during rapid interphase death and the comparative neglect of this subject by radiation biologists. This area clearly offers much uncharted territory for exploration.

Acknowledgements

The author is grateful to Dr. George Hodgson for helpful comments and to Sarah Ellis for preparing the electron micrographs. This work was supported in part by grants from the National Health and Medical Research Council and the Anti-Cancer Council of Victoria.

References

Afanas'ev VN, Korol BA, Mantsygin YA, Nelipovich PA, Pechatnikov VA, and Umansky SR (1986) Flow cytometry and biochemical analysis of DNA degradation characteristic of two types of cell death. FEBS Letters 194: 347-350

Bryant PE (1984) Enzymatic restriction of mammalian cell DNA using Pvu II and Bam H1: evidence for the double-strand break origin of chromosomal aberrations. Int. J. Radiat. Biol. 46: 57-65

Carson DA, Seto S, Wasson DB, and Carrera CJ (1986) DNA strand breaks, NAD metabolism, and programmed cell death. Exp Cell Res 164: 273-281

Dean PN, Dolbeare F, Gratzner H, Rice GC, and Gray JW (1984) Cell-cycle analysis using a monoclonal antibody to BrdUrd. Cell Tissue Kinet 17: 427-436

Del Rio LA, Ortega MG, Lopez AL, and Gorge JL (1977) A more sensitive modification of the catalase assay with the Clark oxygen electrode. Anal Biochem 80: 409-415

Harris AW, and Lowenthal JW (1982) Cells of some cultured lymphoma lines are killed rapidly by X-rays and by bleomycin. Int J Radiat Biol 42: 111-116

Hendry JH (1988) Survival of cells in mammalian tissues after low doses of irradiation: a short review. Int J Radiat Biol 53: 89-94

Hopwood LE, and Tolmach LJ (1979). Manifestations of damage from ionizing radiation in mammalian cells in the post irradiation generations. Adv Radiat Biol 8: 317-362

Joshi GP, Nelson WJ, Revell SH, and Shaw CA (1982) X-ray-induced chromosome damage in live mammalian cells, and improved measurements of its effects on their colony-forming ability. Int J Radiat Biol 41: 161-181

Kondo S (1988) Altruistic cell suicide in relation to radiation hormesis. Int J Radiat Biol 53: 95-102

Lee SK, Singh J, and Taylor RB (1975) Subclasses of T cells with different sensitivities to cytotoxic antibody in the presence of anesthetics. Eur J Immunol 5: 259-262

Lowenthal JW (1982) Effects of X-rays and cytotoxic drugs on lymphoid cells. Ph.D. Thesis, University of Melbourne

Prise KM, Davies S, and Michael BD (1989) Cell killing and DNA damage in Chinese hamster V79 cells treated with hydrogen peroxide. Int J Radiat Biol 55: 583-592

Radford IR (1985) The level of induced DNA double-strand breakage correlates with cell killing after X-irradiation. Int J Radiat Biol 48: 45-54

Radford IR (1986a) Evidence for a general relationship between the induced level of DNA double-strand breakage and cell killing after X-irradiation of mammalian cells. Int J Radiat Biol 49: 611-620

Radford IR (1986b) Effect of radiomodifying agents on the ratios of X-ray-induced lesions in cellular DNA: use in lethal lesion determination. Int. J. Radiat. Biol. 49: 621-637

Radford IR, and Hodgson GS (1985) ^{125}I-induced DNA double-strand breaks: use in calibration of the neutral filter elution technique and comparison with X-ray-induced breaks. Int J Radiat Biol 48: 555-566

Radford IR, and Hodgson GS (1987) Effect of ploidy on the response of V79 cells to ionizing radiation. Int J Radiat Biol 51: 765-778

Ward JF, Blakely WF, and Joner EI (1985) Mammalian cells are not killed by DNA single-strand breaks caused by hydroxyl radicals from hydrogen peroxide. Radiat Res 103: 383-392

Wyllie AH (1987) Cell death. Int Rev Cytol Supplement 17: 755-785

THE INFLUENCE OF GENOME STRUCTURAL ORGANIZATION ON DNA DAMAGE AND REPAIR IN EUKARYOTIC CELLS EXPOSED TO IONIZING RADIATION

O. Sapora[1], M. Belli [2], B. Maione [1], S. Pazzaglia [2] and M.A. Tabocchini [2]
1) Comparative Toxicology and Ecotoxicology Laboratory
2) Physics Laboratory
Istituto Superiore di Sanità
Via Regina Elena 299
Rome, Italy.

The nucleus of the mammalian cell is an highly organized and complex structure containing a large amount of DNA arranged in different order levels (Wojcierowski 1983, Bouteille 1983, Pienta 1984, Gasser 1987, Smith 1990). The chromatin structure and organization is not fixed, but can change with the expression of cellular pattern and differentiation. The basic structural units of the chromatin, the nucleosomes, are not uniform and important DNA functions are modulated by subtle alterations in selected regions of the genome. These modifications, controlled by several biochemical pathways, can be rapidly reversed and so that the cell, at least partially, is able to change the pattern of its chromatin structure in response to appropriate signals. Generally these modificatios produce either (i) an overall decreased in the affinity of the nucleosome histones towards the DNA molecule, and as a consequence a loose chromatin structure, or (ii) a more compact and less accessible structure with a consequent decrease of the enzyme and activators affinities towards DNA. Some of these postsynthetic modifications involve acethylation, ADP-rybosylation, phosphorylation and ubiquitation of the histones, methylation of the DNA, and are implicated in the control of gene transcription. Moreover, nucleosome positioning might control access of macromolecules to specific DNA sequences as well as of repair enzymes to DNA damaged sites (Hancock 1982, Lewin 1987).

In the last few years experimental evidence has accumulated indicating that distribution of the DNA radiation damage and repair can be affected by the complexity and "fluidity" of the nuclear organitation.

Evidence for different damage distributions

Since there are no reasons to suppose that the spatial distribution of energy deposition events could be affected by the target structural organization, it has been assumed for a long time that radiation induced DNA damage is randomly distributed along the molecule. This assumption has been apparently confirmed by the studies

NATO ASI Series, Vol. H 54
The Early Effects of Radiation on DNA
Edited by E. M. Fielden and P. O'Neill

TABLE I

Influence of chromatine structure on the distribution and repair of several damages introduced by ionizing radiation.

Type of damage	Biological system	Experimental method	Conclusion	Ref.
s.s.b.	CHO and L929 cells	Specific probes for active and non active chromatin	More damage in actively transcribing chromatin	Chiu (1982)
s.s.b.	L9/Ro cells and neurons	Alkaline sedimentation	Less repair in differentiated neurons	Wheeler (1983)
s.s.b.	L929 cells	Alkaline elution and specific probes for quiescent or active DNA	More damage in active DNA eluting in the first fraction	Warters (1987)
s.s.b.	Chromatin from V79	FADU technique	More ssb in relaxed than in normal and condensed chromatin	Heussen (1987)
s.s.b.	V79 cells treated with Butiric acid	Alkaline sedimentation	Increase in ssb and radiosensitivity	Nackerdien (1989)
base damage	Hela cells	T' damage chromatography fractionation	More damage in active chromatin	Warters (1982)
base damage	chromatin from Hela cells	Specific probes for active and quiescent chromatin	More damage in active chromatin	Patil (1985)
DNA-protein cross-links	chromatin from CHO cells	Filter binding assay	More cross-links in the nucleosome core	Mee (1981)
DNA-protein cross-links	V79 cells	Filter binding assay	More damage in the matrix regions	Chiu (1986)

of damage production on cellular DNA, showing that the size distribution of the DNA fragments is nearly ramdom. However the techniques used, such as elution and sucrose gradient sedimentation, are unable to distinguish between random distribution of the damage and different radiosensitive regions randomly distribuited throughout the genome. Warters et al.(1987) using the elution technique in combination with a technique capable of identifying specific regions of the genome such as hybridization with specific probes for actively trascribing genes, demonstrated that actually trascribing DNA elutes more rapidly and contains more ssb than bulk DNA. Other experimental evidence (table I), from molecular and cellular systems, indicates that damages such as single (ssb) and double strand breaks (dsb), base damages and DNA-protein cross-links are not randomly distributed along the DNA molecule.

Generally, transcriptionally active DNA sequences and nascent chromatin are more susceptible to radiation induced ssb and base damage than inactive sequences or mature chromatin. Even if most of these experiments used large doses, some of the reported results were obtained in the same dose-range used for cell killing. Cramp et al.(1984) found a remarkable correlation between the instability of newly synthesized DNA in several unirradiated cell lines and their relative radiation sensitivities. This implies that the amount of newly synthesized DNA, possessing more relaxed structural organization compared to mature chromatin, can influence the cellular response to radiation. Bohm and cooworkers reported two different experiments on the subject. In the first set of experiments, chromatin extracted from V79 cells, relaxed or condensed by different treatments, was irradiated with doses as low as 0.5-3 Gy (Heussen 1987). They found, using a fluorimetric assay, that the amount of damage was proportional to the degree of chromatin relaxation: the more chromatin is relaxed, the more damage is produced by the same dose of radiation. In the second set of experiments, V79 cells were treated with butyric acid, an inhibitor of histone deacethylase (Nackerdien 1989). After such treatment the cells possess a more relaxed chromatin structure, due to the extensive acethylation, are more sensitive to the lethal effects of ionizing radiation and show an increase in the yield of DNA damage compared to that with untreated cells. Similar results have been reported by Smith in 1984. This author has shown, using chemical and physical damaging agents, that the differential access to the target, induced by an increase or decrease of acethylation, produces a decrease or an increase in cell survival.

Polynucleosomes as a model for damage distribution studies

The above mentioned results have been related to changes in the chromatin structure and to nucleosomal "density" indicating a correlation between the amount of damage and the chromatin compactness. However they do not clarify the role of DNA structural organization at the nucleosomal level in the damage distribution. Among the different lesions induced by ionizing radiation both ssb and base damage are easily repaired by cellular repair enzymes and according to the present knowledge it is likely that they do not play an important role in determining long term biological effects. A more important role, at least for cell lethality, is generally attributed to DNA double strand breaks (Chadwick 1981, Frankenberg-Schwager 1982, Blocher 1982). To study the distribution of such damage as a function of nucleosomal structure, we used, as a molecular model, trinucleosomes isolated from hen erythrocyte chromatin (fig.1) (Barone 1989). Trinucleosomes were irradiated and analysis of radiation damage was performed by sedimentation on neutral sucrose gradient. The optical density profiles exhibit a discrete distribution with the production of lower oligomers corresponding to the sedimentation positions of mono- and di-nucleosomes (fig.2a) suggesting a preferential distribution of dsb in the linker regions. However the fragmentation of DNA in the core regions could be hidden by histones holding together the DNA fragments. For this reason, immediately after irradiation, the polynucleosomes were treated to

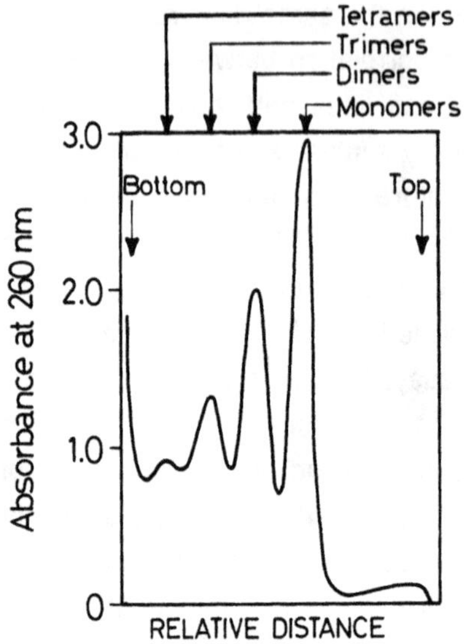

Sedimentation profiles of chromatin from hen erythrocytes nuclei digested with Micrococcus Endonuclease.

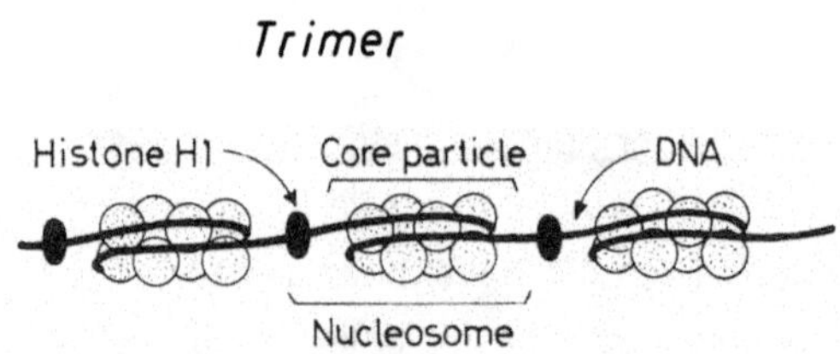

Hen erythrocytes nucleosome:

140 bp core DNA }
~50 bp linker DNA } 130.000 dal

Fig.1. Schematic diagram of nucleosome extraction. Hen erythrocytes chromatin digested with Micrococcus endonuclease was layered on top of 5-30% sucrose neutral gradients and centrifuged using SW28 rotor at 30000 rpm for 14 hours. The peaks corresponding to trinucleosomes were then purified by a further centrifugation and used for the experiments.

remove both proteins and RNA contaminations. Also for the DNA extracted from irradiated trinucleosomes, three peaks can be identified in the sedimentaion pattern (fig.2b), clearly indicating that the dsb are preferentially introduced in the linker regions compared to the core regions.

We compared these results with those obtained upon irradiating DNA isolated from trinucleosomes (fig.2c). In this case a different sedimentation profile is obtained, where the distribution of DNA molecules along the gradient appears to be that expected

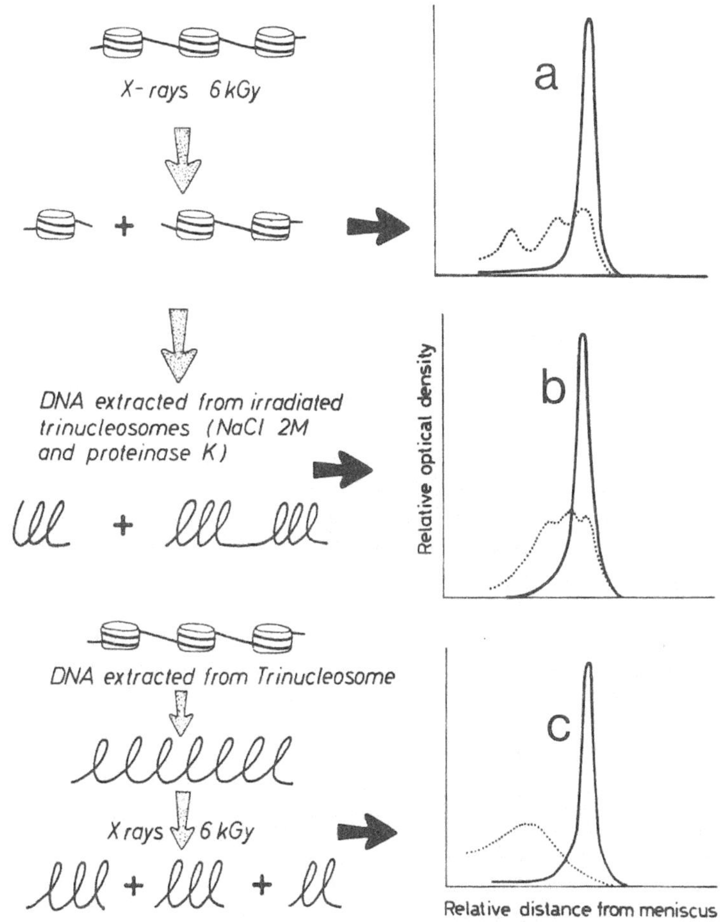

Fig.2. Schematic rapresentation of the experiments carried out with trinucleosomes. The continous lines represent the sedimentation profile of the control while the dotted lines correspond to the sedimentation of the irradiated sample.

expected for randomly induced dsb. From the sedimentation profiles the number of dsb per molecule was evaluated and the results are reported in fig.3.

From figure 2 and 3 it appears that: (i) DNA in the core regions is more resistant than that in the linker regions, and (ii) "naked" DNA is 3-4 fold more sensitive than DNA organized in polinucleosomes. These findings point to the importance of the association of DNA with proteins in determining the amount of damage and strongly suggest a non-random distribution of the damage in the cellular chromatin.

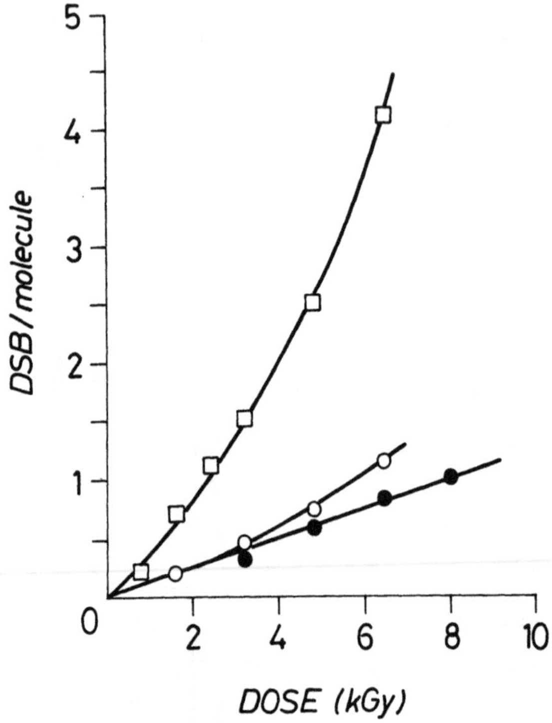

Fig.3. Number of dsb per molecule as a function of radiation dose. Irradiated trinucleosomes (●), DNA extracted from irradiated trinucleosomes (○), DNA extracted from trinucleosomes and then irradiated (□).

Cellular differentiation processes and chromatin structure

The biochemical and structural characteristics of the cell are not constant throught its life span, but they change with the cell cycle, aging and differentiation. In general most cells in a multicellular organism are required to divide or to differentiate and in mammals more than 80 % of them are in differentiated state. The term differentiation can be defined with regard to the cells ability to undergo division and three different classes of cells can be identified: (i) the terminally differentiated cells that have permanently lost the ability to divide, (ii) the reversibly differentiated cells that can be recruited into DNA synthesis and division and (iii) the undifferentiated cells that are continously progressing through the cell cycle.

In terminally differentiated cells, changes involve all the cellular structures, from the membrane (Parasassi 1987) to the nuclear organization (Scher 1978,Groudine 1983, Drzynkiewicz 1985). Part of the genes expressed in actively growing cells are silent while others are active. This means that the enzymatic content of the cell can change drastically.

The nuclear organization of differentiated cells is affected by active biochemical processes such as ADP-ribosylation and DNA methylation.

The nuclear enzyme ADP-ribosyl transferase (ADPRT), NAD^+ dependent, catalyses the formation of ribosylated chromatin proteins and requires nicked DNA as a co-factor (Boulikas 1989). In human lymphocytes an intimate relationship exists between DNA nicks, NAD metabolism, ADP-ribosyl transferase activity and cell differentiation (Carson 1986). Under steady state conditions resting lymphocytes continually break and rejoin their DNA (Johnstone 1982, Francis 1984, Greer 1986). This process is accompanied by a proportional consumption of NAD^+ for poly(ADP-ribose) synthesis. These breaks persist during tissue culture, but are repaired rapidly after exposure of the cells to a mytogenic stimulus. This behaviour has been observed also in other cell types, such as Friend cells, avian mioblasts and rat neurons (McMahon 1984, Farzanesh 1985). Thus, despite their quiescent state, differentiated cells appear to break and rejoin DNA continously and, at any one time, some DNA nicks are present. The average number of nicks per genome depends in part upon the state of activation of the cell. It has been suggested that DNA breaking and rejoining, regulated by ADPRT, is involved in a general mechanism of differentiation acting as an additional safety-catch mechanism that restrains the differentiated cells from replicating their genetic material.

A second important process is DNA methylation (Doerfler 1983, Kautianen 1985, Lyon 1986). The DNA of many higher organisms, including man, contains a small proportion of 5-methyl-cytosine that is formed by enzymatic methylation of some bases present in the DNA. Most of the genes do not code for essential proteins and, in general, DNA of inactive genes tends to be more methylated than DNA of active genes. In several cases an inactive gene that contains methylated DNA has been shown to became undermethylated after gene activation. The DNA methylation, because of the location of the methyl group in the minor grove, can affect the DNA availability to enzymes. In the past few years evidence has been accumulating for the implication of DNA methylation in a number of biological processes. It can affect DNA structure changing DNA-protein interactions, protect DNA against restriction endonucleases, regulate gene expression in eukaryotes, enhance mutation and recombination, and may influence DNA replication, and differentiation (Christman 1980, Sheffery 1982, Tolberg 1984). For some of the listed processes the accumulated evidence begins to be rather convincing, for others it is still tentative. In general DNA in differentiated cells is more methylated than that in actively duplicating cells. It is likely that DNA methylation is one of the several mechanisms that are used to create inherited patterns of gene regulation keeping genes active or repressed.

Cellular systems capable to undergo "in vitro" differentiation as a tool in radiobiological studies

In terminally differentiated cells the overall changes in the nuclear and chromatine structure, the presence of physiologically produced nicks in some of the genome regions, the possible decreased affinity for the repair enzymes, the presence of hypermethylated DNA regions, the absence of DNA duplication which can interfere with DNA repair activities, and the possibility to have more time for such activities and to hold a damage site, in silent genes, with no biological consequences, can affect the rate and the extent of repair and consequently the final biological effects. It is also true that biological end points, such as cell killing and mutation, cannot be studied because their analysis are based on the capacity of cells to duplicate giving visible clones.

To our knowledge very few data have been reported on the use in radiation biology of cellular systems capable to undergo in vitro differentiation (Einspenner 1984). The only systematic work on DNA damage and repair has been published by the Weehler group (Wang 1978, Wheeler 1987). However they used in vivo cellular systems, namely rat neurons from cerebellum and rat intracerebral 9L/Ro tumor cells (gliosarcoma), as models for the differentiated and the undifferentiated status, respectively. They found, using alkaline sedimentation technique, no differences in the initial number of DNA breaks induced by radiation in differentiated and proliferating cells.

We used two different cell lines both able to express in vitro some steps of the differentiation process characteristic of the original tissues: the human erythroleukemic K562 cell line clone S and the rat myoblastoid L5 established cell line. These lines can be stimulated to differentiate by chemical treatment or by lowering the amount of foetal calf serum in the medium, respectively. In this way it was possible to study and compare the radiation induced DNA damage in the same cell type in different physiological conditions. These two lines have been previously characterized for the respective differentiation markers (Cioe 1981, Hartwig 1984, Scarpa 1985, Parasassi 1987). A common characteristic of both cell lines is the presence of nicks in their DNA when the cells are in differentiated conditions, in agreement with the previous reported evidence. The presence of such nicks leeds to the loss of 20-30% of the DNA when the elution technique is used. However, from the experiments carried out using alkaline sedimentation technique and from the elution kinetics (fig.4), it is clear that the nicks are not randomly distributed but probably localized in specific regions of the genome. In fact while the gradient profiles show that a large number of breaks are present in differentiated cells, the elution kinetics do not show the expected corresponding dramatic changes in the slopes. It must be considered that the elution kinetics are dependent on both DNA size and compactness of chromatin structure. In our case the lack of change in the slope, despite nicks characteristic of the differentiated status, is an indication of more compact structure of DNA in differentiated cells.

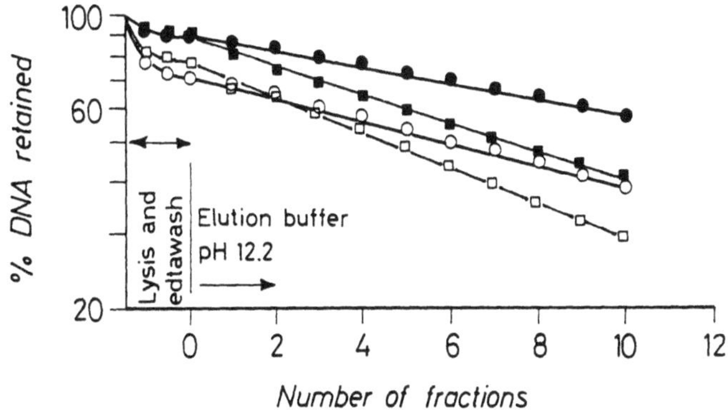

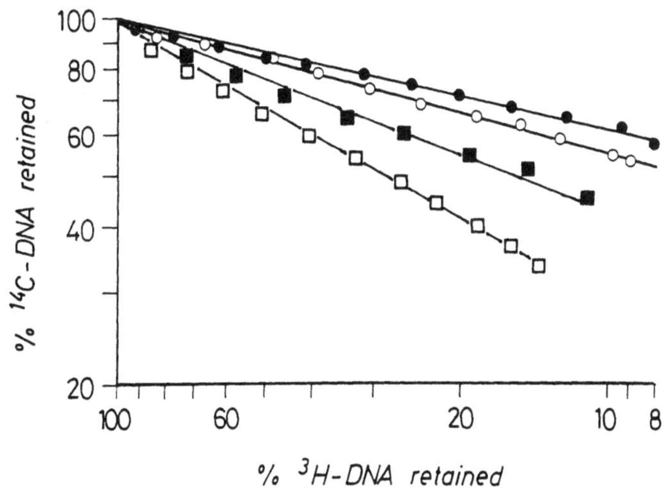

Fig.4. Elution kinetics for unirradiated K562 (●○) and L5 (■□) before and after differentiation induction as a function of fraction number (top) or as a function of an internal standard (bottom) rapresented by the corresponding actively proliferating cells labelled with 3H-tymidine for 48 hours and irradiated with a dose of 4 Gy. The closed symbols represent the elution kinetics of actively proliferating cells while the open ones rapresent those of differentiated cells. The cells to be induced were labelled for 48 hours with 14C-tymidine, then ceased, washed and resuspended in fresh medium containing 2 mM butyric acid (K562) or 1% FCS (L5). The K562 and the L5 cells were irradiated 2 days and 7 days after induction respectively.

For the L5 cells differentiation takes same consistent time. After induction, the committed cells continue to divide for at least three cycles. On the fourth day they reach confluency and, from then on, start to fuse giving, at the sixth seventh day, the characteristic syncytium containing fibres and several nuclei. Instead, the differentiation of K562 cells presents a different pattern. When differentiated, the cells do not loose their individuality and they undergo one division only. 48 hours after induction all changes in the membrane and in the chromatin are complete.

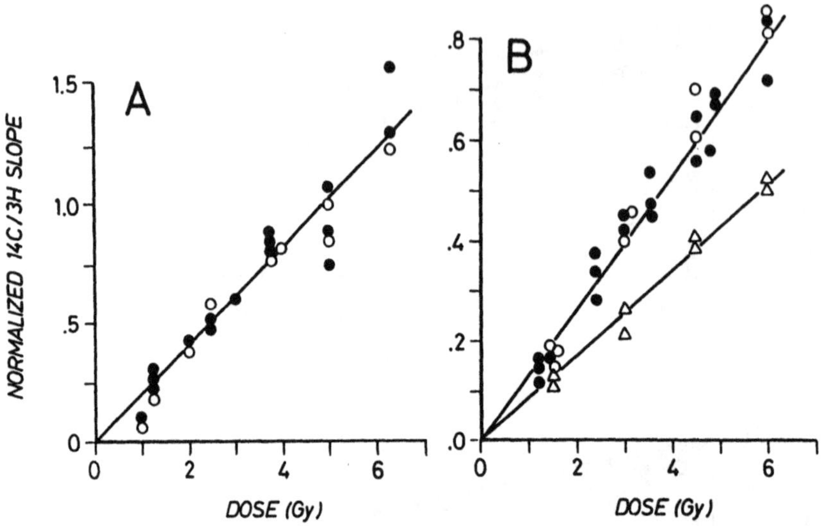

Fig.5 . Dose-response curves for K562 (A) and L5 (B) cells. The normalized 14C/3H slope represent the value of elution slope of irradiated cells minus the value of the corresponding unirradiated control. Panel A: actively proliferating (●) and 48 hours differentiated (○) K562 cells. Panel B: L5 cells before (●) or 4 (○) or 7 (△) days after differentiation induction.

Moreover the differentiation is induced by chemical treatment, which becomes toxic after 96 hours, while L5 cells are still alive after 8 day treatment. For these reasons the production of ssb as a function of dose has been measured in K562 cells, 48 hours after differentiation and in L5 cells 4 and 7 days after differentiation. The results are shown in fig. 5. For both K562 and L5 cells, the initial yield of ssb is similar, before and after 2 and 4 day differentiation, respectively. Different results are obtained with L5 cells seven days after differentiation. In this case a lower slope value is obtained, indicating less damage produced per unit dose. This decrease of sensitivity can be explained assuming that more compact regions, influencing the amount of radiation induced damage, exist in the genome of L5 cells after seven days differentiation.

DNA accessibility and repair

Although in all cell the DNA is freely available to small molecules, its availability to larger molecules is restricted by the chromatin organization. Therefore the nucleosome packing can limit the access of regulatory proteins affecting the gene expression and other processes. Some regions of chromatin might be expected to show a non random positioning of the nucleosomes, leaving relatively exposed those DNA sequences that must be recognized by other proteins.

There is experimental evidence that cellular chromatin is divided into regions more or less accessible to endonuclease enzymes (Jalouzot 1980) and these regions can be classified, according to their sensitivities, as hypersensitive, sensitive, and insensitive to enzymatic action. The hypersensitive sites have been hypothesized to be important in gene regulation and represent a region from which nucleosomal proteins have been removed to allow regulatory molecules to bind. An other indication of differential enzyme accessibility to chromatin regions, comes from the experiments carried out with isolated nuclei from L9/Ro and cerebellar neurons cells (Wheeler 1983/a, Wheeler 1983/b). The DNA of these nuclei were digested with micrococcus endonuclease and DNAase I. The results show that the DNA from neuronal nuclei is digested more slowly than that from tumour nuclei. The genome of neurons contains approximately three times more of less accessible regions than that of the genome of actively duplicating tumour cells. It is also possible that a portion of the neuron genome could be completely inaccessible. These data suggest that the genome of actively proliferating cells may be more accessible to repair enzymes, than the genome of terminally differentiated cells.

Using the K562 cellular system we have studied and compared the repair kinetics of differentiated and actively proliferating cells in the attempt to verify the DNA accessibility

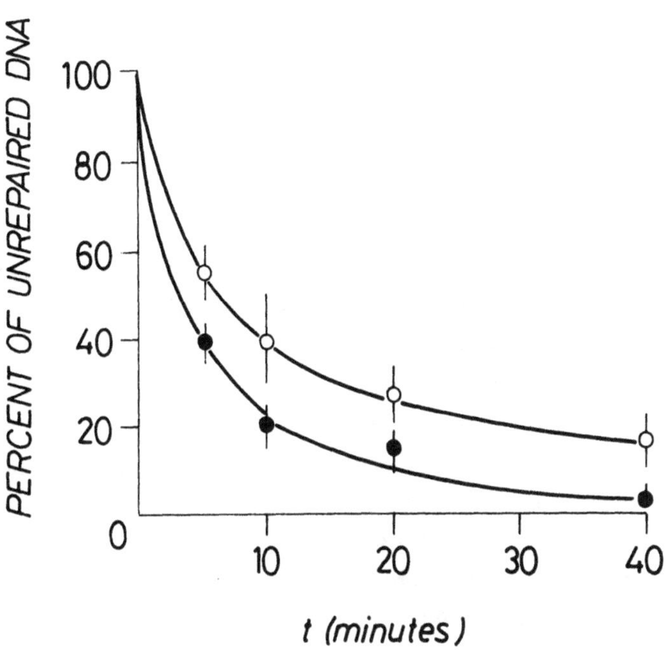

Fig.6 . Repair kinetics for actively proliferating (●) or 48 hours differentiated (○) K562 cells. The cells were irradiated with the dose of 4 Gy, incubated at 37 C for the reported length of time, chilled and then processed for elution.

compared to the proliferating ones. After 60 minutes repair the amount of residual damage in both conditions is similar and this occurs up to a dose of 40 gray (fig.7).

The most important factors that can influence the repair kinetics are: the amount of lesions, the saturation of the repair systems and the accessibility of damaged site to repair enzymes. We have evidence that the initial number of breaks is similar on both proliferating and differentiated K562 cells. Moreover the results obtained after 60 minutes repair as a function of the dose, clearly show that there is no evidence of repair systems saturation up to a dose of 40 Gy. Then, the best explanation of the results obtained is that different kinetics represent repair of damage located in regions of the genome of different accessibility.

Two forms of DNA repair

Since the discovery of DNA repair mechanisms in bacteria different classification schemes have been proposed. Some of these were based on the repair efficiency (error-free or error-prone repair), or on the possibility that they can act in non-growth and in growth conditions, or according to the enzyme categories involved in the repair (excision repair, recombination repair etc.) (Fielden 1982). Reagan and Setlow (1974), using human cells and different chemical carcinogens and mutagens, proposed two classes of DNA repair conveniently described as "short" and "long" repair. In "short" repair the number of nucleotides involved in repairing the lesion is small and the repair process is completed within 60 minutes. In "long" repair there is an extensive run of nucleotide excision and replacement, and repair continues up to 20-24 hours. The typical insult that results in "short" repair is ionizing radiation, while "long" repair is typified by the process that follows damage by UV radiation.

With our cells, which are of human origin, we have evidence of the existence of both "short" and "long" repair acting differently in proliferating and differentiated cells. As reported above, the kinetics of the "short" component is faster in the former cells compared to the latter and after 60 minutes repair no differences in the amount of residual damage can be detected in both cell types. The repair at longer time has been also investigated and the results are shown in figure 7, where the residual damage after 24 hours incubation is plotted as a function of the dose. We used the term "residual damage" because we believe that the ssb detected after 24 hours repair are mainly unrepaired double strand breaks or single large gaps produced by enzymatic action in the attempt to repair DNA damage. After 24 hours repair the amount of residual damage is higher in actively proliferating cells compared to the differentiated ones. If the hypothesis that unrepaired damage is correlated with the cell sensitivity is correct (Frankenberg-Schwager 1982, Blocher 1982) the lower amount of damage seems to indicate a higher radioresistance of differentiated cells. Moreover it must be stressed that in actively replicating cells the DNA synthesis processes can interfere with the

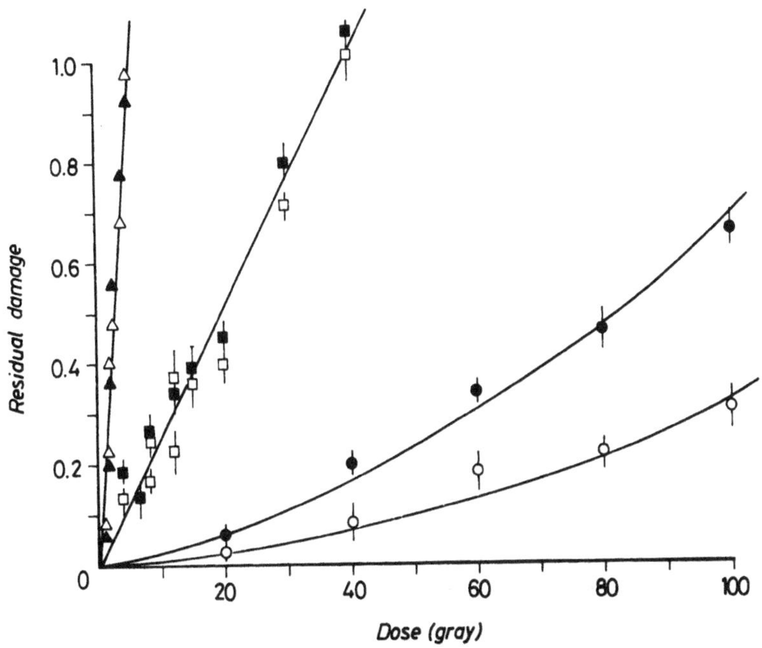

Fig.7 . Dose response curves after no repair (▲△), 60 minutes repair (■□) or 24 hours repair (●○) for actively proliferating cells (closed symbols) or 48 hours differentiated (open symbols) K562 cells. The cells after irradiation were incubated at 37 C for the corresponding lengthof time and then the amount of residual damage tested using elution technique. The residual damage is expressed as the value of the slope of irradiated sample minus the value of the slope of unirradiated control hold in the same conditions as the irradiated sample.

repair process while, in differentiated cells, damage occurring in non trascribing genes presumably can be held without biological consequences.

In conclusion experimental evidence has been reported suggesting a non random distribution of several types of DNA damage due to differences in the nucleosomal compactness. Double strand breaks, the most relevant damage introduced by ionizing radiation, are preferentially introduced in the linker DNA rather than in the core DNA in a in vitro nucleosomal model. The amount of damage can also be different in cells before and after differentiation. In a human erythroid cellular system two repair pathways are present ("short" and "long") with clear differences in the kinetics and in the total amount of repair when actively proliferating or differentiated cells are compared. This is likely to be a consequence, in differentiated cells, of limited accessibility of the repair enzymes to the DNA and of lack of replication activities.

The data show that chromatin structural organization has important implications in molecular and cellular radiobiology and point out to the need of better clarifying the role of chromatin domains in determining biological effects such as cell killing and mutation.

Acknowledgements

Thanks are due to F. Barone and L.A. Serpietri for their skillful technical assistence. Work partially supported by the special Project "Oncologia" of the Italian National Council of Research (CNR).

References

Barone F, Belli M, Pazzaglia S, Sapora O, Tabocchini MA (1989) Radiation damage and chromatin structure. Ann. Ist. Super. Sanita, 25: 59-68.

Blocher D, Pohlit W (1982) DNA double strand breaks in Ehrlich ascites tomour cells at low doses of X-rays. II Can cell death beattributed to double strand breaks. Int. J. Radiat. Biol. 42: 329-338.

Boulikas T (1989) DNA strand breaks alter histone ADP-ribosylation. Proc. Natl. Acad. Sci. USA, 86: 3499-3503.

Bouteille M, Bouvier D, Seve AP (1983) Heterogeneity and territorial organitation of the nuclear matrix and related structures. Int. Rew. Cytol., 83: 135-182.

Carson DA, Seto S, Wasson DB, Carrera CJ (1986) DNA strand breaks, NAD metabolism and programmed cell death. Exp. Cell Res. 164: 273-281.

Chadwick KH, Leenhouts HP (1981) Molecular theory of radiation biology. Springer-Verlag, Berlin, Heidelberg, New York.

Chiu SM, Friedman LR, Sokany NM, Xue LY, Oleinick NL (1986) Nuclear matrix proteins are crosslinked to transcriptionally active gene sequences by ionizing radiation. Radiat. Res. 107: 24-38.

Chiu SM, Oleinick NL (1982) The sensitivity of active and inactive chromatin to ionizing radiation-induced DNA strand breakage. Int. J. Radiat. Biol. 41: 71-77.

Christman JK, Weich N, Schoenbrun B, Schneiderman N, Acs G (1980) Hypomethylation of DNA during differentiation of Friend erythroleukemia cells. J. Cell Biol. 86: 366-370.

Cioe L, McNab A, Hubbel HR, Meo P, Curtis P, Rovera G (1981) Differential expression of the globin genes in human leukemia K562(s) cells induced to differentiate by hemin or butyric acid. Cancer Res. 41: 233-243.

Cramp WA, Edwards JC, George AM, Sabovljev SA, (1984) Subcellular lesions: the current position. Brit. J. Cancer 49, suppl. VI: 7-11.

Doerfler W (1983) Dna methylation and gene activity. Ann. Rev. Biochem., 52: 93-124.

Drzynkiewicz Z, Traganos F, Kapuscinski J Melamed MR (1985) Denaturation and condensation of DNA in situ induced by acridine orange in relation to chromatin changes during growth and differentiation of erythroleukemia cells. Cytometry 6: 195-207.

Einspenner M, Boulton JE, Borsa J (1984) Variation of radiation sensitivity of friend erythroleukemia cells cultured in the presence of differentiation inducer DMSO. Radiat.Res. 97:55-63.

Farzanesh F, Shall S, Johnstone AP (1985) The dynamic nature of DNA strand breaks present in differentiating muscle cells and quiescent lymphocytes. FEBS 189: 62-66.

Fielden EM, Sapora O (1982) Radiation induced cellular DNA damage and repair, and the effect of hypoxic cell radiosensitizers. In: Advanced topics on radiosensitizers on hypoxic cells, Breccia A, Rimondi C and Adams GE (eds), Plenum PC, 105-117.

Francis GE, Gray DA, Ho AD, Berney JJ, Wing MA,Yaxley JJ,Ma DDF, Hoffbrand AV (1984) DNA strand breakage and ADP-ribosyl trans-ferase mediated DNA ligation during stimulation of human bone marrow colony stimulating activity. Leukemia Res. 8: 407-415.

Frankenberg-Schwager M, Frankenberg D, Blocher D, Harbish R, Adamczyk C (1982) Irreparable DNA double strand breaks induced in eukaryotic cells by sparsely or densely ionizing radiation and their importance for cell killing. Mutat. Res. 96: 132-146.

Gasser SM, Laemmli UK (1987) A glimpse at chromosomal order. Trends Genet., 3: 16-22.

Greer WL, Kaplan JG (1986) Early nuclear events in lymphocyte proliferation. Exp. Cell Res. 166: 399-415.

Groudine M, Kohwi-Shigematsu T, Gelinas R, Stamatoyannopoulos G, Papayannopoulou T (1983) Human fetal to adult hemoglobin switching: changes in chromatin structure of the beta-globin gene locus. Proc. Natl. Acad. Sci. USA, 80: 7551-7555.

Hancock R, Boulikas T (1982) Functional organization in the nucleus. Int. Rew. Cytol. 79: 165-214.

Hartwig M, Matthes E,Kopp J, Jantscheff P, Milleck J (1984) DNA organitation changes after butyrate induced differentiation of leukemic K562 cells. Studia Biophys. 99: 185-192.

Heussen C, Neckerdien Z, Smit BJ, Bohm L (1987) Irradiation damage in chromatin isolated from V-79 chinese hamster lung fibroblasts. Radiat. Res. 110: 84-94.

Jalouzot R, Briane D, Ohlenbrusch HH, Wilhelm ML, Wilhelm FX (1980) Kinetics of nuclease digestion of Physarum polycephalum nuclei at different stages of the cell cycle. Eur. J. Biochem. 104: 423-431.

Johnstone AP, Williams GT (1982) Role of DNA breaks and ADP-ribosyl transferase activity in eukaryotic differentiation demonstrated in human lymphocytes. Nature 300: 368-370.

Kautianen TL, Jones PA (1985) DNA methylation in mammalian nuclei. Biochemistry 24: 5575-5581. Lewin B (1987) Genes III. John Wiley & Sons Inc., New York.

Lyon SB, Link H (1986) DNA methylation: The enzymes and a naturally occurring inhibitor. Fed. Proc. 45: 19-27.

McMahon G, Alsina JL, Levy SB (1984) Induction of a Ca^{+}, Mg^{2+} dependent endonuclease activity during the early stages of murine erythroleukemic cell differentiation. Proc. Natl. Acad. Sci. USA 81: 7461-7465.

Mee LK, Adelstein SJ (1981) Predominance of core histones in formation of DNA-protein crosslinks in irradiated chromatin. Proc. Natl. Acad. Sci. USA 78:2194-2198.

Nackerdien Z, Michie J, Bohm L (1989) Chromatin decondensed by acethylation shows an elevated radiation response. Radiat. Res. 117: 234-244.

Parasassi T, Conti F, Gratton E, Sapora O (1987) Membranes modification if differentiating proerythroblasts. Variation of 1,6-diphenyl-1,3,5-hexatriene lifetime distributions by multifrequency phase and modulation fluorometry. Biochim. Biophys. Acta 898: 196-201.

Patil MS, Locher SE, Hariharan PV (1985) Radiation-induced thymine base damage and its excision repair in active and inactive chromatin of HeLa cells. Int. J. Radiat. Biol. 48: 691-700.

Pienta KJ, Coffey DS (1984) A structural analysis of the role of nuclear matrix and DNA loops in the organization of the nucleus and chromosome. J.Cell Sci. (Suppl.1): 123-135.

Regan JD, Setlow RB (1974) Two forms of repair in the DNA of human cells damaged by chemical carcinogens and mutagens. Cancer Res. 34: 3318-3325.

Scarpa S, Uhlendorf BW, Cantoni GL (1985) The differentiation of L5/A10 myoblast cell line (a subclone of L5 line) is controlled by changes of culture conditions. Cell Differ. 17: 105-114.

Scher W, Friend C. (1978) Breakage of DNA and alterations in folded genomes by inducers of differentiation in friend erythroleukemic cells. Cancer Res. 38: 841-849.

Sheffery M, Rifkind RA, and Marks PA (1982) Murine erythroleukemia cell differentiation: DNase I hypersensitivity and DNA methylation near the globin genes. Proc. Natl. Acad. Sci. USA, 79: 1180-1184.

Smith MF, Athey Bd, Williams SP, Langmore JP (1990) Radial density distribution of chromatin: evidence that chromatin fibers have solid centers. J. Cell Biol. 110: 245-254.

Smith PJ (1986) n-butyrate alters chromatin accessibility to DNA repair enzymes. Carcinogenesis 7:423-429.

Tolberg ME, Smith SS (1984) Methylation of middle repetitive DNA sequence class during differentiation in Friend erythroleukemia cells. FEBS 176: 250-254.

Wang TS, Wheeler KT (1978) Repair of X-ray-induced DNA damage in rat cerebellar neurons and brain tumor cells. Radiat. Res. 73: 464-475.

Warters RL, Chielders TJ (1982) Radiation-induced base damage in replicating chromatin. Radiat. Res. 90: 564-574.

Warters RL, Lyons BW, Chiu SM, Oleinick NL (1987) Induction of DNA strand breaks in trascriptionally active DNA sequences of mouse cells by low doses of ionizing radiation. Mutat. Res. 180: 21-29.

Wheeeler KT, Wierowski JV (1983a) DNA repair kinetics in irradiated undifferentiated and terminally differentiated cells. Radiat. Environ. Biophys., 22: 3-19.

Wheeler KT, Wierowski JV (1983b) DNA accessibility: a determinant of mammalian cell differentiation ? Radiat. Res. 93: 312-318.

Wheeeler KT, Nelson GB (1987) Saturation of a DNA repair process in dividing and nondividing mammalian cells. Radiat. Res. 109: 109-117.

Wojcierowski JK (1983) Eukariotic genome: model consideration. J. Theor. Biol., 103: 379-403.

THE LETHALITY OF RADIATION-INDUCED DNA DOUBLE-STRAND BREAKS FOR RADIATIONS OF DIFFERING LET

Kevin M. Prise, Melvyn Folkard, Susan Davies and Barry D. Michael
Cancer Research Campaign Gray Laboratory,
PO Box 100,
Mount Vernon Hospital,
Northwood, Middlesex,
HA6 2JR, UK.

Using the filter elution technique it is possible to measure DNA double-strand breaks (dsb) in the same dose range as cell survival measurements. Unlike other techniques for measuring dsb, the dose-effect relationship for dsb induction with low LET radiation is non-linear with dose. The shape of the induction curve mirrors that of the cell survival curve such that a plot of lethal lesions (-*ln*[surviving fraction]) versus dsb produces a linear relationship (Radford, 1985). We have repeated these measurements for radiations of differing linear energy transfer (LET) and for breaks induced chemically by H_2O_2 treatment. The yield of DNA dsb and cell survival in Chinese hamster V79 cells were measured after irradiation at 4°C by 400 keV electrons, 250 kVp X-rays, 1.90 MeV, 1.15 MeV, 0.76 MeV protons, 2.3 MeV neutrons and 3.8 MeV ^{238}Pu α-particles. For H_2O_2 treatment, cells were incubated with the drug (10^{-5}–10^{-2}M) for 20 minutes also at 4°C. Depending on the source used, cells were irradiated in suspension (X-rays and neutrons), attached on 13 mm diameter Thermanox coverslips (electrons) or collected onto polyvinylidine difluoride membrane filters (protons, α-particles and H_2O_2). Immediately after irradiation, cells were assayed for DNA dsb using the filter elution technique (pH 9.6) or assayed for their colony-forming ability.

RESULTS AND DISCUSSION

DNA dsb induction curves were similar for all radiations studied and all RBE's were close to unity. For cell survival with increasing LET of radiation an increasing RBE was measured up to a value of 4.23 (10% survival) for α-particles (110 keV/μm). Plots of lethal lesions (-*ln*[surviving fraction]) versus dsb were approximately linear with increasing slope with increasing LET. The slope values obtained from this plot are shown versus LET in figure 1. These slope values are a measure of the relative probability of a dsb producing a lethal event.

NATO ASI Series, Vol. H 54
The Early Effects of Radiation on DNA
Edited by E. M. Fielden and P. O'Neill

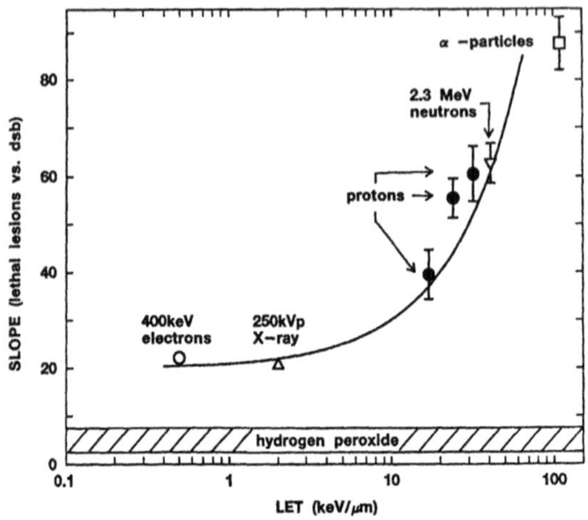

FIGURE 1. Relative probability of a dsb leading to a lethal event *vs.* LET

Thus with increasing LET of radiation there is an increased probability of an induced dsb leading to a lethal event. Dsb induced by α-particles had 4-5 times higher lethal probability than those induced by low LET radiations. Also shown in figure 1 is the slope value obtained for chemically induced dsb (H_2O_2). These had 5 times lower lethal probability than those induced by low LET radiations. This apparent 20-fold range in the lethality of a dsb may be explained in terms of differences in the types of dsb being induced and/or differences in their reparabilty. DNA dsb produced by H_2O_2 are thought to be due to the action of a few OH radicals whereas low LET breaks may be associated with locally multiply-damaged sites (LMDS) (Ward, 1985). We have other evidence, from plasmid DNA studies, which suggests differences in the multiplicity of damaged sites (see Michael *et al.* these proceedings). Differences in the multiplicity of damaged sites with increasing LET of radiation may explain differences in the lethality of the breaks measured. Evidence also suggests differences in both the rate and extent of repair of dsb with LET (Fox and McNally, 1988) which could also be linked to differences in the types of lesions being scored as DNA dsb.

REFERENCES

Fox JC,McNally NJ (1988) Cell survival and DNA double-strand break repair following X-ray or neutron irradiation of V79 cells. Int J Radiat Biol 54: 1021-1030

Radford IR (1985) The level of induced DNA double-strand breaks correlates with cell killing after X-irradiation. Int J Radiat Biol 48:45-54

Ward JF (1985) Biochemistry of DNA lesions. Rad Res 104:S103-S111

SINGLE-STRAND BREAKS AND BASE DAMAGE IN DNA OF HUMAN WHITE BLOOD CELLS IN FULL BLOOD EXPOSED TO IONIZING RADIATION DETECTED AT BIOLOGICALLY RELEVANT DOSES

A.A.W.M. van Loon*, G.P. van der Schans, A.J. Timmerman, F.J.A. Kouwenberg, R.H. Groenendijk, P.H.M. Lohman* and R.A. Baan
TNO Medical Biological Laboratory
Lange Kleiweg 139
2288 GJ Rijswijk
The Netherlands

Both physical (UV- and X-irradiation) and chemical agents can damage cells. Of special interest are the DNA molecules, the carriers of the genetic information. Damage inflicted on DNA, unless repaired, may kill cells by interference with transcription and replication or by the induction of mutations. Non-lethal mutations may result in cells with altered properties which may contribute to ageing, heart diseases and cancer.
Exposure of cells to ionizing radiation damages the DNA. This damage comprises strand breaks and base modifications. It is important to quantify the various damages to get information about their relative contribution and persistency. In case of (accidentally) irradiated persons, analysis of DNA damage in, *e.g.*, blood cells could reveal exposure dose and recovery perspectives. In this context we are developing sensitive immunochemical and biochemical methods to quantify single-strand breaks, alkali-labile sites and base damages.

The immunochemical method is based on the binding of a monoclonal antibody to single-stranded regions in cellular DNA that result from partial unwinding under strictly controlled conditions. Strand breaks and alkali-labile sites form initiation points for the unwinding. The extent of the unwinding (% single-strandedness) is a measure of the number of such sites. The method is simple and fast; it takes only 4 h after collection of the blood to get an answer. For human blood *in vitro* exposured to ^{60}Co-γ-rays it has been shown that the relation between radiation dose and the extent of induced single-strandedness is linear; the induction of breaks is detectable at a dose as low as 0.5 Gy.

*Laboratory for Radiation Genetics and Chemical Mutagenesis, Sylvius Laboratory, State University of Leiden, Netherlands.

NATO ASI Series, Vol. H 54
The Early Effects of Radiation on DNA
Edited by E. M. Fielden and P. O'Neill

By incubation of the irradiated blood at 37°C, the repair of the DNA damage was studied. A fast repair was observed in the first 15 min followed by a much slower repair. The kinetics of the repair are not significantly different from those previously assayed with the alkaline elution (Van Loon *et al.* 1990) or competitive ELISA (Van der Schans *et al.* 1989).

The method was also applied on white blood cells (WBC) from a leukemia patient treated with a cytostatic drug (Endoxan) followed by a total-body irradiation before receiving an autologic or allogenic bone marrow transplantation. We were able to detect single-strand breaks after *in vivo* exposure (4.5-8.6 Gy) to 18-MV photons. After a dose of 4.5 Gy it was still possible to detect DNA damage 30 min after the total-body irradiation.

Base damages can be quantified with alkaline elution if preceded by treatment of the DNA with damage-specific endonucleases (*i.c.* an extract of *Micrococcus luteus* bacteria). *M. luteus*-extracts contain endonucleases and glycosylases acting on specific base damages induced into DNA by ionizing radiation and convert these into single-strand breaks or alkali-labile sites. Such *M. luteus*-extracts have been employed to quantitate DNA-base damage present in nucleated cells of human blood exposed *in vitro* or *in vivo* to ionizing radiation. Subsequently, the additional breaks induced in the DNA by the action of base-damage specific enzymes in the extract, were determined by alkaline elution. With this method, single-strand breaks and base damages have been assayed after doses as low as 1.5 Gy. Radiation-induced base damage in DNA of WBC in human blood exposed to ionizing radiation, increased linearly with dose. The removal of base damage is slower than that of single-strand breaks. Even at 90 min after irradiation, base damage is still detectable after a dose of 3 Gy. The method was also applied on WBC from leukemia patients receiving total body irradiation as described above. It was possible to detect radiation-induced base damage after *in vivo* exposure to ionizing radiation. At 105 min after the total-body irradiation with a dose of 7.8 Gy, the directly induced single-strand breaks had disappeared; however, base damage could still be detected.

Van der Schans GP, Van Loon AAWM, Groenendijk RH, Baan RA (1989) Int J Radiat Biol 55:747-760

Van Loon AAWM, Groenendijk RH, Van der Schans GP, Lohman PHM, Baan RA (1990) (submitted to Int J Radiat Biol)

DISCUSSION: CELLULAR DNA STRAND BREAKAGE

P.L. Olive
British Columbia Cancer Research Centre
601 W. 10th Ave.
Vancouver, B.C.
Canada V5Z 1L3

Many techniques are now available for measuring DNA single- and double-strand breaks by ionizing radiation. As we are reminded by the results of Van Loon and co-workers, single-strand break measurements are far simpler to perform than double-strand break assays and are sensitive to damage by lower radiation doses. However, while the number of single-strand breaks is linearly dependent on dose, it is not predictive for cell killing by ionizing radiation. Conversely, the number of initial double-strand breaks has been shown to correlate with cell killing. Double-strand break damage, if unrepaired or inaccurately repaired, is believed to be the likely "lethal lesion", leading first to chromosome breakage and then to loss of reproductive capacity, mutation and perhaps transformation. While there is little doubt that initial DNA damage is responsible for the final outcome of cell death, time and multiple DNA repair events separate these two processes. It is therefore somewhat unexpected that such good predictions for cell killing across a variety of cell lines can be obtained with the filter elution assay based only on initial numbers of DNA double-strand breaks.

At various times during the Workshop, the concept of lesion subsets was addressed. All DNA double-strand breaks are not likely to be equal in terms of their chemical nature or biological effect; some double-strand breaks are more likely to lead to chromosome damage and cell killing perhaps because they are less able to be repaired. Methods for measuring double-strand breaks probably do not distinguish between these subsets. In addition, measurement of initial damage does not take into account differences in rates or accuracy of repair of lesions, factors which are known to play a major role in cell survival and mutation.

The relevance of DNA damage measurements in predicting cell response to radiation damage was discussed at length, especially with regard to the neutral filter

NATO ASI Series, Vol. H 54
The Early Effects of Radiation on DNA
Edited by E. M. Fielden and P. O'Neill

elution method (or neutral "illusion" method, as one participant quipped) now in widespread use. Radford's excellent correlation between DNA damage and cell killing in fibroblasts (Radford et al., 1986) promises, within certain limitations, a method for predicting cell viability after radiation exposure. As shown by Prise and Michael, mammalian cell killing also correlates well with DNA double-strand break induction for high LET radiations, although the slope of the curve is increased due to the increased killing efficiency of the DNA damage. For low LET radiation, Radford showed that the correlation between initial DNA damage and cell killing is altered for ataxia telangectasia cells, similar to results reported by Iliakis (1988) for xrs-5 cells. It is not surprizing that the correlation breaks down for these cells which are deficient in repair. But perhaps this is telling us that the variety of radiation responses observed for different cell lines is not a result of relative repair deficiency or proficiency, but instead reflects some physical property of the cell which the neutral filter elution assay is able to detect.

The linear correlation between initial double-strand breaks and cell killing is of particular importance since conclusions about the mechanism of radiation killing have been drawn based on the presence of a shoulder on the dose-response curve for double-strand break induction (Radford et al., 1988). The mechanism for such a non-linear induction of double-strand breaks by low LET radiation remains an area of considerable controversy. Could the linear-quadratic dependency indicate one and two-track action as initially proposed by the Chadwick-Leenhouts model? Or perhaps the non-linearity reflects inherent differences in cell lines (e.g., levels of radioprotective molecules). On a statistical basis, the probability of two independent single-strand breaks interacting to produce a double-strand break is highly unlikely. The basis for a possible chemical saturation process needs to be defined. As shown by several investigators at this Workshop, other methods for measuring DNA double-strand breaks, such as gel electrophoresis, neutral DNA precipitation and neutral sucrose sedimentation generally demonstrate a linear dependency between DNA damage and radiation dose. Regardless of the reason for the non-linearity of double-strand break induction measured with the filter elution method, the lesion (or subset of lesions) measured by this method appears to correlate very well with cell killing. Understanding why non-linearity is detected in initial events could tell us much more about what factors influence radiation-induced cell killing.

While the majority of laboratories report non-linear induction of DNA double-strand breaks with dose using the filter elution method, it is not a universal observation. Furthermore, significant variations in the shape of the dose-response curve have been observed for different lysing conditions (e.g., detergent type, pH, proteinase K digestion) suggesting that factors other than initial damage contribute to the shape of the dose-response curve. Results by Iliakis indicating less rapid elution of replicating DNA also need to be considered when applying this method. Inhibition of elution by protein binding to DNA (perhaps an explanation for the slow elution of replicating DNA) cannot be discounted as we were reminded that some DNA binding proteins can only be removed by boiling (Bodner et al., 1983). Certainly a biological basis for these non-linear kinetics remains a viable explanation especially considering the complexities of DNA conformation described by Sapora which might influence the rate and extent of DNA elution. Even some single-strand break assays can be influenced by the "packaging" of the DNA molecule (Olive et al., 1986).

Another possible (but less likely) explanation for the non-linear kinetics observed using the elution method is the way in which data are collected and analyzed. Some investigators base their calculations on the logarithm of the fraction of DNA eluted, while others use a linear relationship. Using the logarithm was considered to be more accurate, but the differences in results using the two methods was not thought to be large. Radford cautioned that adequate calibration of the neutral filter elution method, using ^{125}I, was mandatory.

As proposed by Iliakis, one approach to overcome the limitations of the neutral elution method is to use a second method to examine double-strand breaks. The pulsed-field gel electrophoresis assay was suggested by Iliakis to be technically superior to the filter elution method for measuring DNA double-strand breaks. Specifically, more samples can be handled with greater ease, and the assay appears to be equally sensitive in detecting DNA double-strand damage. However, as with the filter elution method, there are problems with variability due to lysis conditions and cell cycle effects (replicating DNA migrates slower in the gel). In addition, the physical basis for the method has not been adequately established and molecular weights of DNA molecules cannot be unequivocally determined.

A comment made later in the meeting by John Warman is perhaps an appropriate summary for this session - that chemists use well-defined systems and treat radiation as the variable, while biologists seem to treat radiation as the constant and examine how the DNA responds. This insight emphasizes the uncertainty biologists face when dealing with complex cellular systems. Certainly a better understanding of the factors which influence the response of DNA in our assays is essential. While we risk studying assays rather than questions, until we understand all of the variables which influence our measurements, it will be difficult to defend specific mechanistic interpretations of our data.

Bodner JW, Jones CJ, Coombs DH, Peason GD and Ward DC (1983) Proteins tightly bound to HeLa cell DNA at nuclear matrix attachment sites. Mol Cell Biol 3: 1567-1579

Iliakis G, Okayasu R and Seaner R (1988) Radiosensitive xrs-5 asnd parental CHO cells show identical DNA neutral filter elution dose-response: implications for a relationship between cell radiosensitivity and induction of DNA double-strand breaks. Int J Radiat Biol 54: 55-62

Olive PL, Hilton J and Durand RE (1986) DNA conformation of Chinese hamster V79 cells and sensitivity to ionizing radiation. Radiat Res 107:115-124

Radford IR (1986) Evidence for a general relationship between the induced level of DNA double-strand breakage and cell killing after X-irradiation of mammalian cells. Int J Radiat Biol 49: 611-620

Radford IR, Hodgson GS and Matthews JP (1988) Critical DNA target size model of ionizing radiation-induced mammalian cell death. Int J Radiat Biol 54: 63-79

THE ROLE OF RADIATION INDUCED CHARGE MIGRATION WITH DNA: ESR STUDIES

MARTYN C. R. SYMONS

[Department of Chemistry, The University, Leicester, LE1 7RH, U.K.]

E.s.r. spectroscopy is useful for studying primary processes in irradiated DNA. From the results it is argued that primary radical-cations migrate into the DNA base system, becoming trapped at guanine by proton loss (G^{+}-H^{+}). Similarly, ejected electrons migrate to, and within, the base system, becoming trapped at cytidine ($C\cdot^{-}H^{+}$) or thymine ($T\cdot^{-}H^{+}$) by proton attachment. When dilute frozen aqueous systems are used, ($T\cdot^{-}H^{+}$) seems to be a major product and is well separated from the ($G\cdot^{+}$-H^{+}) units. It is stressed that trapping is probably controlled by the rates of protonation and that these vary greatly with temperature and environment. Sites of such protonation are considered, and these are also likely to be very sensitive to the conditions used.

In frozen aqueous systems, the effective target is glassy DNA, crystalline water being a separate phase whose damage is not relevant to the DNA ions. Various additives can change the volume of the glassy region and hence change the yields of DNA radicals.

Finally, some outstanding problems in this field are discussed.

INTRODUCTION

For dilute aqueous DNA exposed to ionizing radiation, most ionizing events will comprise electron ejection from water [1]. The two primary ions will rapidly reach 'stable' solvated states but, prior to this, migration must be extensive. For $H_2O\cdot^{+}$, this occurs by electron-transfer [2], which is an extremely fast process, and will continue until process [3] occurs. Thus, solvated electrons $(e^-)_{aq}$ and solvated $\cdot OH$ radicals are not expected to be close to each other in pairs at this stage.

$$H_2O \rightarrow H_2O\cdot^{+} + e^- \quad .... [1]$$

$$H_2O\cdot^{+} + H_2O \rightleftharpoons H_2O + H_2O\cdot^{+} \quad .... [2]$$

$$H_2O\cdot^{+} + H_2O \rightarrow HO\cdot + H_3O^{+} \quad .... [3]$$

NATO ASI Series, Vol. H 54
The Early Effects of Radiation on DNA
Edited by E. M. Fielden and P. O'Neill

Most discussions centre attention on attack by ·OH radicals on DNA. Possibly this is because the 'shell' of negative charge around DNA hinders electron addition to the base system? Certainly, the major products can be ascribed to the results of attack by ·OH radicals. It might be supposed that a similar sequence of events would apply to frozen dilute aqueous DNA solutions, but this is not the case, as is shown below.

I start by considering different systems and methods of study. This is followed by a consideration of e^- and H^+ movement and of average distances travelled by electrons and holes prior to trapping. Some aspects of target volume are discussed followed by a brief consideration of the sites at which electrons are trapped. Finally, aspects of secondary reactions are considered, together with some speculation regarding certain outstanding problems.

Types of DNA Systems

These include dilute aqueous solutions at ambient temperatures, similar frozen solutions, dry DNA, minimally hydrated DNA (including oriented fibres or ribbons), and DNA in water containing a range of additives. Also, DNA from various sources have been studied, as well as single-stranded (SS) DNA, various synthetic oligonucleotides, etc. More recently, studies have been extended to include DNA-histone complexes, chromatin and cell nuclei, also in the presence of various additives. These studies are beginning to bridge the gap between fundamental work on the various components of DNA and the equally extensive work on cell-lines and on whole bodies. One rather negative aim of the present work is to show that results may vary widely using DNA in different situations, and that extrapolation from one system to another may be dangerous.

In our own work, solid systems have been used in order to facilitate the application of e.s.r. spectroscopy [Boon 1984; Boon 1985; Cullis 1985; Cullis 1986; Cullis 1987; Symons 1987]. Very recently, e.s.r. signals have been obtained from DNA in dilute aqueous solutions at room temperature [Hildenbrand 1989, 1990] but the significance of these results is not yet clear. Obviously, it is necessary to know how the e.s.r. results tie in with ultimate damage. We have made the crude assumption that a study of single and double strand breaks (SSB and DSB) give us a measure of overall damage, but this is only a first approximation. Nevertheless, there is considerable opinion that DSB's are a major source of chromosome aberrations, of cell mutation and of cell death. We therefore look especially at the effects of additives on yields of DSBs.

Movements of Electrons, Holes and Protons

Our results leave little doubt that there is an extremely rapid stage in which electrons and holes migrate into the stacked base system. Infrared studies suggest that our systems comprise two phases, pure crystalline ice, and a glassy DNA phase. This result depends upon the use of HOD, D_2O solutions and studying the isolated O-H stretching band [Blackburn, unpublished results]. This is a sharp symmetrical feature for ice, and a broader asymmetric feature for the O-H oscillators in the DNA-glass phase. We suggest that electrons and holes generated in this glassy region migrate efficiently to DNA.

All additives either accumulate in the glass phase, or at high concentrations, produce a third phase, also detectable by infrared spectroscopy (Fig. 1). The results strongly suggest that most electrons

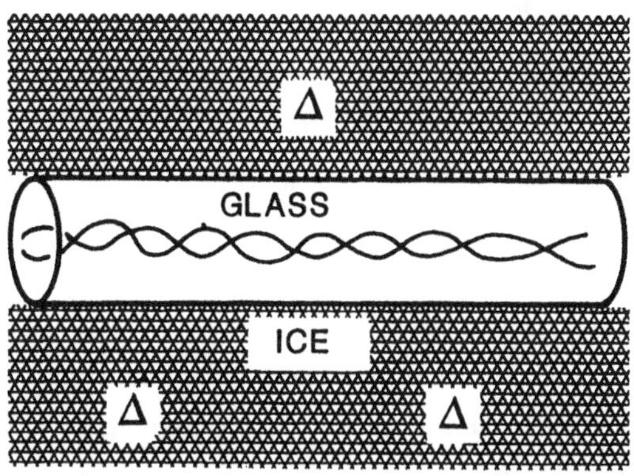

Fig. 1

Schematic representation of a DNA duplex ion in a frozen aqueous system. The inner cylinder is thought to be glassy water which will contain any additives besides the DNA anions and sodium cations. The outer regions represent pure ice crystallites together (Δ) with crystallites of a third phase possibly formed from an additive such as NaCl.

and holes formed in the glass-phase migrate to the bases before any of the other potential chemical reactions can occur. So, for example, reaction [2] is fast relative to [3], thereby facilitating hole migration. Also, the potentially stable $>P-\dot{O}$ electron loss centres (characterised by a ^{31}P doublet and relatively large *g*-shift [Hughes 1963; Subramanian 1970]) are not detected. Similarly, the three possible reactions of electrons with phosphate [4-6] should all be detectable, but are not seen.

$$(RO)_2PO_2^- + e^- \rightarrow (RO)_2PO_2^{2-} \quad .... [4]$$
$$(RO)_2PO_2^- + e^- \rightarrow RO^- + \cdot PO_2(OR)^- \quad .... [5]$$
$$(RO)_2PO_2^- + e^- \rightarrow R\cdot + (RO)PO_2^{2-} \quad .... [6]$$

The phosphoranyl radicals in [4] and the phosphoryl radicals in [5] should both be characterised by well defined ^{31}P doublets with splittings in the region of 600-1,000 G. In our studies of electron capture by phosphate esters [Nelson 1975] one or both of these species were always detected together with alkyl radicals [6] on electron capture. Hence their absence in the DNA spectra argues against reaction [6]. These radicals (I) should

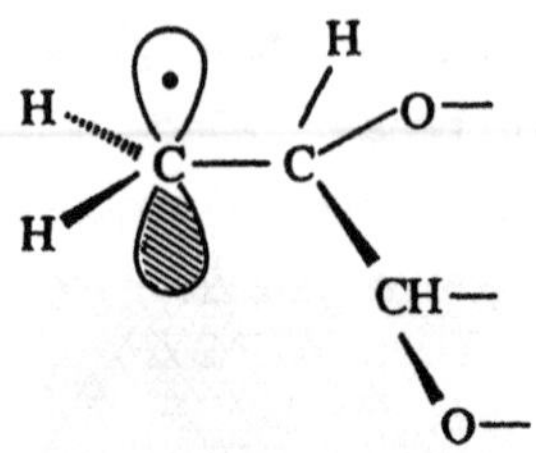

(I)

be characterised by an anisotropic $-CH_2$ triplet (a ~22 G) together with an isotropic doublet of unpredictable magnitude. Such features are not evident in the normal spectra obtained from frozen aqueous DNA. Also, trapped electrons are not formed in these glasses.

Once in the base-stacks, there is good evidence for large relative migrations of holes and electrons prior to trapping. In our earlier studies of non-interacting electron-capture agents [Boon 1985; Cullis 1985], we estimated average separations of the order of 30 base-pairs (b-p). However, in our more recent studies of intercalators (I), we found much greater sensitivity, with clear reduction in DNA radicals for b-p/I ratios of 150 [Cullis 1990]. In fact, there was a *ca.* 50% reduction for a ratio of *ca.* 110 (Fig. 2). These results can only be explained on the assumption that there is extensive charge migration. Because of the more diffuse nature of the SOMO for base radical-anions compared with cations, we suggest that the major migration is of electrons rather than holes, but this problem is still being investigated. However, it is the relative migration that is important. As has already been stressed [Rackovsky 1989], because of the screening effect of large concentrations of phosphate anions and sodium cations, electron return *via* coulombic attraction falls off rapidly with separation and would be insignificant under our

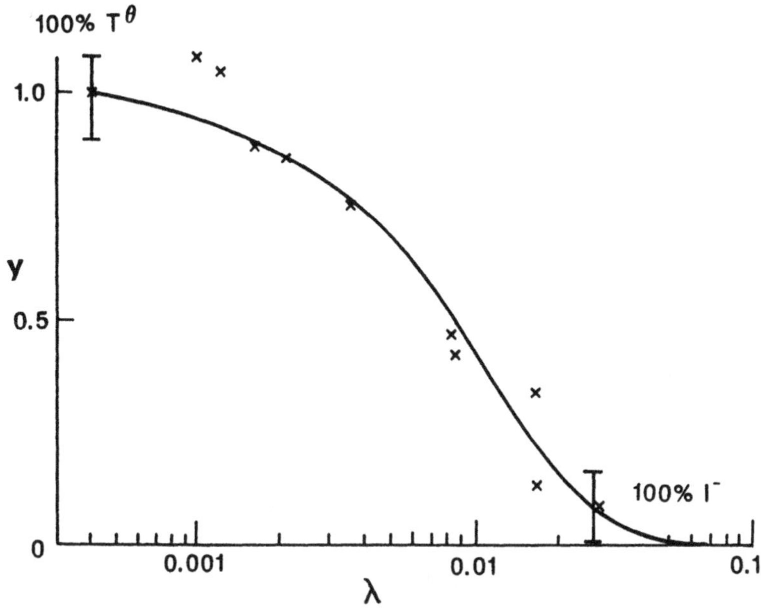

Fig. 2

Trend in the yield of ($T\cdot^-H^+$) and/or ($C\cdot^-H^+$) as the concentration of intercalator (I) is increased; $y = [T\cdot^-]/([T\cdot^-] + [I\cdot^-])$ and λ is the ratio of the number of I molecules to the number of base-pairs.

conditions. Conductivity, however, must be extremely short-lived, since it is most probable that the electron- and hole-centres are rapidly stabilised by proton transfer [Steenken 1989; Cullis 1986]. This, unless it is reversible, should completely quench further charge migration.

There is much evidence, especially from studies using oriented DNA [Gräslund 1981], in favour of the postulate that the main hole centre is confined to guanine bases. This accords with the fact that, for the isolated bases, the ionization potential for guanine is the lowest (8.2 eV). There is some controversy regarding which proton is involved in stabilising the $G\cdot^+$ centre to give $G^+(-H^+)$ [Hüttermann 1987; Hole 1987], but there is little doubt that it is the (N_1) proton that is lost for the monomer in water [Steenken 1989]. The constraints of crystal ctructure can impose themselves just as can the constraints within the DNA duplex.

For many years it was accepted that electrons were captured largely at thymine sites. Competition experiments, if they are to be trusted, imply that capture at C or T should be favoured, but do not lead to a clear distinction between $C\cdot^-$ and $T\cdot^-$ centres. In our own studies, using water or methanol glasses, these are formed in about equal yield [Hole 1989]. However, it is clear that C·⁻ is protonated extremely rapidly (to give C·⁻(H⁺) units, characterised by an e.s.r. triplet [Cullis 1989]) and most probable that T·⁻ is also protonated, to give T·⁻(H⁺) [Sagstuen 1989]. In that case, the electron-transfer [7] which would lead to the most stable

$$C^{\cdot -} + T \leftrightharpoons C + T^{\cdot -} \qquad \text{.... [7]}$$

which would lead to the most stable anion, is inhibited at a very early stage, and the 50% result cannot be taken as being significant.

The reason why T·⁻ was favoured in earlier studies was because, on annealing, ·TH is formed (II). The new results of Hüttermann and his co-

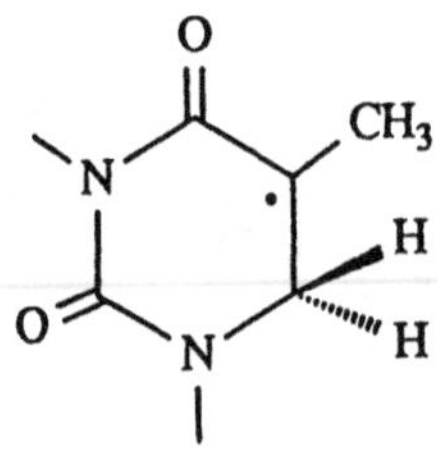

(II) (·TH)

workers throws doubt on this conclusion, since they claim that there is only *ca.* 10% conversion [Hüttermann, personal communication]. Taken together with their other evidence using oriented solid DNA, which suggests that C·⁻ centres predominate strongly over T·⁻ centres on initial electron-capture, this result looks compelling. One aim of the present work is to stress that the nature of the species formed is strongly controlled by protonation. This, in turn, is a subtle function of environment. This is nicely illustrated by the recent results for the protonation of C in frozen glasses [Cullis 1989], and of T in crystalline thymidine [Sagstuen 1989]. These are remarkable in that, despite the large amount of preceding work, they reveal for the first time that protonation is rapid, even at 4 K, and that, under these particular conditions, there are large extra doublet splittings from the added protons [see structures (III) and (IV)]. [This was not previously

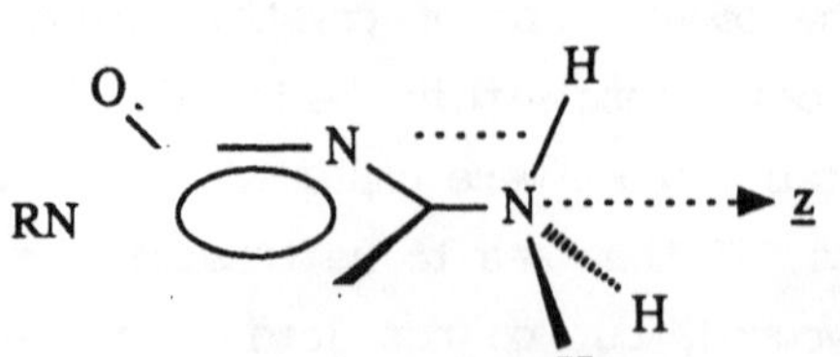

(III)

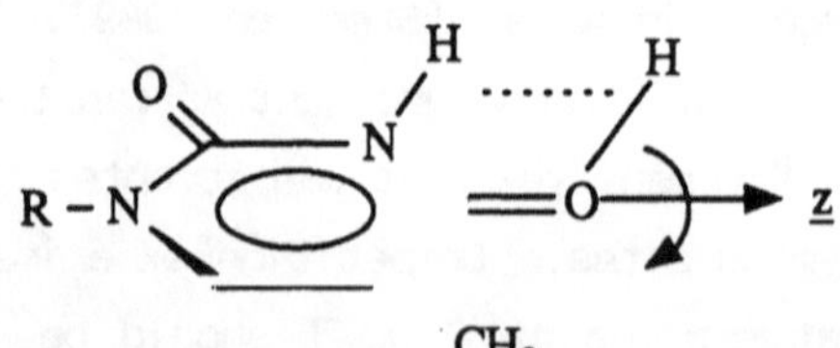

(IV)

appreciated because deuterated samples were always used.] These results are important because they show that protonation is rapid. They also show that, for thymine, the site of protonation is the C_4 carbonyl oxygen and that, for cytosine, it is not the expected ring nitrogen atom (N_3) but probably the NH_2 group (N_4). In the former, the proton is out-of-plane, rather than being in the more stable in-plane site. The large coupling is due to hyperconjugation and would fall to a small value on rotation about the *z*-axis into the molecular plane. Hence, in-plane protonation, expected for $T\cdot^-$ in DNA, would not give rise to a large extra splitting.

Similarly, in DNA, $C\cdot^-$ might be protonated on N_4 rather than on the NH_2 group (though clearly the issue must be a subtle one), in which case again, the extra splitting would probably be small. I suggest therefore, that the site of protonation, and which anion is to be protonated, is controlled by subtle kinetic factors, and hence could well differ on going from "dry" oriented DNA to dilute aqueous DNA; from room temperature to 77 K; or from DNA to chromatin.

Can $\cdot$TH formation (II) be taken as a guide to the presence of $T\cdot^-(H^+)$? It is possible that reaction [7] can occur on annealing, such that <u>some</u> $T\cdot^-$ is formed from $C\cdot^-$ and that rapid conversion of $T\cdot^-$ into $\cdot$TH tips the equilibrium to suit $\cdot$TH formation. If that were correct, then the results from oriented DNA become anomalous. Also, since $C\cdot^-$ and $T\cdot^-$ are almost certainly protonated, it would be necessary to deprotonate prior to electron transfer. It is far simpler to argue that $\cdot$TH comes from $T\cdot^-(H^+)$ without the intervention of $C\cdot^-(H^+)$.

Another important factor is the curious temperature evolution of $\cdot$TH (Fig. 3). Reactions in solids often show this behaviour, in which a certain percentage of a reaction occurs at one temperature, the remainder requiring further heating to be able to proceed [Plonka 1988]. $\cdot$TH radicals grow in and decay in the same temperature range, and hence the maximum in the intensity of the e.s.r. features (occurring at *ca.* 210 K for our systems) cannot be taken as a proper measure of yield. The value can only be taken as a minimum estimate, and the proper yield may be much higher, as indicated in Fig. 3. Our best estimate of this minimum yield for frozen aqueous DNA is *ca.* 40%, which means that $[T\cdot^-(H^+)]$ must be $\gg$40% of the total anion yield. At present, therefore, we conclude that <u>in our systems</u>, the $[C\cdot^-(H^+)]$ is $<$60% and could be small. Possibly the approach taken by Bernhard and coworkers will solve the problem eventually [Bernhard 1989], but even then, the results cannot be safely extrapolated

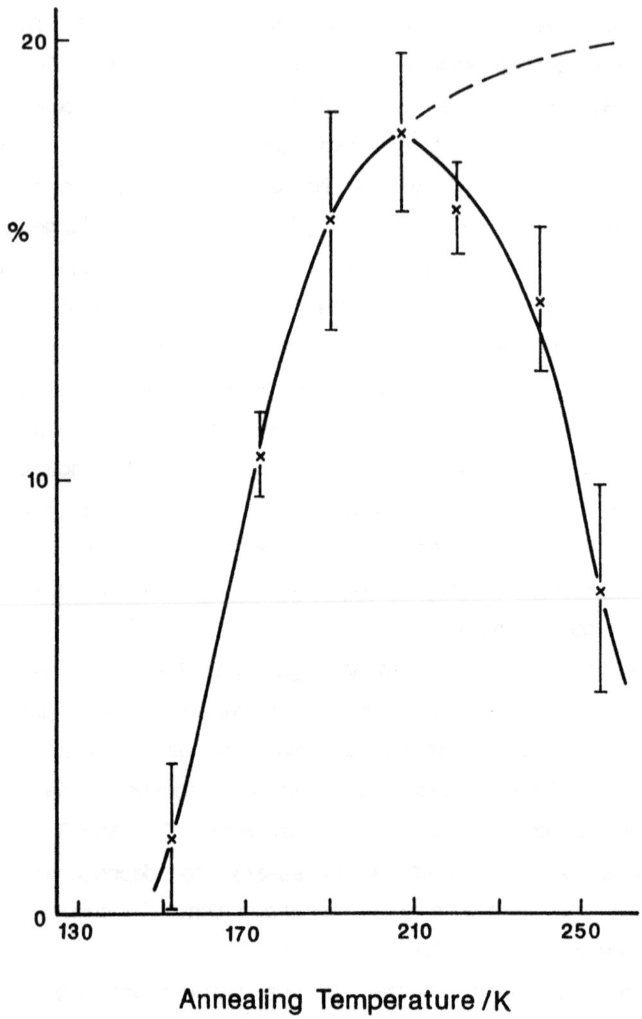

Fig. 3

Temperature profile for the growth and decay of $\cdot$TH radicals formed after exposure of aqueous calf-thymus DNA to ^{60}Co γ-rays at 77 K. The dashed line indicates a possible growth profile in the absence of decay.

to room temperature, nor to DNA in the cell, without much more work. There are, however, some important pointers suggesting that $T\cdot^-$ formation is still important in these more complex systems. One is that there is a marked enhancement in the yield of $\cdot$TH in irradiated chromatin relative to DNA [Cullis 1987]. This may imply migration of electrons from histone proteins onto DNA, as we have tentatively suggested, or it might arise, at least in part, because $C\cdot^-$ formation is less important under these tions. Also, the isolation of dihydrothymine from irradiated cells [Cadet, personal communication] not only supports the direct damage mechanism and the concept of hydrogen atom transfer, but also confirms that $T\cdot^-$ formation is still important.

It is natural to assume that in the DNA duplex, it is the protons of the inter-base hydrogen bonds that move. The alternative, of course, is that water acts as the proton donor or acceptor. In either case, hydrogen bonded protons will surely be involved. In the former case, all that is needed is a minor "vibration" type shift within the bond, as indicated in (V) for the G-C pair. Possibly only the central strong N---H-N unit is

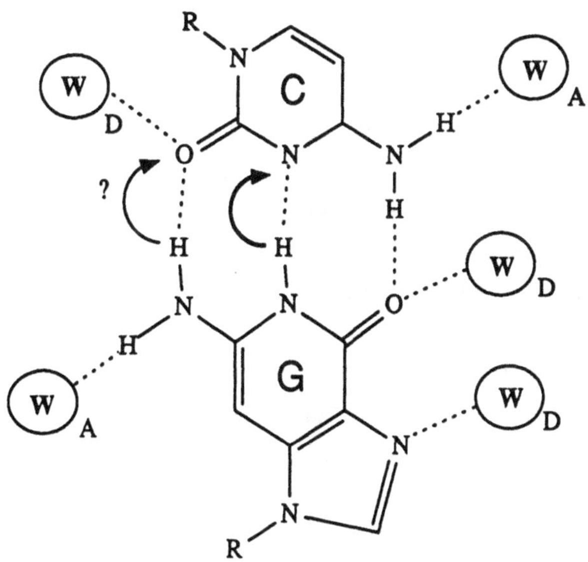

(V)

involved, the proton migrating towards C both for $C\cdot^-$ and for $G\cdot^+$ formation. However, there are expected to be at least five water molecules, the two W_A molecules being potential acceptors and the three W_D molecules potential donors. The potential advantage of water as a mediator is that the resulting H_3O^+ or OH^- ions will rapidly move away and can ultimately neutralise each other. These movements, of course, are themselves rapid proton-transfers rather than being bulk movements of the ions. Implications regarding conductivity changes are worth considering in this context.

<u>Target Volume</u>

It might be thought that estimates of G-values would help in solving some of these problems. This is, however, a difficult task, because of the rôle of the glassy water region discussed above. We have recently found that certain 'inert' cosolvents or salts greatly increase the yields of DNA radicals, for no apparent chemical reasons. These changes can, at least in part, be understood in terms of increases in target volume. These results strongly support our contention that damage to peripheral water can move into the DNA over considerable distances [Blackburn, unpublished results].

Some Outstanding Problems

We are currently extending our studies to include cell nuclei and isolated chromatin [Cullis 1987]. This includes studies of damage to the histone proteins, and to DNA-scaffold protein adducts. However, I stress that many problems remain for the simple DNA systems.

The $C\cdot^-/T\cdot^-$ Problem - As stressed above, the nature of the anion site in DNA remains ambiguous, and it seems certain that some proportion of each of the anions is formed, probably varying with different conditions. This potential variability needs to be borne in mind in all studies, and it is clearly unsafe to use the ·TH yield as a measure of the extent of electron capture. Our results suggest strongly that ·TH sites can lead to strand breaks, possibly *via* the transfer of 'poised' hydrogen atoms from adjacent sugar groups [Symons 1987] (Fig. 4). The fate of $C^-(H^+)$ units still needs to be established. The cytosine analogue of ·TH does not seem to be formed [Gregoli 1979], or if it is, it reacts so rapidly that the signal fails to accumulate. Also, the structure of the species responsible for the triplet detected after electron addition to C, thought to be (III), can hardly be taken as definitive. In particular, we have, so far, been unable to detect rotation about the C-N bond, which is surprising. A remote, but conceivable, alternative is that the species is formed by *ipso* addition to the C_4 carbon to give (VI). It is well established that

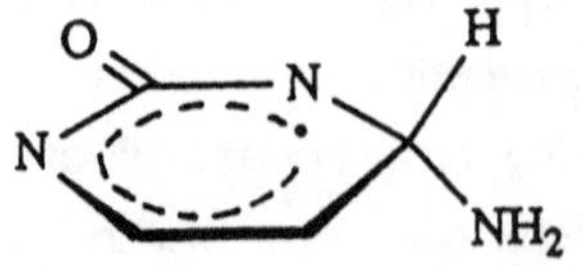

(VI)

deamination to give uracil derivatives occurs during radiolysis and this is a possible intermediate. I would have expected a splitting greater than *ca.* 13 G from the added hydrogen for this structure, but know of no comparable results to draw upon for comparison.

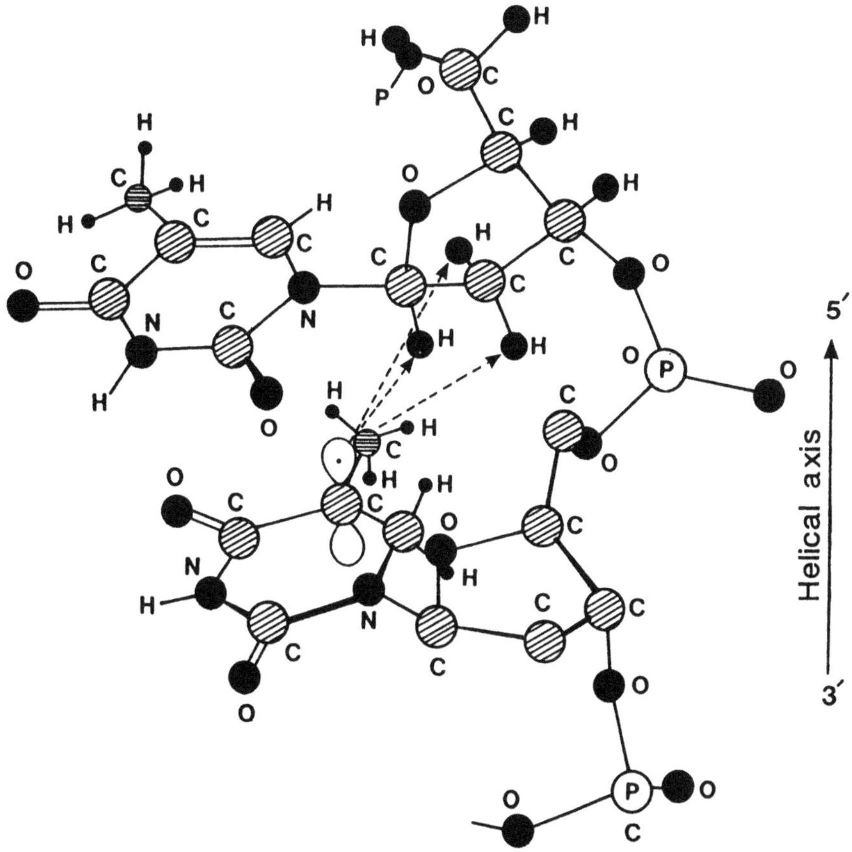

$$TH^{\bullet} + H-C\lessgtr \rightarrow (R)TH_2 + {}^{\bullet}C\lessgtr$$

Fig. 4

Partial structure of DNA, showing a ·TH radical and the nearest C-H units. The arrow indicates transfer of one of these 'poised' hydrogen atoms to give TH_2 and a deoxyribose radical.

Radical Intermediates - Having been remarkably kind in giving us only two or three primary radical centres, nature has been less kind to the e.s.r. spectroscopist with secondary species. The formation of ·TH is clear, but no other intermediate can be seen in the absence of additives. This almost certainly arises because they are too short-lived. For frozen aqueous systems, DNA radical signals have fallen to *ca.* 10% of the initial value by 250 K, which is well below the melting point of the systems. Radicals can only be lost by radical-radical interactions. Given that the radical centres are well separated and remain linked to DNA, how do they manage to encounter each other? We are hoping that a combined NMR-IR study may help to resolve this problem.

Rôle of Additives - In most of our work, electron-scavengers reduce the concentrations of DNA radical-anions, whilst hole scavengers reduce the concentration of $G^{\cdot+}$ centres. As expected, there is a concomitant reduction in the yields of SSBs and a relatively large reduction in DSBs [Boon 1984; Boon 1985; Cullis 1985; Cullis 1986; Cullis 1987; Symons 1987]. However, in our study of intercalators [Cullis 1990], electron and possibly hole scavenging is extremely efficient, but the numbers of SSBs is hardly affected. Whilst possible explanations suggest themselves, this still remains an unsolved paradox.

Acknowledgements

I thank Dr. Paul Cullis and our various collaborators for doing so much over the past few years, and also Dr. Peter O'Neill for very helpful discussions. We thank the Cancer Research Campaign (CRC), the Association for International Cancer Research (AICR) and the Ministry of Defence (MOD) for supporting these studies.

REFERENCES

Bernhard WA (1989) Sites of electron trapping in DNA. J Phys Chem 93:2187-2194.

Bernhard WA (1989) Free radicals formed by electron gain in oligomers of DNA. Free Rad Res Commun 6:15-18.

Blackburn C, Cullis PM, Elsy D, Davies A, Symons MCR (unpublished results).

Blackburn C, Symons MCR (unpublished results).

Boon PJ, Cullis PM, Symons MCR, Wren BW (1984) Effects of ionizing radiation on deoxyribonucleic acid and related systems, Part 1 - The role of oxygen. J Chem Soc Perkin Trans 2 1393-1399.

Boon PJ, Cullis PM, Symons MCR, Wren BW (1985) Effects of ionizing radiation on deoxyribonucleic acid, Part 2 - The influence of nitroimidazole drugs on the course of radiation damage to aqueous deoxyribonucleic acid. J Chem Soc Perkin Trans 2 1057-1061.

Cadet J (personal communication).

Cullis PM, Jones GDD, Symons MCR, Lea JS (1987) Electron transfer from protein to DNA in irradiated chromatin. Nature 330:773-774.

Cullis PM, McClymont JD, Symons MCR (1990) Electron conduction and trapping in DNA: an electron spin resonance study. J Chem Soc Faraday Trans 1 86:591-2.

Cullis PM, Podmore ID, Lawson M, Symons MCR, Dalgarno BG, McClymont JD (1989) The site of electron capture in irradiated deoxyribonucleic acid: cytosine vs thymine. J Chem Soc Chem Commun 1003-1005.

Cullis PM, Symons MCR (1986) Effects of direct radiation on deoxyribonucleic acid. Radiat Phys Chem 27:93-100.

Cullis PM, Symons MCR, Jones GDD, Lea JS, Sweeney MC (1987) The effects of ionizing radiation on deoxyribonucleic acid, Part 5 - The role of thiols in chemical repair. J Chem Soc Perkin Trans 2 1907-1914.

Cullis PM, Symons MCR, Sweeney MC, Jones GDD, McClymont JD (1986) The effects of ionizing radiation on deoxyribonucleic acid, Part 4 - The role of hydrogen peroxide. J Chem Soc Perkin Trans 2 1671-1676.

Cullis PM, Symons, MCR, Wren BW (1985) Effects of ionizing radiation on deoxyribonucleic acid, Part 3 - The effect of iodoacetamide. J Chem Soc Perkin Trans 2 1819-1825.

Gräslund A, Rupprecht A, Köhnlein W, Hüttermann J (1981) Specific formation of electron gain and loss centres in X-irradiated oriented fibres of DNA. Radiat Res 88:1-5.

Gregoli S, Olast M, Bertinchamp A (1979) ESR spectra for the radical anion of cytosine and its protonated product. Radiat Res 77:417-425.

Hildenbrand K, Schulte-Frohlinde D (1989) Electron spin resonance studies of radicals formed during photolysis of aqueous nucleic acid derivatives. Int J Radiat Biol 55:725-730.

Hildenbrand K, Schulte-Frohlinde D (1990) ESR spectra of radicals of single and double stranded DNA in aqueous solution. In: 23 Ann Int Conf Electron Spin Resonance. 26-30 March 1990. Royal Holloway and Bedford New College, Egham.

Hole EO, Nelson WH, Close DM, Sagstuen E (1987) EPR and ENDOR study of irradiated single crystals of purines at 4.2 K. J Chem Phys 86:5218-5223.

Hughes E, Moulton WG (1963) EPR spectra of radicals in irradiated phosphates. J Chem Phys 39:1359-1364.

Hüttermann J, personal communication and this volume p.....

Hüttermann J, Voit K (1987) Free radicals from direct action of ionizing radiation in DNA. In: Weil JA (ed) Electronic magnetic resonance of the solid state (Canadian Soc Chem) 267-275.

Nelson DJ, Symons MCR (1975) ESR spectra and structure of dichlorophoranyl radicals. J Chem Soc Dalton Trans 1164-1167.

Plonka A (1988) Dispersive kinetics in condensed phases. Ann Rep Chem Soc 85C:47-69.

Rackovsky S, Bernhard WA (1989) Electron trapping in double stranded DNA: implications for formation of double strand breaks. J Phys Chem 83: 5006-5012.

Sagstuen E, Hole ED, Nelson WH, Close DM (1989) Structure of the primary reduction product of thymine after X-irradiation at 10 K. J Phys Chem 93:5974-5979.

Steenken S (1989) Electron and proton gain and loss by DNA bases. Chem Rev 89:503-509.

Subramanian S, Symons MCR, Wardale HW (1970) Oxides and oxyions of the non-metals. Part XIII - Electron spin resonance studies of the PO_4^{2-} radical and related species in γ-irradiated phosphates. J Chem Soc (A) 1239-1242.

Symons MCR (1987) Application of ESR spectroscopy to the study of the effects of ionizing radiation on DNA and DNA complexes [The Bruker Lecture]. J Chem Soc Faraday Trans 1 83:1-11.

EARLY CHEMICAL EVENTS IN THE DEVELOPMENT OF RADIATION DAMAGE OF DNA - NOVEL APPROACHES

P. O'Neill, A.T. Al-Kazwini, E.J. Land* and E.M. Fielden
MRC Radiobiology Unit
Chilton
Didcot
Oxon OX11 ORD
U.K.

The various effects of ionising radiation in biological systems are generally thought to be a consequence of chemical modifications induced within DNA (Painter, 1979). Indeed DNA is generally accepted to be the major target which is sensitive to ionising radiation resulting in the various biological effects determined. Following the initial energy deposition events, the radiation-induced damages to DNA most generally measured in biological systems are single and double strand breaks and modified bases. At present, there are little experimental data on the time course of the early processes which occur following the initial energy deposition events ($\sim 10^{-15}$ s). The present biophysical models (Goodhead, this volume) generally relate the distribution of energy deposition events to the biological effects of radiation quality. These models do not consider the effects of dose modifiers, e.g. OER. Therefore, more detailed information of the chemical processes occurring at early times is required.

Up to date the majority of the time-resolved studies on radiation-induced changes in DNA at room temperature have concentrated on the interactions of water radicals generated in dilute, aqueous solution (O'Neill & Fielden, 1986; Steenken, 1989; Von Sonntag, 1987). These studies have provided essential information on the properties and reactivities of radicals formed within DNA (see Fielden *et al.*, this volume). The few time-resolved studies which have been performed using solid DNA have concentrated on radiation-induced electrical conductivity and luminescence

* CRC Paterson Institute for Cancer Research, Department of Biophysical Chemistry, Christie Hospital and Holt Radium Institute, Manchester M20 9BX, U.K.

NATO ASI Series, Vol. H 54
The Early Effects of Radiation on DNA
Edited by E. M. Fielden and P. O'Neill

(Van Lith *et al*., 1983, 1986). In fact, the majority of the information obtained on characterisation of the species produced upon irradiation of DNA has been obtained using the technique of esr whereby the radicals are generally produced at low temperature in ice, glasses or single crystals (Bernhard, 1989; Symons, 1987; Hüttermann *et al*., 1984; Boon *et al*., 1984; Close *et al*., 1985). These systems minimise the effects of water radicals diffusing considerable distances to the DNA so that the damage is initiated by species produced in the vicinity of the DNA. In general the radiolysis of DNA in these systems has identified guanine and thymine/cytosine as the major electron loss and gain centres respectively and rather surprisingly sugar radicals are not detected (Symons, 1987) Of interest in these observations is the fact that charge migration is invoked to explain the localisation of the charge at the identified centres. Recently the use of electron traps to intercept the e^- have shown that a 50% probability of electron capture by the trap occurs at a ratio of 110:1 base pairs to trap (Cullis *et al*., 1990). Interestingly, the yield of positive centres does not significantly change with added trap (Boon *et al*., 1985; Cullis *et al*., 1985). These latter findings are in contrast with those of Gräslund *et al* (1977). Incorporation of misonidazole in orientated solid DNA did cause an increase in positive centres following irradiation at 77K. The DNA radicals reported following irradiation and detected by esr are discussed in more detail in this volume.

In an endeavour to study time-resolved changes in DNA induced by ionising radiation, two techniques have recently been utilised, namely luminescence (Al-Kazwini *et al*., 1988) and diffuse reflectance spectroscopy (O´Neill *et al*., 1989) to gain an insight into the early processes occurring within the DNA environment. These studies have concentrated on (i) the role of radiation-induced energy migration within DNA following the initial energy deposition event and (ii) the role of water of hydration on the properties of the DNA radicals observed. In this paper we discuss some of the conclusions drawn from the earlier studies and present some further evidence for positive charge migration based upon recent diffuse reflectance studies using DNA and polynucleotides.

Role of water of hydration of DNA

In the "dry" state, DNA exists predominantly in the A-form. Addition of water results in conformational changes of the DNA giving the B- and Z- forms. In order to fully hydrate DNA, about 20 water molecules per nucleotide are required (Saenger, 1984).

Many biological studies have demonstrated that addition of radical scavengers such as DMSO and ethylene glycol afford ~ 60-80% protection under aerobic conditions when considering survival (Roots et al., 1982), single strand breakage (Chapman et al., 1975) and chromosomal aberrations (Littlefield et al., 1988). Such protection has been related to dehydration of the DNA or to scavenging of reactive water species such as the hydroxyl radical which diffuses to the target. These species are thought to be produced in water associated with or near to the DNA, since high concentrations of scavenger are required to afford protection. A strong argument against significant perturbation of the water of hydration by such additives has been presented recently (Ward, 1988). This protective effect resulted in the evolution of the terms "direct" and "indirect" effects for radiation action. The use of these terms has led to confusion and is at least misleading. Furthermore, energy deposition within DNA and associated water occur simultaneously so that it is not possible to separate the effects into rigid compartments. Indeed, ionisation of DNA and the interaction of OH with DNA may ultimately yield the same DNA radical (see Fielden et al., this volume).

The studies on radiation-induced conductance in DNA have elegantly demonstrated (Van Lith et al., 1986) the role of water of hydration in influencing the migration of electrons as shown in Figure 1. Radiation-induced mobile charge carriers in the DNA were only observed once the DNA was fully hydrated. It was suggested that the electrons move in the

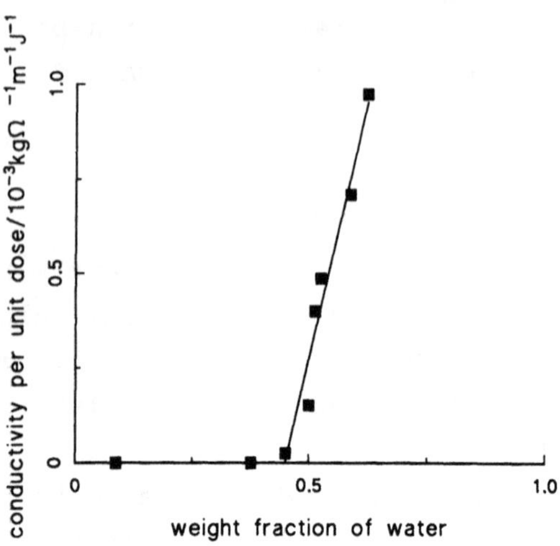

Figure 1. Dependence of the radiation-induced conductance in DNA upon water content (reproduced with permission of J. Warman *et al.*, Delft)

Table 1.
Some species involved in early processes

Water Radicals

electrophiles	- $^{\bullet}OH$, $H_2O^{\bullet +}$
nucleophiles	- e^-, e^-aq
	- H - atoms

DNA Species

Species from ionisation	- $DNA^{\bullet +}$ +e^-
excited states (directly formed or via ion recombination)	- $^1DNA^*$, $^3DNA^*$

ice sheath around the DNA. At 95K the charge migration distance was estimated to be of the order of 10^4 base pai whereas in dried DNA the average migration distance is restricted to ≤ 30 base pairs.

The energy is deposited within the systems as discre energy packages resulting in clusters of radical which ha been described by Ward (1981) as Locally Multiply Damaged Sites (LMDS). Indeed, these radical clusters which, for typical energy deposition in water, may contain 2.27 wate dissociations represent an energy loss of 38 eV (Pimblott LaVerne, 1990). In Table 1 are presented some of the spec which are involved in the early processes leading to DNA damage. Many of these species will be present within 10^{-1} 10^{-12} sec. For example, electron thermalisation occurs wi times $<10^{-13}$s corresponding to distances of <7 nm (Goulet Jay-Gerin, 1988). Only some of these species will be for near to or within the DNA. From estimations of the scavenging efficiency in cellular systems, it is inferred that only those radicals produced within 3-4 nm of the D will be involved in chemical modification of the DNA. If $H_2O^{\cdot+}$ is produced sufficiently close to the DNA, e.g. wit the structured water, it is thought that charge transfer the DNA will occur as opposed to the ion molecule reactic which occurs in bulk water.

$$H_2O + H_2O^{\cdot+} \longrightarrow H_3O^+ + {}^{\cdot}OH$$

Indeed, those water species with electrophilic properties thought to be more important in modifying DNA to produce lesions which are biologically significant. It is propos that when we consider energy deposition within DNA the following processes should be included in models:

(i) ionisation of DNA
(ii) ionisation of water of hydration
(iii) radicals produced within bulk water ≤ 4nm from the

In order to gain information on these early processes occurring within solid, hydrated DNA, two novel approache have been used whereby time resolved changes in irradiate DNA may be followed. These approaches are described belc

Radiation-induced changes in solid DNA - nanosecond timescale

Since solid DNA is opaque, conventional optical detection with pulse radiolysis may not be used to study the early processes occurring in solid DNA following irradiation. Two detection techniques, namely luminescence and diffuse reflectance, have been utilised to study the early radiation chemical effects of solid DNA. The apparatus for both these techniques have been fully described previously (Al-Kazwini et al., 1988; Wilkinson et al., 1984; O´Neill et al., 1989).

The earlier luminescence studies using electrons whose energies produced significant Cerenkov emission showed that energy may migrate up to distances of 100 base pairs in DNA containing bromo-deoxyuridine (Fielden et al., 1971). The present system (Al-Kazwini et al., 1988) utilises electrons whose energies ($\leq$200 keV) are sufficiently low so that Cerenkov radiation is insignificant and therefore allow observation of luminescence during the pulse (pulse width 30ns). From previous observations with this system on irradiation of "dry" DNA samples with different A+T/G+C content (Al-Kazwini et al., 1988), the "in-pulse" luminescence reflects the base composition of the DNA. The luminescence spectra and intensity of luminescence are independent of temperature (77-293 K). Further, hydration of the DNA (1:1 w/w DNA:H_2O) does not significantly modify the luminescence spectrum. Virtually no "post pulse" luminescence was observed in these studies. It was inferred (Al-Kazwini et al., 1988) that the short-lived luminescence induced by the radiation is due to singlet excited states of DNA which may arise from ion recombination and/or by direct excitation. It was previously shown that the emission profiles follow the time profile of the electron pulse from which it is estimated that the processes leading to luminescence occur within times $\leq$ 1ns. This lifetime is to be compared to that estimated for the lifetime of mobile charge carriers in irradiated dry DNA of a few picoseconds. That the emission spectra of different DNA samples reflect the A+T/G+C ratio, it is proposed that on the timescale of ion recombination the electron returns to the positive centre prior to development of the chemistry of the positive centre. If positive hole migration, prior to recombination, had occurred with localisation of the elec loss centre at guanine, as inferred from esr studies (Sym 1987), then the luminescence spectra from irradiated DNA samples should have been essentially independent of the b composition of these samples. This finding questions the importance of separation of spin and charge (proton shift for the electron loss centres of DNA at these early times (Steenken, 1989). Such processes may play a more prominen role for those electron loss centres which survive ion recombination within a few ps.

In order to assess the ability of an electron trap t interfere with the ion recombination processes leading to luminescence, "dry" samples of DNA doped with misonidazol an electron trap were pulse irradiated (see Al-Kazwini et al., 1990 for details). If the charge is prevented from returning to the electron loss centre by the trap, the luminescence from DNA should be reduced. The time profil the "in-pulse" radiation-induced luminescence from DNA al was not affected by the presence of misonidazole, however intensity of the luminescence was reduced.

The dependence of the luminescence intensity at 550 on misonidazole content of the DNA (Al-Kazwini et al., 19 is shown in Figure 2. The luminescence intensity relative

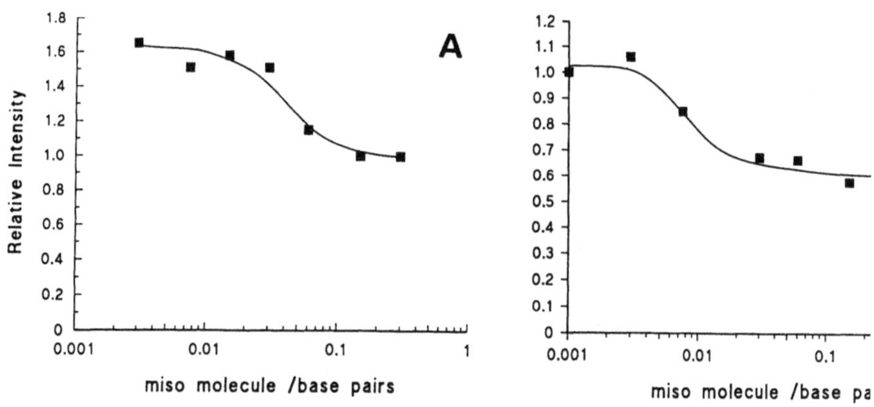

Figure 2. Dependence of the luminescence intensity at 550nm (A) and the absorbance at 600nm determin with diffuse reflectance (B) on misonidazole content of "dry" DNA.

that for DNA decreases with increasing concentration of misonidazole and at the higher concentrations appears to approach a limiting value. This value represents ~50% quenching of the total luminescence. Since <10% of the energy is deposited within the misonidazole, this dependence cannot be understood in terms of energy partition between DNA and the trap. From Figure 2, a 50% probability for electron capture occurs at a ratio of 25:1 base pairs to traps. This ratio is less than that determined for electron trapping (110:1) in esr studies (Cullis et al., 1990) and the ratio determined from diffuse reflectance (see later).

The following scheme is proposed to account for these observations at times ≤1 nanosecond and to rationalise these findings with those from esr studies.

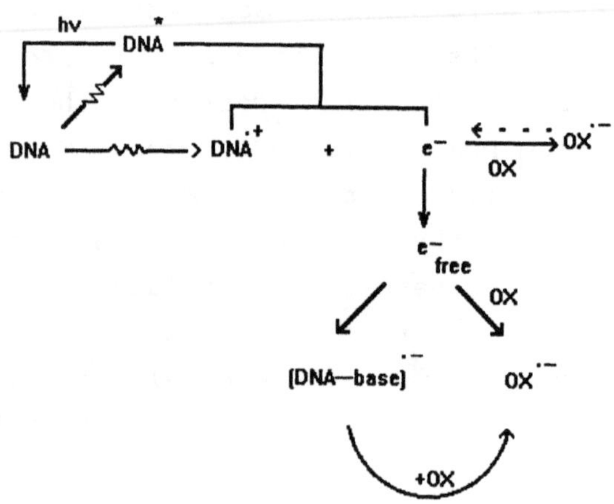

Irradiation of DNA initially results in ionisation and charge separation. In the absence of an added electron trap, charge recombination may occur in competition with localisation of the electron on the bases, which from esr studies have been identified as the pyrimidine bases. Charge recombination results in formation of a singlet excited state of DNA which fluoresces. From variation of the radiation-induced luminescence spectra on base composition of the DNA, it is inferred that insignificant hole migration occurs on the timescale for these ion recombination processes. It is suggested that the electron gain centres observed by esr at low temperatures reflect those electrons which escape ion recombination and become localised upon the bases, e.g. $T^{\cdot -}$ and/or $C^{\cdot -}$. From the esr studies, localisation of the electron allows the electron loss centre to migrate and localise on guanine.

Addition of an electron trap (Figure 2) has been show (Al-Kazwini et al., 1990) to influence the formation of radiation-induced excited states of DNA, from which it is suggested that the trap interferes with ion recombination. The trap may also capture "free" electrons either by competing with the base moieties for localisation of the electrons or by subsequent electron transfer from the base radical anion. Interference with ion recombination should lead to an increase in electron loss centres ($DNA^{\cdot +}$) and a reduction in luminescence intensity, whereas trapping of electrons which escape their positive partners should not modify the yield of electron loss centres or the intensity luminescence. Indeed, the latter effect was inferred from esr studies for DNA containing electron traps (Cullis et a 1990), reinforcing the view that the majority of the elect gain centres originate from trapping of "free" electrons. discussed later, these differences in electron trapping ma result in different concentration dependences for the effe of electron traps.

A further consequence of interference with ion recombination by an added electron trap is an enhancement the local density of radicals produced within the cluster shown below.

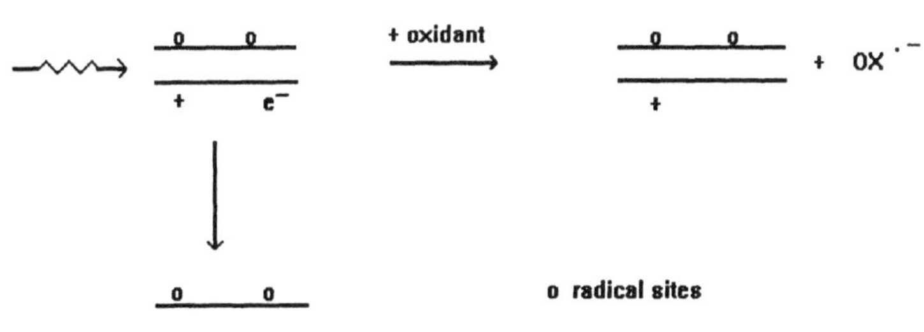

If, for example, 4 radical centres are produced in the DNA by the energy deposition event for low LET radiation and one pair of radicals recombine as shown, these processes would result in the existence of 2 lesions. However, if a trap interferes with recombination, the number of possible lesions increases from 2 to 3, assuming that hole migration is limited. Using the arguments presented by Ward, the number of radicals (lesions) within the LMDS could be significantly increased by the presence of electron traps. Since the number of lesions within a cluster (LMDS) also depends upon LET, it is interesting to speculate that the enhancement of biological effectiveness by oxygen/sensitisers may become less significant as the LET increases, since the effects of electron trapping in increasing the local concentration of lesions becomes less significant. Contrastingly, scavenging of electrons which do not undergo ion recombination on the ps timescale may result in the formation of fewer DNA lesions, assuming that $T^{\bullet -}/C^{\bullet -}$ lead ultimately to ssb as previously suggested. Indeed, electron traps have been shown to decrease the yield of ssb and dsb produced in irradiated ice containing DNA (Boon et al., 1985). Such protection has been related to a decrease in the yield of $T^{\bullet -}/C^{\bullet -}$ in the presence of a trap with a corresponding decrease in the yield of ssb.

Radiation-induced changes in solid DNA - microsecond timescale

The technique of diffuse reflectance has been utilised (O´Neill et al., 1989) to investigate the processes occurring on a longer timescale following the very early processes discussed above on the sub-nanosecond timescale. The apparatus has been fully described previously whereby a 5 µs pulse of electrons is delivered to a solid/hydrated sample of DNA in which changes in the diffusely reflected light are induced.

The effect of water of hydration upon the optical absorption spectrum of DNA measured 0.5 ms after irradiation (O´Neill et al., 1989) is shown in Figure 3. It is assumed that the absorption spectra are predominantly due to base

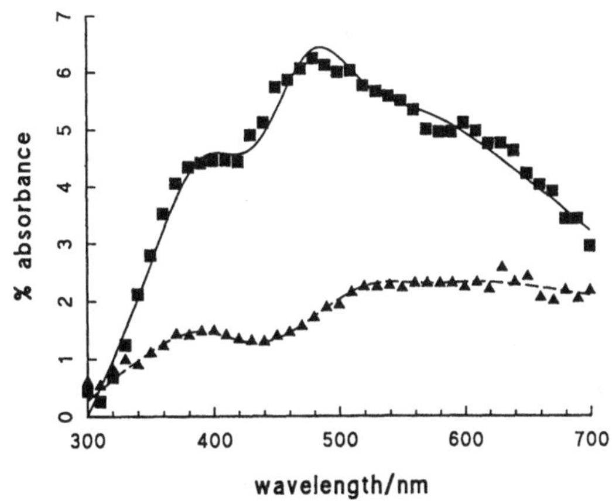

Figure 3. The effect of water content upon the diffusely reflected optical absorption spectrum of DNA determined 0.5 ms after pulse irradiation. (Δ) "dry" DNA, (■) 1:1 w/w DNA/water

radicals. Furthermore, whereas with "dry" DNA the transie species are long-lived ($t_{1/2}$ > 30 ms), upon hydration (1:1 w/w), shorter-lived species were observed ($t_{1/2}$ ~ 0.6 ms) wi ~ 30-50% loss of the total transient absorption. It is apparent from these observations that hydration of DNA pla an important role in modifying the chemical events followi irradiation. Further hydration of the DNA (e.g. 3:1 w/w H_2O/DNA) results in an absorption spectrum which resembles that observed following pulse irradiation of DNA in dilute aqueous solution. Of the many roles of water, the followi are considered to be of major importance:

(i) Water radicals, formed within the hydration structur DNA, interact with the DNA (e.g. $H_2O^{\cdot+}$, e^-), the yield depending upon the partition of absorbed energy between DN and its associated water;

(ii) The interaction of DNA radicals with water, e.g. prototropic equilibria, hydration reactions, and

(iii) Charge migration within the DNA.

Whether this latter point is a reflection of the different conformations of DNA remains an open question. This influence of hydration upon long-lived DNA species contras with the minor effects of hydration upon the much faster processes leading to light emission. Indeed it has been suggested that ionisation of crystalline water in single crystals of purines yields OH radicals which interact with the purine (Sagstuen et al., 1988). However, the site of addition is dependent upon the position of the water molec relative to the base.

In an attempt to characterise the DNA radicals presen within ~ 200 μs of irradiation, the transient spectra were determined upon pulse irradiation of "dry" samples of poly G, U and C and are shown in Figure 4. From a comparison o these spectra with that from "dry" DNA, it is apparent tha at λ >400 nm the absorption of purine radicals predominate Hydration of the polynucleotides (1:1 w/w) results in significant changes of the absorption spectra of the intermediates with the exception of that from poly G. The absorption spectra of hydrated poly G and poly A following irradiation are also shown in Figure 4 (cf. hydrated DNA in Figure 3). From a comparison of these spectra and the similarity between that of poly G and DNA, it is tempting suggest that guanine is one of the major sites for localisation of the radical centres on these timescales an that such localisation is facilitated by hydration. Such localisation would be consistent with the esr studies in which the major electron loss centre has been identified a the guanine base moiety. It is assumed that we are observi those DNA species which result from separation of ion pair and/or unstable intermediates produced as a result of radiationless decay processes of the excited states of DNA produced as a result of ion recombination. The occurrence of fast processes occurring after the pulse (within 50 μs) are presently under investigation.

The ability of an electron trap to modify the transie absorption spectrum of "dry" DNA was assessed following pu irradiation of a "dry" DNA sample containing misonidazole. At a concentration of 10 base pairs per misonidazole molec a reduction was observed in the absorption at all waveleng monitored (440-700 nm). Since this quenching occurs durin the 5 μs pulse, it is assumed that a precursor species is trapped prior to localisation upon the base moieties.

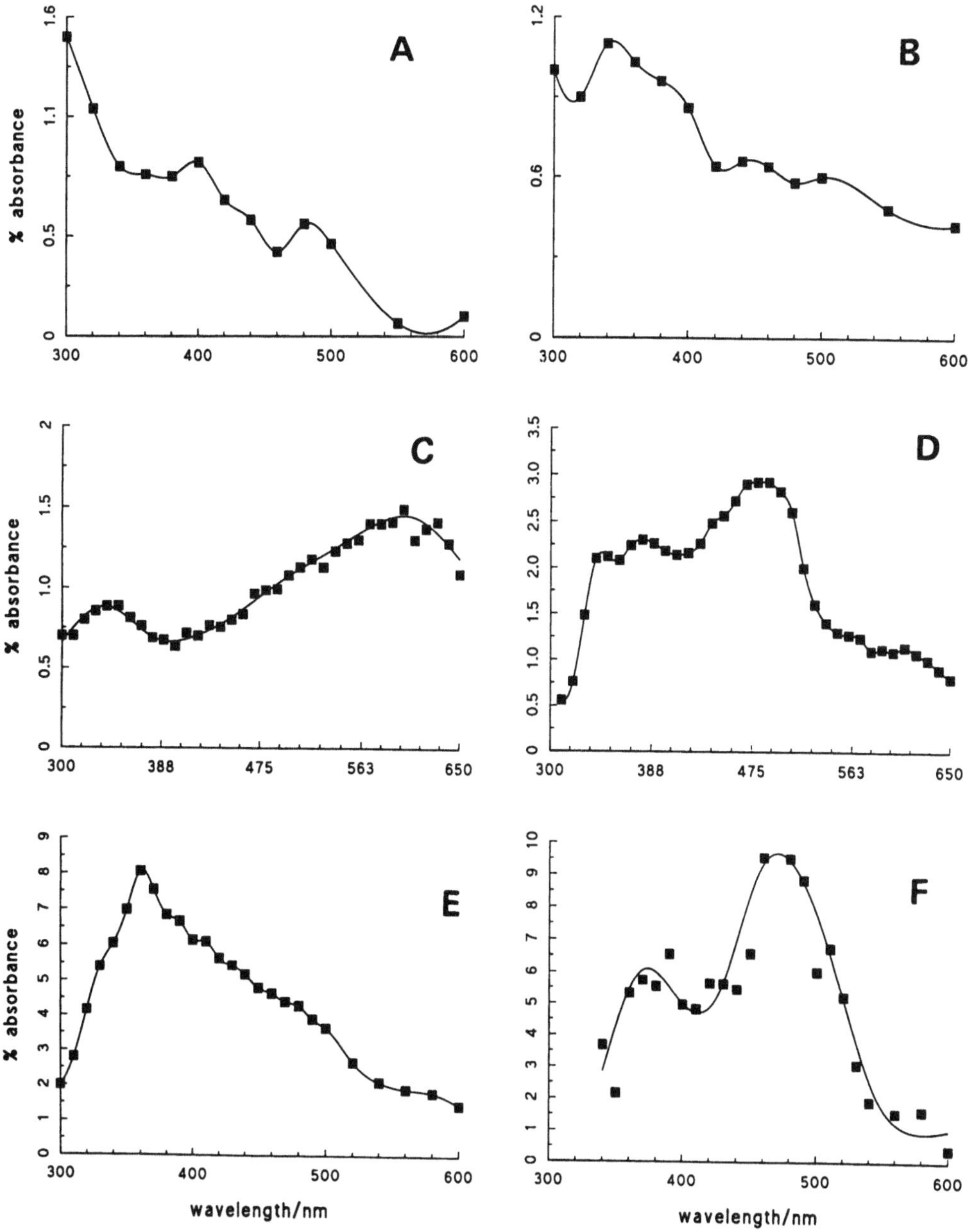

Figure 4. The optical absorption spectra determined 0.5 ms after pulse irradiation of either "dry" samples of poly U (A), poly C (B), poly A (C) and poly G (D) or hydrated samples (1:1 w/w) of poly A (E) and poly G (F).

Since electron traps are known to reduce the yield of pyrimidine electron gain centres, it is tentatively proposed that the precursor species is an electron which has escaped the coulombic attraction of its positive partner. Indeed in "dry" DNA, the spectrum may contain contributions from purine radical anions. Therefore misonidazole competes with the localisation of the e^- on DNA bases. The dependence of the scavenging effect of misonidazole as determined at 600 nm on its concentration is shown in Figure 2. The 50% probability of charge trapping occurs at a concentration of base pairs to misonidazole molecules of ~ 90:1, a value which is greater than that determined from luminescence changes. This dependence, as mentioned previously, may predominantly reflect the distribution for electrons which escape their germinate partner and is similar to that recently determined by esr for DNA containing an intercalator.

Conclusions

From these novel approaches of investigating the time-resolved chemical changes in DNA induced by ionising radiation, the following conclusions are drawn:

(i) radiation-induced charge migration may occur over several base pairs in "dry" DNA,

(ii) electron traps may act by increasing the local radical density at the site of energy deposition by interfering with ion recombination,

(iii) water radicals produced within or close to the DNA influence the types of DNA radicals produced, and

(iv) hydration may facilitate positive charge migration within DNA and result in modification of the DNA radicals.

When considering models of radiation action, the role of water associated with or near to the DNA should be taken into account as it influences the ensuing chemistry and the ultimate "stable" chemical products existing at longer times. Finally, it should be emphasised that the use of the terms "direct" and "indirect" effects are confusing when discussing the events occurring at or near to the DNA.

References

Al-Kazwini AT, O'Neill P, Fielden EM, Adams GE (1988) Radiation-induced luminescence from dry and hydrated DNA and related macromolecules. Radiat.Phys.Chem. 32: 385-389

Al-Kazwini AT, O'Neill P, Adams GE, Fielden EM (1990) Radiation-induced energy migration within solid DNA: the role of misonidazole as an electron trap. Radiat.Res. 121: 149-153

Bernhard WA (1989) Sites of electron trapping in DNA. J.Phys.Chem. 93: 2187-2194

Boon PJ, Cullis PM, Symons MCR, Wren BW (1984) Effects of ionizing radiation on deoxyribonucleic acid and related systems. Part I. The role of oxygen. J.Chem.Soc. Perkin Trans. II: 1393-1399

Boon PJ, Cullis PM, Symons MCR, Wren BW (1985) Effects of ionizing radiation on deoxyribonucleic acid, Part II - The influence of nitroimidazole drugs on the course of radiation damage to aqueous deoxyribonucleic acid. J.Chem.Soc. Perkin Trans. II: 1057-1061

Chapman JD, Gillespie CJ, Reuvers AP, Dugle DL (1975) The inactivation of Chinese hamster cells by X-rays: The effects of chemical modifiers on single- and double-events. Radiat.Res. 64: 365-375

Close DM, Sagstuen E, Nelson WH (1985) ESR study of the guanine cation. J.Chem.Phys. 82: 4386-4388

Cullis PM, Symons MCR, Wren BW (1985) Effects of ionizing radiation on deoxyribonucleic acid, Part III - The effect of iodoacetamide. J.Chem.Soc. Perkin Trans. II: 1819-1825

Cullis PM, McClymont JD, Symons MCR (1990) Electron conduction and trapping in DNA: an electron spin resonance study. J.Chem.Soc. Faraday Trans. 1, 86: 591-2

Fielden EM, Lillicrap SC, Robbins AB (1971) The effect of 5-bromouracil on energy transfer in DNA and related systems: DNA with incorporated 5-BUDr. Radiat.Res. 48: 421-431

Gräslund A, Ehrenberg A, Rupprecht A, Strom G, Crespi H (1975) Ionic base radicals in γ-irradiated oriented non-deuterated and fully deuterated DNA. Int.J.Radiat. Biol. 28: 313-323

Gräslund A, Ehrenberg , Rupprecht A. (1977) Free-radical formation in γ-irradiated oriented DNA containing electron-affinic radiosensitizers. Int.J.Radiat.Biol. 31: 145-152

Goulet T, Jay-Gerin J-P (1988) Thermalization distances and times for subexcitation electrons in solid water. J.Phys.Chem. 92: 6871-6873

Littlefield LG, Joiner EE, Colyer SP, Sayer AM, Frome EL (1988) Modulation of radiation-induced chromosome aberrations by DMSO, an OH radical scavenger. 1: Dose-response studies in human lymphocytes exposed to 220 kV X-rays. Int.J.Radiat.Biol. 53: 875-890

Hüttermann J, Voit K, Oloff H, Köhnlein W, Gräslund A, Rupprecht A (1984) Specific formation of electron gain and loss centres in X-irradiated oriented fibres of DNA at low temperature. Faraday Disc. of the Chemical Society, 78: 135-149

O'Neill P, Fielden EM (1986) Molecular aspects of DNA damage and its modification. NATO ASI series, Series A, Life Sciences 124: 425-438

O'Neill P, Al-Kazwini AT, Land EJ, Fielden EM (1989) Diffuse reflectance pulse radiolysis of solid DNA: the effect of hydration. Int.J.Radiat.Biol. 55: 531-537

Painter RB (1979) The role of DNA damage and repair in cell killing induced by ionizing radiation. In: Meyn RE and Withers HR (eds) Radiation Biology in Cancer Research. Raven Press, New York, pp. 59-68

Pimblott SM, LaVerne JA (1990) Comparison of stochastic and deterministic methods for modelling spur kinetics. Radiat.Res. 122: 12-23

Roots R, Chatterjee A, Blakely E, Chang P, Smith K, Tobias C (1982) Radiation response in air-, nitrous oxide- and nitrogen-saturated mammalian cells. Radiat.Res. 92: 245-254

Saenger W (1984) Principles of nucleic acid structure. Springer, New York, pp. 368-384

Sagstuen E, Hole EO, Nelson WH, Close DM (1988) ESR/ENDOR study of guanosine 5'-monophosphate (free acid). Single crystals X-irradiated at 10K. Radiat.Res. 116: 196-209

Steenken S (1989) Electron and proton gain and loss by DNA bases. Chem.Rev. 89: 503-509

Symons MCR (1987) Application of esr spectroscopy to the study of the effects of ionising radiation on DNA and DNA complexes. J.Chem.Soc. Faraday Trans. 1, 83: 1-11

Van Lith D, de Haas MP, Warman JM, Hummel A (1983) Highly mobile charge carriers in hydrated DNA and collagen formed by pulse ionization. Biopolymers 22: 807-810

Van Lith D, Warman JM, de Haas MP, Hummel A (1986) Electron migration in hydrated DNA and collagen at low temperatures. J.Chem.Soc. Faraday Trans. 1, 82: 2933-2943

Von Sonntag C (1987) The chemical basis of radiation biology. Taylor & Francis, London

Ward JF (1981) Some biochemical consequences of the spatial distribution of ionizing radiation-produced free radicals. Radiat.Res. 86: 185-195

Ward JF (1988) DNA damage produced by ionizing radiation in mammalian cells: Identities, mechanisms of formation and reparability. Prog.in Nucl.Acid Res. 35: 95-125

Wilkinson F, Willsher CJ, Warwick P, Land EJ, Rushton FAP (1984) Diffuse reflectance pulse radiolysis of opaque samples. Nature 311: 40-42

INITIAL SITES OF ONE ELECTRON ATTACHMENT IN DNA

William A. Bernhard
University of Rochester
Rochester, NY 14642
U.S.A.

Introduction

Chemical alterations in DNA, due to the direct effects of ionizing radiation, are initiated by three possible actions: excitation, electron abstraction, or electron addition. It is generally assumed, but unproved, that excitation events are negligible and that 50% of the damage stems from one-electron reduction and 50% stems from one-electron oxidation (Bernhard, 1981). Here the focus will be on the one-electron reduction of DNA, specifically the initial sites of attachment by unsolvated electrons. Ideally the initial distribution of electron attachment should be measured at very short times ($<10^{-12}$ s) under physiological conditions. Instead, we use low temperatures (<4K) to freeze out highly unstable events. This affords the advantage of giving ample time to make measurements but, of course, leaves uncertainties because low temperatures do not necessarily capture the short lived events of high temperatures.

Electrons move through aqueous DNA by at least two different routes. One is through the solvation sheath of DNA (van Lith *et al.*, 1986). These electrons are long lived (below 170K the lifetimes are in the microsecond range) and travel long distances. The other is through the stacked bases. These electrons must be short lived, less than a few picoseconds, according to the experiments of van Lith *et al.* (1986). The distance of travel through the stacked bases is an unsettled question with estimates ranging from ~5 base pairs (Whillans, 1975) to ~100 base pairs (Al-Kazwini *et al.*, 1990).

NATO ASI Series, Vol. H 54
The Early Effects of Radiation on DNA
Edited by E. M. Fielden and P. O'Neill

Electron attachment occurs at the bases of DNA, interaction with the sugar phosphate backbone being negligible. As to the distribution between the four bases, it is known that trapping at the pyrimidines is favored over the purines. ESR experiments on frozen aqueous solutions (Gregoli *et al.*, 1982) and DNA fibers at 77K (Gräslund *et al.*, 1975; Hüttermann *et al.*, 1984) have been interpreted as evidence that thymine is the predominant, or even exclusive, trapping site. Other similar experiments have left open the possibility that cytosine is also a trapping site (Sevilla *et al.*, 1979; Gräslund *et al.*, 1975). Indeed it has been pointed out that protonation of the cytosine anion makes it a probable site of electron attachment (Bernhard, 1981). Recently it has been shown, using oligomers in 12 M LiCl at 4K, that cytosine is the predominant site of electron trapping (Bernhard, 1989).

There are two objectives in our current work. One is to determine the initial distribution of electron attachment in oligodeoxynucleotides of variable base composition and the other is to identify the variables that influence the distribution of electron attachment sites.

Methods

ESR spectra of one-electron reduced deoxyoligonucleotides were obtained using the sample configuration shown in Figure 1. The deoxyoligonucleotides were purchased from Sigma Chemical Co. and Pharmacia Molecular Biologicals and used without further purification. The oligomers were dissolved in water (either HOH or DOD), ~5 OD_{260} units per 5 microliter, and then mixed 1:1 with 24 M LiCl (HOH or DOD). Cooling to 2K, the 12 M LiCl solutions yielded a clear uncracked glass. While at 2K, the samples were pushed through the ESR cavity and out the bottom, X-irradiated at 50 KeV for 15 min to a dose of ~1500 Gray, and then pulled back into the cavity to record the ESR spectra (Mercer and Bernhard, 1987).

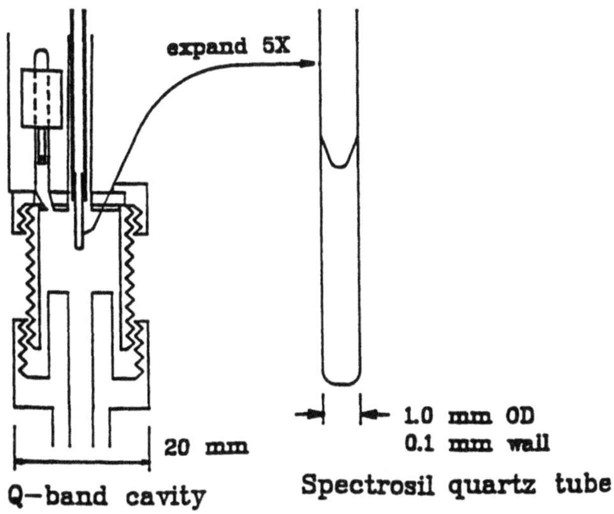

Fig. 1 ESR configuration for studies of 12 M LiCl glasses.

Before analyzing the ESR spectra of the oligomers, the background spectrum due to radicals trapped in the LiCl matrix was subtracted out and baseline drift was corrected using a polynomial fitting routine. The resulting spectrum was then fit using a weighted sum of two or more of the four bench mark spectra calculated for the four one-electron reduced bases: $T^{\cdot -}$, $C^{\cdot -}$, $A^{\cdot -}$, and $G^{\cdot -}$. The goodness-of-fit for the spectral simulations was measured by R, where $R = \Sigma|F_o-F_c|/\Sigma|F_o|$. The F_o are observed data points of a spectral trace and the F_c are calculated points (Bernhard and Patrzalek, 1989). In the fitting routine, the g value of each bench mark spectrum was allowed to vary within the error limits of g, that is ±.0002. When the g values were locked during the fit, the values given in Table 1 were not greatly effected. In the simulations there was no attempt to ascertain a global minimum. On the other hand, when different sets of starting parameters were used comparable solutions were obtained. The weighted sums reported here are, therefore, reasonable but an error analysis still needs to be done.

The four bench mark spectra, for both HOH and DOD glasses, are shown in Figure 2. Each bench mark spectrum is derived from a set of simulation parameters. These parameters are determined by generating a powder spectrum that closely fits ($R<0.05$) the experimental spectrum for each base in a configuration which reflects that expected in duplex DNA. Thus, the spectra are derived from $d(pT)_{10}$ for $T^{\cdot -}$, d(pGpCpGpC) for $C^{\cdot -}$, $d(pA)_{10}$ for $A^{\cdot -}$, and GdR for $G^{\cdot -}$. It is notable that only the monomer of G gives rise to a spectrum equivalent to that observed in the oligomers. In the case of cytosine, the difference in monomer versus oligomer is dramatic.

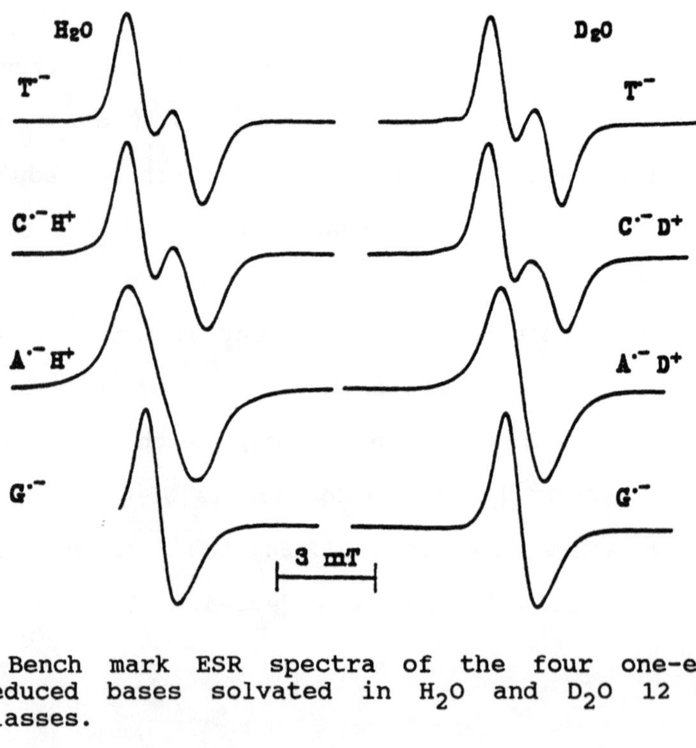

Fig. 2 Bench mark ESR spectra of the four one-electron reduced bases solvated in H_2O and D_2O 12 M LiCl glasses.

It has been suggested that the extra hfc in one-electron reduced deoxycytidine stems from protonation of the exocyclic NH_2 so as to leave the original two protons in equatorial positions and the added proton in an axial position (Cullis *et al.*, 1989). Although this does explain the ESR characteristics, it goes against the expectation that $C^{\cdot -}$ should protonate at N3 not N4 (Bernhard, 1981). There is, however, a good explanation for a departure from protonation at N3 when cytidine is in 12 M LiCl (Bernhard, 1990). Metal cations are known to coordinate preferentially at N3 of cytosine (Saenger, 1984). Thus, as shown in Figure 3, Li^+ most likely blocks N3, redirecting the site of protonation to N4. When C is hydrogen bonded to G, the Li^+ ion is displaced by G:C hydrogen bonding. Even in 12 M LiCl, protonation will occur at N3 of cytosine when C is in a G:C pair. Protonation at N3, in contrast to protonation at N4, has a very small effect on the ESR spectrum. This explains the difference between the monomer and oligomer spectra.

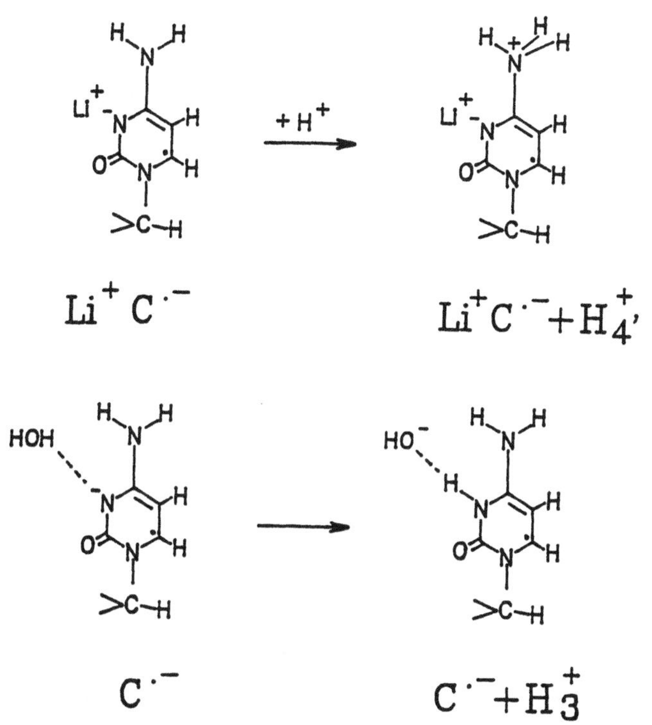

Fig.3 Protonation of $C^{\cdot -}$ in the presence of Li^+ vs. H_2O.

The difference between monomeric and oligomeric cytosine is important for two reasons. One, from an analytical viewpoint, it is important to be cautious in using data from the monomers to draw conclusions about the polymer. Two, the strong tendency for the cytosine radical anion to protonate explains the observed difference between monomer and oligomer; the tendency for the thymine radical anion to protonate is not nearly as strong (Steenken, 1990a & 1990b) explaining why there is very little difference between the monomer and oligomer (Bernhard and Patrzalek, 1989).

Results

Three examples of ESR spectra of one-electron reduced deoxyoligonucleotides are given in Figure 4. The background spectrum has been subtracted out and the baseline corrected for the spectra shown. Simulations, the dashed lines, are superimposed to show the goodness-of-fit associated with the R-values obtained.

The results from 12 deoxyoligonucleotides are summarized in Table 1. Only the spectra corresponding to bases contained in each oligomer were used in the simulation; for example d(pApTpApT) was simulated using only $A^{\cdot-}$ and $T^{\cdot-}$ spectra. For each oligomer there are two simulations given: one for the HOH glass the other for the DOD glass. Looking at the top spectrum in Figure 4, of d(pApTpApT), it is not difficult to see the close similarity to the $T^{\cdot-}$ bench mark spectrum. This is reflected in the simulation results of 94% $T^{\cdot-}$ + 6% $A^{\cdot-}$. The spectrum of d(pTpApTpA), middle of Figure 4, shows less of the doublet due to $T^{\cdot-}$ and more of the singlet due to $A^{\cdot-}$. The simulation gives 63% $T^{\cdot-}$ + 37% $A^{\cdot-}$. Therefore, the distribution of electron attachment is changed when the AT tetramer is changed from the 5′-3′ to the 3′-5′ direction. Also shown in Figure 4, there appears to be a change in the distribution of electron attachment in going from d(pApTpApT) in an HOH glass to a DOD glass. As discussed later, there are difficulties in interpreting the DOD effect.

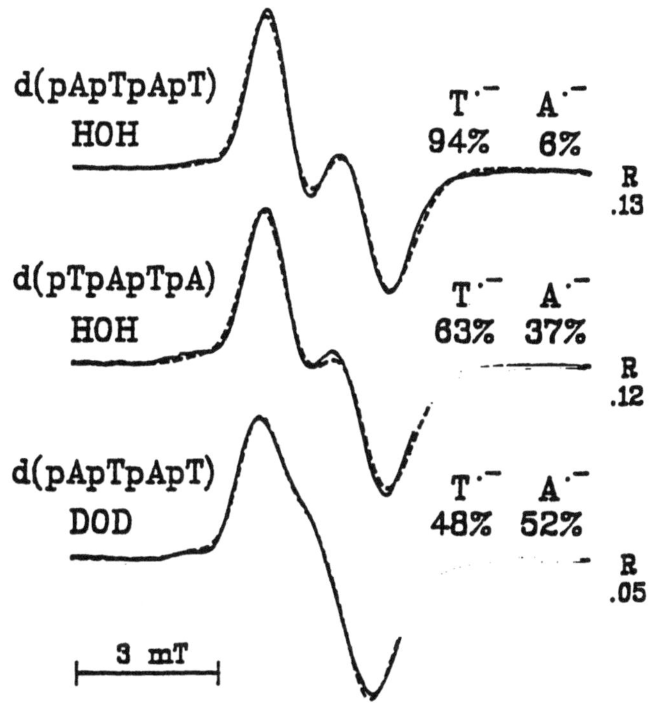

Fig. 4 ESR spectra of one-electron reduced tetramers of AT showing dependence on sequence and HOH versus DOD 12 M LiCl glass.

The top two entries in Table 1 are tetramers that are not self complementary. In both cases C is the better electron trap. The probability of electron attachment declines in the order C > T > A. In the HOH glasses the differences are not very large. The $[C^{\cdot -}]:[T^{\cdot -}]:[A^{\cdot -}]$ ratios are about 10:8:6.

Table 1. Percent of bench mark spectra used to simulate the esr spectra of one-electron reduced oligomers.

	HOH					DOD				
oligomer	T %	C %	A %	G %	R	T %	C %	A %	G %	R
$d(pCpT)_5$	44	56			.06	24	76			.09
$d(pApC)_5$		63	37		.17		94	6		.07
d(pGpCpGpC)		99		1	.03		99		1	.03
d(pCpGpCpG)		88		12	.09		100		0	.05
d(pGpGpCpC)		91		9	.16		94		6	.09
$d(pGpC)_5$		92		8	.13		95		5	.06
d(pApTpApT)	94		6		.13	48		52		.05
d(pTpApTpA)	63		37		.12	36		64		.03
d(pApApTpT)	94		6		.13	94		6		.11
$d(pApT)_5$	99		1		.18	91		9		.12
d(pApGpCpT)	18	78	4	0	.07	3	91	5	0	.07
d(pTpCpGpA-pCpG):d(pA-pGpCpTpGpC)	3	87	10	0	.04	7	89	4	0	.05

The GC oligomers, the next four entries in Table 1, are self complimentary. Electron tapping at C dominates over G by a ratio of 10:1 or more. The following four entries are also self complimentary, the AT oligomers. Still looking at just the HOH glasses the ratio of $[T^{\cdot-}]:[A^{\cdot-}]$ is >10:1 in three out of the four. In the fourth, d(pTpApTpA), the $[T^{\cdot-}]:[A^{\cdot-}]$ ratio drops to 2:1.

Of the bottom two entries, the first is a self complimentary tetramer with GC/AT=1 and the second is an equal molar mixture of complimentary hexamers with GC/AT=2. The tetramer gives ratios $[C^{\cdot-}]:[T^{\cdot-}]:[A^{\cdot-}]:[G^{\cdot-}]$ of 20:5:1:0. For the hexamer the ratios are 29:1:3:0; if the trapping ratios in the tetramer held true for the hexamer then the hexamer ratios should have been 40:5:1:0.

In the DOD glasses the electron attachment distribution appears to change for some of the oligomers but not others. In both $d(pCpT)_5$ and $d(pApC)_5$, trapping at C appears to increase relative to either T or A. In both d(pApTpApT) and d(pTpApTpA), trapping at A appears to increase relative to T; in contrast, there is essentially no change in d(pApApTpT) and $d(pApT)_5$.

Discussion

Interpretation of these results requires some knowledge of the conformation of the oligomers in 12 M LiCl at 4K. There are no structural studies, under these conditions, in the literature. It is reasonable , however, to extrapolate from lower salt concentrations and higher temperatures. It is known that base stacking is enhanced by either increasing ionic strength or decreasing temperature (Saenger, 1984). It seems a good operating assumption that the conformations of the oligomers at lower salt and higher temperatures will be good predictors of conformation at high salt and low temperatures. This assumption is supported by observations on $C^{\cdot -}$. As pointed out above, the $C^{\cdot -}$ spectrum changes substantially and predictably when going from the monomer to duplexed oligomers.

The structural information encapsulated in this paragraph is from the review article by van de Ven and Hilbers (1988), the book by Saenger (1984), and from references therein. The entries in Table 1 that are known to form duplexes are $d(pGpC)_5$ and $d(pApT)_5$; those that can be predicted, from the sequence dependence of duplex formation, to form duplexes are d(pGpCpGpC), d(pGpGpCpC), d(pApApTpT), and d(pTpCpGpApCpG): d(pApGpCpTpGpC). The entries known to be single stranded are d(pTpApTpA) and d(pApGpCpT); those that can be predicted to be single stranded are $d(pCpT)_5$ and $d(pApC)_5$. The d(pCpGpCpG) and d(ApTpApT) oligomers are likely to be in both single strand and duplex forms; in addition, d(ApTpApT) forms frame shifted duplexes. Stacking interactions in the AT oligomers are greatest in d(pApT)$_5$ and d(pApApTpT), are somewhat less in d(pApTpApT), and drop off significantly in d(pTpApTpA). Among the GC oligomers, it is anticipated that d(pCpGpCpG) will have the poorest stacking.

Under the assumption that these structures apply in our system the following correlations can be made.

1. In single stranded oligomers, electron trapping is probable at C, T, and A, with the probability decreasing in that order, and improbable at G. (Consider d(pCpT)$_5$, d(pApC)$_5$, and d(pApGpCpT).)
2. In going from a single stranded conformation to a duplex, the probability of trapping at T increases relative to A and the probability of trapping at C increases relative to T. (Compare d(pApTpApT) with d(pApT)$_5$ and d(pCpT)$_5$ with the hexamer duplex.)
3. In duplexes containing all four bases, C is the major site of electron attachment while attachment at G is negligible. (This is based on the hexamer results.)

Interpretation of the changes in going from HOH to DOD will require additional work. It is possible that this is a true isotope effect but it seems more likely to be an anomaly introduced by the choice of bench mark spectra. That is the apparent changes could be the result of having chosen an incorrect bench mark spectrum for $A^{\cdot-}$ in HOH. This would be the case if $A^{\cdot-}$ behaves the same way as $C^{\cdot-}$ with regard to displacement of the protonation site by Li^+, as discussed above.

The observed correlations can be explained by drawing on some key observations concerning the radiation chemistry of DNA and related systems. The calculated electron affinities of the bases decrease in the order T > C >> A > G (Bodor *et al.*, 1970; Berthod *et al.*, 1966). Sevilla *et al.* (1976) and Gregoli *et al.* (1977) have shown that base stacking promotes electron mobility between the bases. Steenken has stressed the importance of protonation in influencing electron attachment (Steenken, 1989). It has been pointed out that not only is the propensity of the radical anion to <u>add</u> a proton important but equally important is the propensity of its complementary base to <u>donate</u> a proton. Thus, not just the pK_a's of the radical anions are relevant but also the pK_a's of the complementary bases (Steenken 1989, 1990a, 1990b).

These observations can be condensed into the following three relationships governing electron attachment in single stranded (ss) DNA:

electron affinity................ T > C >> A > G [1]
addition of a proton............. $C^{\cdot -} > A^{\cdot -} > G^{\cdot -} >> T^{\cdot -}$ [2]
ss, abstraction of a proton by... $C^{\cdot -} = A^{\cdot -} = G^{\cdot -} = T^{\cdot -}$ [3a]

The change in distribution, in going from single stranded to double stranded (ds) DNA, is explained in the change from relation [3a] to [3b]

ds, abstraction of a proton by... $C^{\cdot -} > A^{\cdot -} > G^{\cdot -} >> T^{\cdot -}$ [3b]

Equation [1] dictates the distribution of electron attachment for stacked bases in an aprotic apolar environment. Stacking of the bases provides for electron mobility between the bases while the absence of proton transfer leaves the relative electron affinities to dictate the probability of trapping. The aprotic condition has not been experimentally tested and is not attainable in double stranded DNA.

Equations [1], [2], and [3a] apply for single stranded DNA in an aqueous medium. In the absence of base pairing, HOH is the potential proton donor for each of the radical anions. Thus, equations [1] and [2] together dictate the sites of electron trapping. Equation [2] asserts that the probability of trapping at C and A is increased relative to T due to the higher propensity for $C^{\cdot -}$ and $A^{\cdot -}$ to protonate. The positions of $C^{\cdot -}$, $A^{\cdot -}$, and $T^{\cdot -}$ in equation [2] are taken from Steenken (1990b) and the position of $G^{\cdot -}$ is my speculation.

Equations [1], [2], and [3b] apply for duplex DNA in an aqueous medium. The propensity of the hydrogen bonded donor to deprotonate affects the relative probability for protonation of the anion. Consequently, this influences the electron trapping. Since A is a much poorer proton donor than G, it is more costly for $T^{\cdot -}$ to obtain a proton than $C^{\cdot -}$.

Even though T has a slight advantage over C with regard to electron affinity, $C^{\cdot -}$ has the dual advantage of having a higher proton affinity and a more facile proton donor. In the duplex, one would then predict the following relative concentrations

$$[C^{\cdot -}H^{+}] \gg [A^{\cdot -}H^{+}] \sim [T^{\cdot -}] \gg [G^{\cdot -}] \qquad [4]$$

The probable absence of protonation of $T^{\cdot -}$ has been discussed above and fits well with recent measurements by Steenken. As Steenken has pointed out (Steenken 1989, 1990a), it is the irreversible protonation of a carbon position that will stabilize a particular one-electron reduced base as a reduction site. As can be seen from the distribution predicted in equation [4], thymine is the only base that is both likely to form an electron adduct and unlikely to protonate reversibly; it is, thereby, the best candidate for protonation irreversibly. Upon warming, the increased vibrational and rotation modes make it possible to reach the higher activation energy needed to protonate carbon positions, in this case C6 of $T^{\cdot -}$.

Conclusions

1. Initially, the electron adds to the stack of bases, as opposed to individual bases.
2. The initial sites of electron trapping are determined by:
   a. the electron affinity of the bases
   b. proton transfer to the radical anion
   c. resolvation of the reduced base.
3. In duplex DNA we expect the initial distribution to be
   $[\ C^{\cdot -}H^{+}\ ] \gg [\ A^{\cdot -}H^{+}\ ] \sim [\ T^{\cdot -}\ ] \gg [\ G^{\cdot -}\ ]$
4. Going from single stranded to double stranded DNA results in:
   a. decreased trapping at A
      (because T is a poor proton donor relative to HOH)

b. decreased trapping at T
(because A destabilizes $T^{\cdot-}$ relative to HOH)

c. increased trapping at C relative to T and A
(because C is paired with the best proton donor, G).

5. The unprotonated $T^{\cdot-}$ is positioned to acquire, upon warming, a proton at C6 and thereby locking in the one electron reduction.

Acknowledgement

The assistance of Andrew Patrzalek and Kermit Mercer in performing the ESR experiments is gratefully acknowledged. This investigation was supported by PHS Grant R01-CA32546, awarded by the National Cancer Institute, DHHS.

References

Al-Kazwini AT, O'Neill P, Adams GE, Fielden EM (1990) Radiation-induced energy migration within solid DNA: the role of mizonidazole as an electron trap. Radiat Res 121:149-153

Bernhard WA (1981) Solid state radiation chemistry of DNA: the bases. Adv Radiat Biol 9:199-280

Bernhard WA (1989) Sites of electron trapping in DNA as determined by ESR of one-electron-reduced oligonucleotides. J Phys Chem 93:2187-2189

Bernhard WA (1990) Protonation of the cytosine radical anion. Radiat Res, submitted

Bernhard WA and Patrzalek AZ (1989) ESR characteristics of one-electron reduced thymine in monomer, oligomer, and polymer derivatives. Radiat Res 117:370-394

Berthod H, Gressner-Prettre C, Pullman A (1966) Theoretical study of the electronic properties of the purine and pyrimidine components of the nucleic acids. I. A semi-empirical sef-consistent-field calculation. Theor Chim Acta 5:53-68

Bodor N, Dewar MJS, Harget AJ (1970) Ground states of conjugated molecules. XIX. Tautomerism of heteroaromatic hydroxy and amino derivatives and nucleotide bases. J Am Chem Soc 92:2929-2936

Cullis PM, Podmore I, Lawson M, Symons MCR, Dalgarno B, McClymont J (1989) The site of electron capture in irradiated deoxyribonucleic acid: cytosine vs. thymine. J Chem Soc Chem Commun 1003-1005

Gräslund A, Ehrenberg A, Rupprect A, Ström G (1975) Ionic base radicals in gamma-irradiated oriented non-deuterated and fully deuterated DNA. Int J Radiat Biol 28:313-323

Gregoli S, Olast M, Bertinchamps A (1977) Charge migration phenomena in gamma-irradiated costacking complexes of DNA nucleotides. I. A computer-assisted ESR analysis of dAMP:dTMP complexes in frozen solution. Radiat Res 70:255-274

Gregoli S, Olast J, Bertinchamps AJ (1982) Radiolytic Pathways in gamma-irradiated DNA: influence of chemical and conformational factors. Radiat Res 89:238-254

Hüttermann J, Voit K, Oloff H, Köhnlein W, Gräslund A, Rupprecht A (1984) Specific formation of electron gain and loss centres in x-irradiated oriented fibres of DNA at low temperatures. Faraday Discuss Chem Soc 78:135-149

Mercer K and Bernhard WA (1987) Design and operation of a variable temperature accessory for Q-band ESR. J Mag Res 74:66-71

Saenger W (1984) Principles of Nucleic Acid Structure, Springer-Verlag, New York Berlin Heidelberg Tokyo

Sevilla MD, Fallor R, Clark C, Holroyd RA, Pettel M (1976) Electron transfer in dinucleoside phosphate anions. J Phys Chem 80:353-358

Steenken S (1989) Purine bases, nucleosides and nucleotides: aqueous solution redox chemistry and transformation reactions of their radical cations, e-, and OH adducts. Chem Rev 89:503-520

Steenken S (1990a) Coupled electron and proton transfer reactions of nucleic acid bases. Abst 38th Ann Meet Radiat Res Soc p55, ch-3

Steenken S (1990b) private communication.

van Lith D, Warman JM, de Haas MP, Hummel A (1986) Electron migration in hydrated DNA and collagen at low temperatures. Part 1. effect of water concentration. J Chem Soc Faraday Trans 1, 82:2933-2943

van de Ven JM and Hilbers CW (1988) Nucleic acids and nuclear magnetic resonance. J Biochem 178:1-38

Whillans DW (1975) Studies of electron migration in DNA in aqueous solution using intercalating dyes. Biochim Biophys Acta 414:193-205

MOLECULAR ORBITAL CALCULATIONS ON DNA BASE PAIRS

D.M. Close
Dept. of Physics
East Tennessee State University
Johnson City, TN 37614
USA

INDO calculations have been performed on entire G:C and A:T base pairs. For the G:C pair the calculations identify cytosine as the site of reduction and guanine as the site of oxidation. For the A:T pair one finds that thymine is the site of reduction while adenine is the site of oxidation. Calculations have also been performed on all protonated anions and deprotonated cations for both base pairs.

Maps of the spin densities and charge densities have been made. From these maps one can see what are the most electronegative sites of the anions and the expected influence protonation has on neutralizing the excess negative charge. In most cases a proton has little influence on the calculated spin densities.

These results can be compared with trends suggested by Steenken (1989). First for the reduction products, the MO calculations predict that it is energetically favorable for the cytosine anion (in C:G) to protonate (either at N3 or C2=O), while it is energetically unfavorable for the thymine anion (in T:A) to protonate at C4=O. This agrees with Steenken's guidelines since $C^{\cdot -}$ is a strong base, whereas $T^{\cdot -}$ is a rather weak base.

On the oxidation side, the MO calculations predict that it is not energetically favorable for the guanine cation (in G:C) to deprotonate either at N1 or C2-NH_2, while the calculations find little difference in energy between the adenine cation (in A:T) and the deprotonated cation. In Steenken's view there would be little tendency for $A^{\cdot +}$ to deprotonate, while $G^{\cdot +}$ should transfer a proton to cytosine (a "good base").

NATO ASI Series, Vol. H 54
The Early Effects of Radiation on DNA
Edited by E. M. Fielden and P. O'Neill

Analysis of the results show some interesting similarities between the HO4-C4< protonated anion in thymine and the HO2-C2< protonated anion in cytosine. It has recently been claimed that in cytidine the anion protonates at the exocyclic NH_2 (Cullis et al. 1989). Based on the results presented here, and on a recent study of the thymine anion observed in thymidine (Sagstuen et al. 1989) it can be argued that the cytidine ESR spectra (Cullis et al. 1989) could equally well be explained by the HO2-C2< protonated cytosine anion.

Bernhard (1989) has shown that in oligomers, although the thymine and cytosine anion have very similar ESR spectra, it is possible to show that a large fraction of the reduction events involve the cytosine base. These results suggest that one needs to know in more detail just what are the exact characteristics of the DNA reduction product.

Bernhard WA (1989) Sites of electron trapping in DNA as determined by ESR of one-electron-reduced oligonucleotides. J Chem Phys **93**: 2187-2189

Cullis PM, Podmore I, Lawson M, Symons MCR, Dalgarno B, and McClymont J (1989) The site of electron capture in irradiated deoxyribonucleic acid; cytosine vs. thymine. J Chem Soc, Chem Commun: 1003-1005

Sagstuen E, Hole EO, Nelson WH, Close DM (1989) Structure of the primary reduction product of thymidine after x-irradiation at 10 K. J Phys Chem **93**: 5974-5977

Steenken S (1989) Purine bases, nucleosides, and nucleotides: Aqueous solution redox chemistry and transformation reactions of their radical cations and e- and OH adducts. Chem Rev **89**: 503-520

This work is supported by NIH Grant CA36810 and by NATO Travel grant RG 0426/88.

ROLE OF ENERGY AND CHARGE TRANSFER IN DNA DAMAGE BY DENSELY IONIZING RADIATION*

John H. Miller
Biology and Chemistry Department
Pacific Northwest Laboratory
Richland, WA 99352

Experiments by Arroyo et al. (1986) with neutrons indicate that radical yields in oriented DNA at 77 K are both qualitatively and quantitatively different when the secondary proton flux traverses the sample in a direction that is approximately parallel to the orientation of DNA molecules. These findings suggest that energy and/or charge transfer influences the production and/or decay of free radicals when multiple excitations are produced in the same DNA chain. Singlet and triplet excitons can migrate along DNA for a few nanometers before they are trapped, but this type of energy transfer should be equally effective in coupling events within the small clusters of excitation (spurs) that are present in both irradiation geometries.

Mechanisms of energy or charge transfer in DNA that have sufficient range to produce interspur reactions are less firmly established than exciton migration. The effect of ligands with high electron affinity on DNA luminescence produced by pulsed radiolysis with electrons has been used to deduce electron transfer distances of the order of 100 base pairs (Fielden et al 1971). However, similar studies with photoexcitation placed an upper limit of 5 base pairs on electron migration in DNA (Whillans, 1975). Transient conductivity observed in pulsed radiolysis studies of solid DNA at low temperatures suggest that charge transfer is associated with water of hydration (van Lith et al. 1986). Above a critical water content, excess electrons appeared to have a mobility comparable to that observed in ice (de Hass et al. 1983) and a lifetime that was limited by homogeneous recombination. Under their irradiation conditions (van Lith et al. 1986), migration distances of 100 nm were typical of the high-mobility phase. Self-cohering vibrational excitations (solitons) that move long DNA chains with a velocity of about 100 nm/ns (Yomosa, 1984) are an even more speculative mode of long-range energy transfer.

*Work supported by the Office of Health and Environmental Research (OHER), U. S. Department of Energy under contract DE-AC06-76RLO 1830.

NATO ASI Series, Vol. H 54
The Early Effects of Radiation on DNA
Edited by E. M. Fielden and P. O'Neill

We are developing methods to calculate the effects of long-range energy and charge transfer on radical yields in oriented DNA and experimental techniques to test our results by direct proton beam irradiation. A macroscopic model that approximates intramolecular energy/charge transfer by a large asymmetry in thermal diffusion parallel and perpendicular to the molecular orientation (Miller et al. 1988) predicts a significantly greater conversion of primary radical anions to 5,6-dihydrothymin-5-yl (TH) radicals in the parallel irradiation geometry. However, unlike the findings reported by Arroyo et al. (1986) which suggest negligible conversion of primary radical anions to TH in the perpendicular case and complete conversion in the parallel case, this thermal-spike model predicts a mixture of primary anions and TH radicals under both irradiation conditions. The difference between the TH yields calculated for the parallel and perpendicular cases depends upon the magnitude of the asymmetry in thermal diffusion, the energy barrier for radical conversion, and the LET of the radiation. With an asymmetry of 1000 and an activation energy of 0.2 eV, we find about 60% conversion of primary anions to TH in 300 keV track segments that are parallel to the helical axis and 10% conversion when the track segment is perpendicular to DNA. Our preliminary experimental data with direct proton irradiation also suggest a mixture of primary anions and TH radicals in the perpendicular case.

References

Arroyo CM, Carmichael AJ, Swenberg CE, and Myers LS (1986) Neutron-induced free radicals in oriented DNA. Int J Radiat Bio 50:789-793

de Hass MP, Kunst M, Warman JM, and Verberne JB (1983) Nanosecond time-resolved conductivity studies of pulse-ionized ice. 1. The mobility and trapping of conduction-band electron in H_2O and D_2O ice. J Phys Chem 87:4089-4092

Fielden EM, Lillicarp SC, and Robins AB (1971) The effect of 5-bromouracil on energy transfer in DNA and related systems: DNA with incorporated 5-BUDR. Radiat Res 48:421-431

Miller JH, Wilson WE, Swenberg CE, Myers LS, and Charlton DE (1988) Stochastic model of free radical yields in oriented DNA at 77 K. Int J Radiat Bio 53:901-907

van Lith D, Warman JM, de Hass MP, and Hummel A (1986) Electron migration in hydrated DNA and collagen at low temperatures. Part 1. Effect of water concentration. J Chem Soc Faraday Trans 1 82:2933-2943

Whillans DW (1975) Studies of electron migration in DNA in aqueous solutions using intercalating dyes. Biochim Biophys Acta 414:193-205

Yomosa S (1984) Solitary excitations in deoxyribonucleic acid (DNA) double helices. Phys Rev A 30:474-480

EARLY CEMICAL EVENTS IN DNA DAMAGE

Jürgen Hüttermann
Fachrichtung Biophysik
und Phys. Grundlagen der Medizin
Universität des Saarlandes
6650 Homburg/Saar
Federal Republic of Germany

The lectures given on tuesday morning's session dealt with the distribution of damage sites over DNA. These were studied either by involving specific chemical reactions in LiCL-glasses of oligonucleotides model compounds (Bernhard) or in frozen aqueous solutions containing DNA (Symons) and utilizing electron spin resonance spectroscopy as tool applicable to extract free radical structures. Another, novel approach was presented by O'Neill who used diffuse reflectance and time resolved optical spectroscopy to study energy migration and sites of trapping. It was common to all lectures that, even when using LiCl-glasses or frozen systems, the direct effect of radiation action and its modification by hydration and additives was at the center of the investigation.

The chairman provided a short introduction which meant to indicate the various stages of development of knowledge which the history of more than three decades of this kind of research has run through starting from the early ESR-studies of DNA [Shields and Gordy, Boag and Müller]. While the identification of the famous TH˙ octet species occurred in 1963 [Ehrenberg et al.] it was not before the seventies that the approach by Gräslund and co-workers employing oriented fibers of DNA succeded in obtaining information about other radical components. It was proposed that only two primary radical patterns (plus an undentified singlet) were formed at 77 K and assigned to G^+ and T^-, a cation on the guanine and an anion on the thymine base [Gräslund et al.]. Using another approach, Gregoli and co-workers, starting from building-block patterns from monomer nucleotides in frozen glasses in the seventies came to the same conclusion for the primary species in DNA and developed a scheme for the secondary reactions upon thermal annealing in 1982 [Gregoli et al. (1976), Gregoli et al.

NATO ASI Series, Vol. H 54
The Early Effects of Radiation on DNA
Edited by E. M. Fielden and P. O'Neill

(1982)]. This experimental system was later pursued and extended by Symons and co-workers who related biological end-points of damage to initial radical formation [Boon et al., Cullis et al.]. Work from the chairman's laboratory, employing oriented DNA-fibers in cooperation with Gräslund and Köhnlein (Münster) has initially confirmed the view of G^{+} and T^{-} as the only primary radicals [Hüttermann et al.] and has given the structure of a successor radical to G^{+}, a radical formed by deprotonation at N_1, GN· [Hüttermann and Voit]. In recent studies, it was shown, that T^{-} was indeed present, but only as a minority component; the pattern which was initially proposed to be due to T^{-} had to be reassigned with the cytosine anion, C^{-}, as a feasable candidate on account of specific deuteration of thymine [Zell et al.].

In view of this intricate development it was not surprising that the speakers came up with different conclusions about primary radical sites in DNA. A controversial issue appears to concern the electron addition site with thymine and cytosine being the proposed candidates. In LiCl-glasses, the tendency of electron addition in oligonucleotides with mixed bases is in favour of C vs. T with duplex formation even enhancing that trend (Bernhard). In frozen aqueous systems containing DNA, T^{-} appears to be favoured at least when judged by TH· formation at higher temperatures and working on the assumption that there is no crossover from C^{-} to T^{-} to TH· (Symons).

In the chairman's view there are two main questions connected with the issue. One concerns the predictive value of the LiCl studies to another environment and to DNA. Can it be assumed that the C^{-}-benchmark from oligonucleotides which differs only slightly but significantly from the T^{-}-benchmark in the same LiCl-glass retains that feature under any condition and, if so, also in DNA? This is the basic problem with all synthetic approaches working from building blocks of subunits to DNA. The problem is presently even enhanced by the fact that C^{-} exhibits a polymorphic ESR-spectrum yielding a triplet in monomers and a doublet in oligomers (Symons, Bernhard). Although the explanation for this feature given by Bernhard appears to be elegant, the question as to the benchmark pattern of C^{-} to be applied at solving the DNA case remains relevant.

This relates directly to the second question which concerns the influence of physico-chemical parameters like hydration structure, base composition, oxygen contents and such on the distribution of (primary) radical sites in DNA. Can we expect under all circumstances G being the primary hole trap? How degenerate are C, which appears to attract electrons in DNA equilibrated in 66 % relative humidity [Zell et al.] and T, which may be the favoured electron addition site in fully hydrated DNA and in the frozen aqueous phase work of Symons' in their electron trapping efficiency?

In this context it is worth mentioning the poster by Close which deals with theoretical calculations on the feasability of reduction and oxidation from partners in base-pairs as well as on their protonation and deprotonation properties. Another aspect treated in the poster by Miller again relates to physico-chemical or better physical parameters governing the distribution of radical sites over DNA which is the energy deposition pattern by densely ionizing radiation.

Energy and charge transfer within dry DNA was the main topic of the paper by O'Neil in which it was shown that migration can take place at a considerable range (20 - 30 base puiss) making trapping of the defect at long distances from energy deposition a reasonable assumption. Also, it was concluded from the spectral characteristics that direct damage in DNA appears to involve purine centered radicals and that hydration influenced to site of localisation of the defect. It is apparent, that the technique of diffuse reflectance measurement has quite a potential, especially due to its time resolution capacity.

In summing up it appears that, despite considerable progress over the past two decades leading, among others, to more and more refined methods of analysis and computer treatment, conventional ESR-spectroscopy has yet to live up to its potential promise to identify structurally and to quantify free radical damage in environments close to the cellular situation. There is a need for spectroscopic probes with a better resolution of radical structures and there is a demand for better definitions of DNA-containing systems under a wide range of modifying parameters (electron- and hole-traps, SH-groups, hydration variation, base compositon, protein-DNA association etc.). It is not before the whole gamut of possible primary and secondary free radicals from DNA in various environments is understood fully that attempts can be made to correlate free radical formation with biological damage in a detailed, mechanistic approach.

Boag JW, Müller A (1959) Electron spin resonance in irradiated deoxyribonucleic acid. Nature 183: 831-832

Boon P, Cullis PM, Symons MCR, Wren BW (1984) Effects of ionizing radiation on deoxyribonucleic acid and related systems. Part 1. The role of oxygen. J. Chem. Soc. Perkin Trans. II: 1393-1399

Cullis PM, Symons MCR, Wren BW (1985) Effects of ionizing radiation on deoxyribonucleic acid. Part 3. The effect of iodoacetanide. J. Chem. Soc. Perkin Trans. II: 1819-1825

Ehrenberg A, Ehrenberg L, Löfroth G (1963) Thymidine-like electron spin resonance spectra in gamma-irradiated deoxyribonucleic acid. Nature 200: 376-377

Gräslund A, Ehrenberg A, Rupprecht A, Ström G (1971) Ionic base radicals in γ-irradiated DNA. Biochem. Biophys. Acta 254: 172-186

Gregoli S, Olast M, Bertinchamps A (1976) Free radical formation in deoxythymidine-5'-monophosphate γ-irradiated in frozen solution. A computer-assisted analysis of temperature dependent ESR-spectra. Radiat. Res. 60: 388-404

Gregoli S, Olast M, Bertinchamps A (1982) Radiolytic pathways in γ-irradiated DNA: Influence of chemical and conformational factors. Radiat. Res. 89: 238-254

Hüttermann J, Voit K (1987) Free radicals from direct action of ionizing radiation in DNA: Structural assignments from EPR spectroscopy in: Electronic Magnetic Resonance of the solid state (Weil J, ed) Canadian Society for Chemistry (Ottawa), 267-279

Hüttermann J, Voit K, Oloff H, Köhnlein W, Gräslund A, Rupprecht A (1984) Specific formation of electron gain and loss centers in X-irradiated oriented fibers of DNA at low temperatures. Faraday Discuss. Chem. Soc. 78: 135-149

Shields H, Gordy W (1959) Electron spin resonance studies of radiation damage in nucleic acids and their constituents. Proc. Natl. Acad. of Sci. 45: 269-281

Zell I, Hüttermann, J, Gräslund A, Rupprecht A, Köhnlein W (1989) Free radicals in irradiated oriented DNA fibers: Results from B-form DNA and from deuterated DNA samples. Free Rad. Res. Comms. 6: 105-106

CONCEPTS OF MICRODOSIMETRY AND THEIR APPLICABILITY TO DNA STUDIES

Joachim Breckow and Albrecht M. Kellerer
Institut für Medizinische Strahlenkunde
Universität Würzburg
Versbacher Str. 5
D-8700 Würzburg
Federal Republic of Germany

Abstract

Ionizing radiations differ from other genotoxic agents by producing, in addition to point mutations, a disproportionate frequency of deletions and other complex DNA lesions. This special characteristic reflects the spatial correlation of energy deposits within the tracks of ionizing charged particles. In the tracks, even of sparsely ionizing particles, clusters of ionizations occur and local concentrations of free radicals that would be extremely unlikely for a uniform random distribution of energy and that could be expected only at doses that are far too large to be of biological relevance. The microscopic correlations of energy deposits are also the reason for the fact that the yield of DNA lesions is proportional to dose - except at very high doses - and that any deviation from linearity must be the expression of energy transport or lesion interaction over comparatively large distances of fractions of a micrometer or of a dose dependent change of repair efficiency.

Microdosimetry provides the concepts that are required to quantify the microscopic spatial correlation of energy deposits for radiations of different quality and to link them to the spatial distribution of the resulting DNA lesions. The basic concepts and quantities are explained, and examples are given that can be related to the size of the DNA structures and to critical distances for the production of DNA lesions.

NATO ASI Series, Vol. H 54
The Early Effects of Radiation on DNA
Edited by E. M. Fielden and P. O'Neill

Introduction

The study of dose-effect relations has always been a central topic of radiation biophysics. One of the principal aims was, to infer the underlying molecular mechanisms from the observed relations. It was soon realized, however, that in radiation studies one deals with a distinct and characteristic feature, namely the statistical fluctuations of energy deposition on a microscopic and molecular scale. Such fluctuations were treated in a variety of approximations: Poisson statistics and the concept of Linear Energy Transfer (LET) were utilized in target theory (Timofeeff-Ressowsky and Zimmer, 1947). Later, a more subtle description was developed by Lea (1955) who introduced a detailed description of the random configurations of energy deposition in the tracks of charged particles. However, a systematic approach towards the quantification of the microscopic distribution of energy deposition was introduced much later, when Rossi developed the concepts and quantities of microdosimetry (Rossi, 1960; Rossi, 1967; Rossi, 1968).

Microdosimetry was designed to further the biophysical analysis of radiation studies (Kellerer and Chmelevsky, 1975; ICRU, 1983; Kellerer, 1985). Somewhat against original expectations it has by now found its most important applications in radiation protection and radiation therapy, where it is used as a means to quantify radiation quality. Mechanistic studies of cellular radiation action have not, as yet, been given rigorous microdosimetric formulations.

The failure of microdosimetry to become a routine tool for mechanistic studies in radiation biology may merely reflect the fact that such studies are still largely unquantitative. However, new techniques will permit more quantitative approaches, and there will thus be a need for rigorous treatments in terms of microdosimetry. It is therefore desirable to summarize the definitions and concepts of microdosimetry and to indicate possible extensions of present concepts that may be particularily suitable for DNA studies.

Basic Concepts and Quantities

Ionizing radiation transfers energy to the exposed materials in discrete random events. The *energy imparted*, ϵ, to the matter in a volume is the difference between the energy of ionizing radiation entering the volume and that emerging from it. The *specific energy*, z, is the quotient of the energy imparted and the mass of the volume. Specific energy in a volume may be due to one or more *energy deposition events*. An event is the energy deposition by an ionizing particle and/or its secondaries, i.e., by energy deposits that are statistically dependent.

The specific energy is a random variable. It has the same unit, Gy, as the *absorbed dose*, D, which is merely the statistical mean of the stochastically deposited energy per unit mass, i.e., of the specific energy. The absorbed dose, therefore, loses its meaning when applied to microscopic regions where the statistical fluctuations of imparted energy can be large.

The *lineal energy*, y, is the microdosimetric analogue of the *Linear Energy Transfer* (LET). It is the quotient of the energy imparted, ϵ, to the volume of interest and the *mean chord length*, $\bar{l}$, in that volume. The mean chord length is the average length of the straight-line segments that result when the reference volume is randomly traversed by straight particle tracks from a uniform, isotropic field. For a convex volume the mean chord length is equal to 4 times its volume divided by its surface (for spherical volumes with diameter d: $\bar{l}=2/3 \cdot d$). The distributions of lineal energy are analogeous to the distributions of specific energy, but they relate only to energy increments produced by *single* events.

The quantity lineal energy has been conceived as the random analogue to the LET, and both quantities are conventionally expressed in the same unit, keV/μm. Measurements that aim at the determination of LET distributions determine, in actuality, the distribution of lineal energy. Due to energy-loss straggling and other stochastic factors these distributions can differ substantially from the distributions of LET. For electrons, for example, the application of the LET concept never permits a direct prediction of energy deposition on a microscopic scale.

The Meaning of Microdosimetric Quantities

Microdosimetric quantities, such as specific energy, are required whenever one considers regions small enough that the relative fluctuations of energy deposition are substantial. Fig.1 gives two scatter diagrams that were obtained by the simulation of energy depositions in small volumes. The diagrams indicate the distribution of specific energies that result at various levels of absorbed dose; they refer to spherical microscopic regions of 0.5 μm (left panel) and 6 μm diameter (right panel). One recognizes that there are very substantial deviations of the specific energy from its mean value, the absorbed dose, and, furthermore, that the distributions of specific energy need not be centred at the value D (diagonal of the diagram). At a given absorbed dose the specific energy may be zero or it may exceed the expectation value by orders of magnitude. The stochastic nature of specific energy becomes more prominent with decreasing site size, with decreasing absorbed dose, and with increasing ionization density of the radiation.

Distributions of specific energy, as represented in Fig.1, can be useful tools in the biophysical analysis of cellular radia-

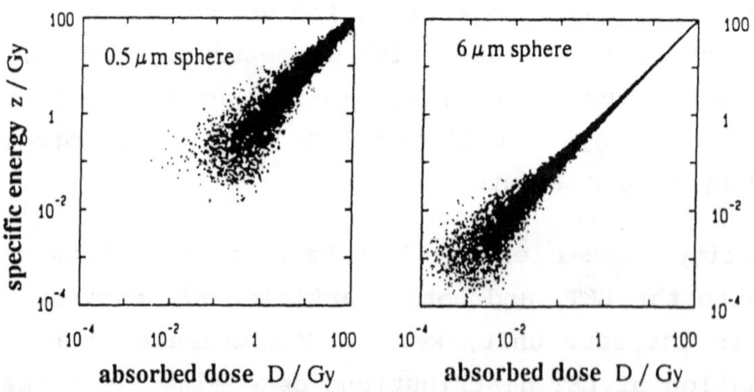

Fig.1.
Scatter diagrams of simulated distributions of specific energy, z, at various levels of absorbed dose, D. The simulations refer to Co-γ-radiation for spherical regions of diameter 0.5 μm (left panel) and 6 μm (right panel). Each panel contains 4000 simulations per decade of D. The number of plotted points is considerably less at low doses due to events with z=0. (After Kellerer, 1985)

tion effects. Specifically, they provide meaningful information on the event frequencies at the stated absorbed dose in specified regions, for example in the nucleus of a cell. They also indicate the possible energy concentrations that can occur in subcellular sites even at very small doses. However, there are definite limits to the applicability of the distributions of specific energy.

There have been various attempts to analyse dose-effect relations in terms of "response functions" for assumed critical sites in the cell which are frequently identified with the cell nucleus. In these approaches one tries to determine the effect probability as a function of the energy concentration, i.e., the specific energy in the nucleus of the cell. However, it was soon realized that the response of the cell or the cellular components is not merely determined by the specific energy in the nucleus or any other assumed target site; it depends on the complex spatial distribution of energy within the totality of its sensitive structures. Microdosimetric distributions of specific energy can not describe such spatial patterns adequately. It is, therefore, desirable to consider concepts that are linked more closely to the structure of charged particle tracks and to the *non-homogeneous spatial distributions* of energy which they produce in the cell.

Characteristics of Track Structure

Fig. 2 exemplifies the spatial patterns of energy deposition by a diagram that superimposes electron and proton tracks with a schematic representation of a chromatin fiber. The dots represent ionizations in the charged particle tracks.

Sparsely ionizing radiation, for instance fast electrons, produce ionizations with fairly wide spacing (Fig.2). However, they produce also a large number of δ-rays (energy clusters) that contribute to a considerable extent to the imparted energy. In effect, one could say that each electron produces not only one electron-track end, but many track ends. A 100 keV photon radiation delivers about 60% of the absorbed dose

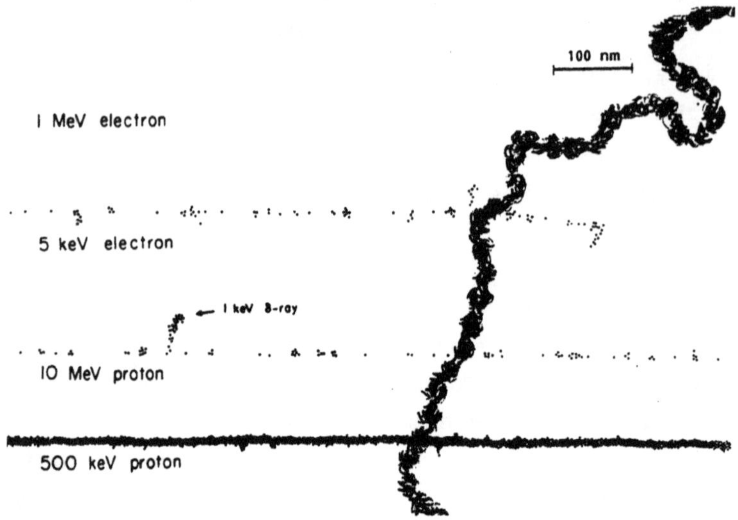

Fig.2.
Schematic representation of electron and proton tracks of the specified energies superimposed on a schematic chromatin fiber that illustrats molecular dimensions.

through electrons of kinetic energies less than 100 eV. Even fast electrons have a high chance to produce local clusters of ionizations which would never occur randomly from independent charged particles, except at enormous doses.

The Proximity Function as a Measure for the Spacing of Energy Deposits

Mutations induced by low-LET radiation are primarily DNA deletions which result from two double-strand breaks (Ward, 1988). This contrasts with the action of chemical mutagens which tend to produce a higher fraction of point mutations. The difference of the interaction mechanisms and the disproportionate frequency of deletions and other complex DNA lesions may be understood in terms of the high local energy concentrations that can occur in the charged particle tracks and that are present even at low absorbed doses. In the following we will consider certain formal concepts which can be utilized in quantitative approaches. They are related to the probability distributions of spatial distances between energy deposits produced in a charged particle track.

The frequency of distances, r, within a given (three-dimensional) point pattern may be represented by the socalled *point-pair distance distribution*. Each point out of an ensemble of N points has N-1 distances to all others. This results in a total of M=N(N-1)/2 distances for a pattern of N points. As an example, three different point patterns and the corresponding

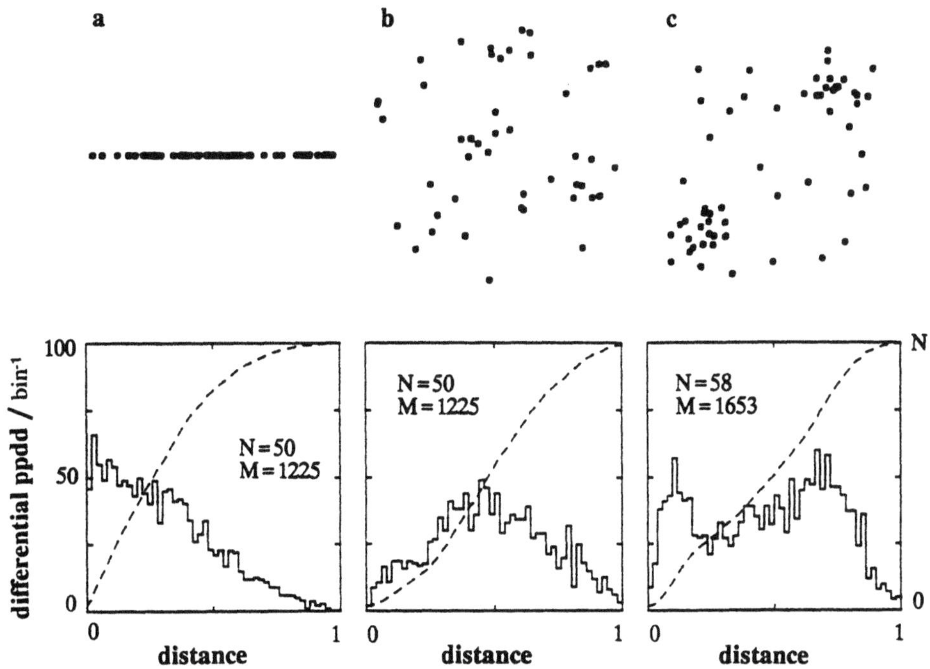

Fig.3.
Examples of point patterns and the corresponding differential (solid line; left ordinate) and integral (dashed line; right ordinate) point-pair distance distributions (ppdd). The points are positioned randomly on a straight-line segment (**a**) or in a plane (**b** and **c**). In **c** two clusters of points are superimposed on the random pattern. The number of points, N, and the number of distances, M, are indicated.

(differential and integral) point-pair distance distributions are shown in Fig.3. In these simple plots the points are positioned randomly on a line (Fig.3a), or in a plane (Fig.3b and c). In Fig.3c two clusters of points are superimposed on the random pattern. Each pattern generates a characteristic point-pair distance distribution. The one-dimensional "track" results in a triangular distance density with the peak value at zero, i.e., there are many short distances and relatively few large distances. For the two-dimensional patterns one obtains maxima at non-zero distances and small or vanishing densities at distance zero. The double peak in the density of the pattern of

Fig.3c reflects the cluster structure of the pattern. In a 3-dimensional pattern one can have an even larger relative contribution of the larger distances.

Fig.4 represents three somewhat more realistic patterns for particle tracks and it can be seen that, here too, the point-pair distance distributions indicate in a general way the nature of the patterns.

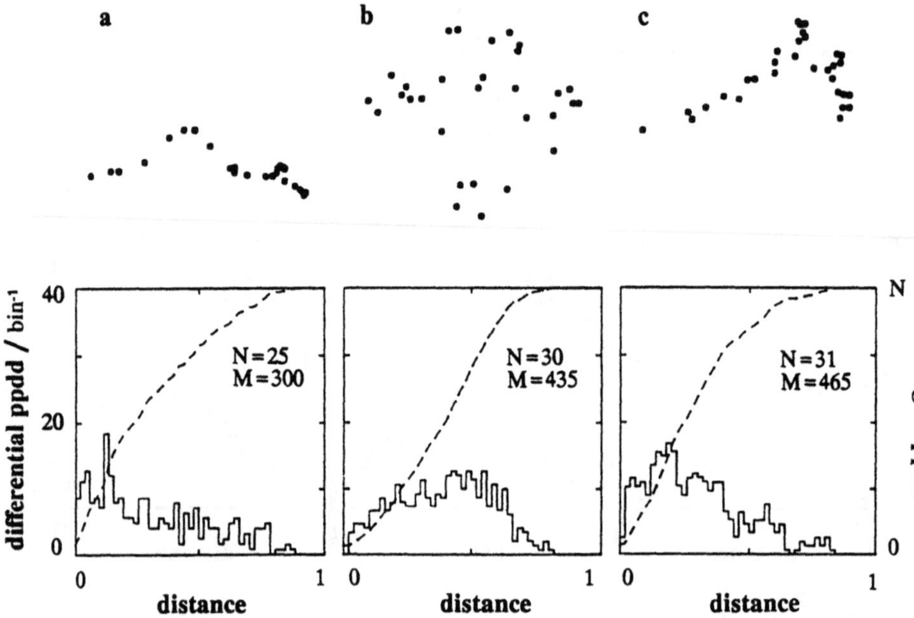

Fig.4.
Examples of point patterns and the corresponding differential (solid line; left ordinate) and integral (dashed line; right ordinate) point-pair distance distributions (ppdd). The number of points, N, and the number of distances, M, are indicated.

In microdosimetry the concept of the socalled *proximity function* is utilized. It is analogous to the point-pair distance distribution. The proximity function, t(r), of a pattern of ionizations in a charged particle track is the energy-weighted distribution of distances between the energy deposits, i.e., it is a weighted point-pair distance distribution, usually expressed in units of keV/μm. More formally, one can state that t(r)dr is the expected energy within the distance interval r to r+dr from a "typical", i.e. randomly selected, energy transfer-point.

Analogous proximity functions can be defined also for geometric objects like volumes, areas, point patterns, or any other configuration, e.g. complex models of molecules. The proximity functions can be a highly valuable mathematical tool in all problems that concern the random intersection of geometric structures. Image analysis is a field where Matheron has introduced important methods based on the *geometric covariograms* (Matheron, 1971) which are closely related to the proximity functions. Another area where the proximity functions provide interesting results on the random intercept of geometric objects is *systematic sectioning* for the estimation of volumes or surfaces (see Cruz-Orive, 1988).

Microdosimetry, too, is concerned with random interception, since energy imparted, ϵ, can be seen as the *random overlap* of the receptor structure (sensitive target) with the charged particle track. One can then use a general formula for the *weighted mean* of the overlap. In terms of the dose-mean event size of energy imparted, $\overline{\epsilon}_D$, it has the form (Kellerer and Chmelevsky, 1975; ICRU, 1983):

$$\overline{\epsilon}_D = \overline{\epsilon^2}/\overline{\epsilon} = \int_0^\infty \frac{s(r)\cdot t(r)}{4\pi r^2}\,dr \qquad (1)$$

where s(r) is the (volume) proximity function of the sensitive site, and t(r) is the (energy) proximity function of the charged particle track. Eq.(1) applies regardless of the complexity of the tracks and the structure of the sites. The site need be neither spherical nor convex, nor even simply connected. For simple geometries the proximity function of the site may be evaluated analytically. For arbitrarily formed configurations it can be derived by means of numerical methods or by Monte-Carlo procedures. The *dose-mean energy imparted*, $\overline{\epsilon}_D$, in a sensitive volume of mass m is a measure for the spatial correlation of energy transfers.

The proximity function makes it possible to determine the dose-mean energy imparted - or corresponding quantities - for any radiation and any geometry of interest. The proximity functions of the radiation and the proximity function of the sensitive sites can be evaluated separately; one obtains then ϵ_D by merely integrating the product of the proximity functions. One single function, t(r), characterizing the radiation field permits thus the computation of ϵ_D for regions of any shape and size.

With this formalism the sensitive site is assumed to be a sharply bounded region. An energy deposit is counted if it is within the volume, otherwise it is not counted. However, the formalism can be extended to apply also "disperse" or "diffuse" regions where energy deposits are given variable weight depending on location (Kellerer, to be published).

Delta-ray, Inter-, and Intratrack Contributions

The proximity function represents the spatial correlation of energy within the particle tracks. The basic features that are involved are the total range and energy of the tracks, and the "pattern" of the energy transfers according to the clustering along the tracks due to *δ-rays*. The proximity function of a charged particle track can be separated into two terms (Chmelevsky et al., 1980):

$$t(r) = t_\delta(r) + t_a(r) \qquad (2)$$

where the term $t_\delta(r)$ represents energy transfers that result from the collisions within the same δ-ray, while the second term is the contribution from energy transfers not on the same δ-ray. Eq.(2) can be a very useful relation in microdosimetric considerations involving heavy ions. However, it is applicable also to electrons. In Fig.5 the separation into two components is illustrated for electrons with initial energies of 5 keV (upper panel) and 1 keV (lower panel). For 1 keV-electrons the peak value at about 5 nm results mainly from non-delta contributions and δ-rays influence the proximity function only well below distances of about 1 nm. This implies that the spatial pattern of energy deposits is essentially determined by LET and

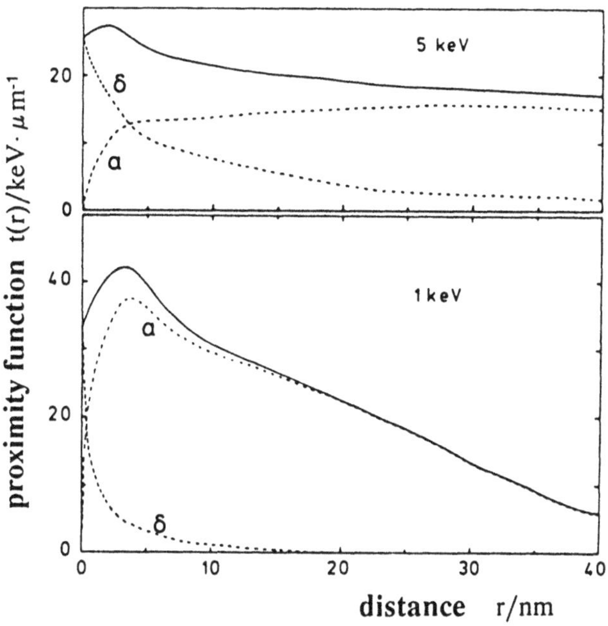

Fig.5.
Proximity functions for simulated electron tracks with initial energies of 5 keV (upper panel) and 1 keV (lower panel). The dashed curves represent the separation into the two components corresponding to the terms $t_\delta(r)$ and $t_a(r)$ in Eq.(2). (After Chmelevsky et al., 1980)

the curling of the electron tracks; the added structuring of the tracks due to δ-rays is not very important. For 5 keV-electrons, on the other hand, the role of the δ-rays predominates below distances of about 3-5 nm; it contributes considerably to the peak value at about 3 nm. For electrons of higher energy the role of the δ-rays becomes even more important; the local concentrations of energy are then largely independent of LET and predominately determined by the clustering of energy in the δ-rays. Thus, even for low-LET radiation, the distribution of the spacing between energy transfers peaks at distances that are in the order of the dimensions of DNA structures. The LET-concept can not directly account for this, and microdosimetric concepts are therefore essential. The frequency of pairs of ionizations only a few nm apart and, therefore, the yield of double-strand breaks increases not only at the ends of electron tracks. It is caused by δ-rays along the entire electron track.

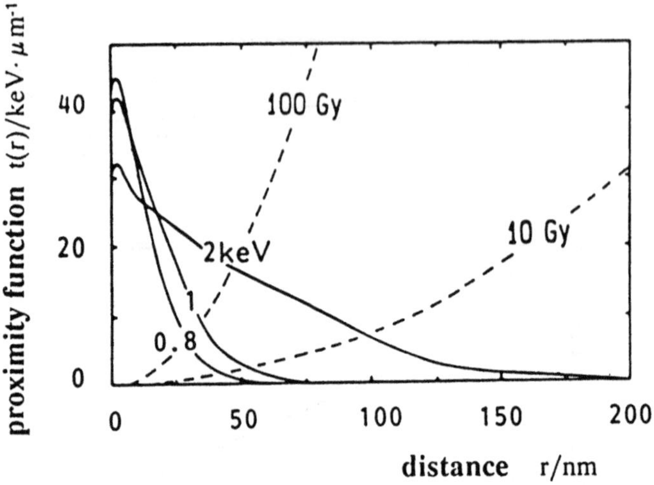

Fig.6.
Proximity functions for simulated electron tracks of the specified energies. The dashed curves represent the dose dependent terms of the intertrack contribution according to Eq.(3) at doses of 10 and 100 Gy. (After Chmelevsky et al., 1980)

The preceding considerations refer to energy deposits within individual particle tracks. Their spatial correlation is, accordingly, independent of dose and the yield of any lesions produced by energy deposits in the same track must be simply proportional to dose. The temporal and spatial proximity of several statistically independent tracks, on the other hand, depends on dose, and associated effects will show non-linear dependence on dose. The concept of the proximity function can be extended to include this aspect. The generalized proximity function becomes then a function of dose, t(D;r). The extension is, however, nearly trivial because t(D;r) differs from the simple proximity function, t(r), the *intratrack* component, merely by an added term that is proportional to the dose and independent of radiation quality (the *intertrack* component) (Chmelevsky et al., 1980):

$$t(D;r) = t(r) + 4\pi\rho_m \cdot r^2 \cdot D \qquad (3)$$

where ρ_m is the density of the medium. This separation is illustrated in Fig.6 for electrons of different initial energies. The figure shows the relative importance of clustering of energy within one electron track (solid lines) in comparison to the accumulation of energy from different tracks (dose dependent term; dashed lines). The energy concentration over short distances and, therefore, the ability to induce DNA deletions is almost entirely determined by the intratrack term, even at high doses.

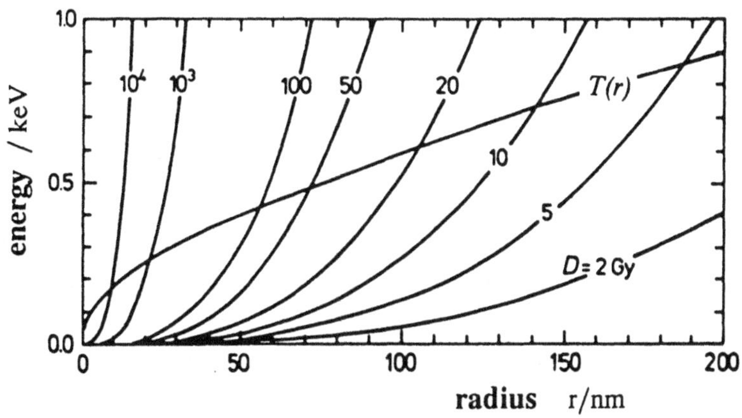

Fig.7.
Comparison of the inter- and intratrack contributions on the clustering of energy deposition for electrons. The line labelled T(r) represents the intratrack contribution in terms of the integral proximity function T(r). The other lines show the intertrack contributions at the specified doses. (After Kellerer, 1976)

A similar picture is given in Fig.7 in terms of the integral proximity function, $T(r) = \int t(r')dr'$. The integral proximity function gives the expected energy on the same particle track within a sphere of radius r around a typical energy transferpoint. For comparison, the expected energy due to independent particles (intertrack contribution) is given for different doses. The figure shows that the energy contribution due to *independent* energy transfers within a sphere of radius about 55 nm equals that due to the *same track* only for doses of at least 100 Gy. One would need a dose in excess of 1 kGy to obtain a larger contribution of other tracks within a sphere of radius 20 nm. For small site sizes that are of interest in DNA studies one is therefore always primarily concerned with the action of individual particles.

Application to DNA Studies

The essentials of the preceding section are summarized in a somewhat more illustrative picture in Fig.8. As has been stated, it is not unlikely, even for fast electrons, that an ionization has in its near neighborhood one or more ionizations produced by the same electron. It is much less likely that it has in its immediate vicinity further ionizations from an independent particle. Equal contributions from independent particles and from the same particle would result for 1 Gy if a large region of 360 nm radius were considered (see spheres in Fig.8). A dose of 10 Gy would be necessary for a region of 140 nm radius, and 100 Gy for 55 nm (cf. Fig.7). The intratrack effect is thus predominant in all studies of multiple DNA lesions, except at very high doses and except for interactions that extend over fairly large distances. Even sparsely ionizing radiations have, therefore, the ability to generate clusters of DNA lesions in close proximity, a feature not common with other

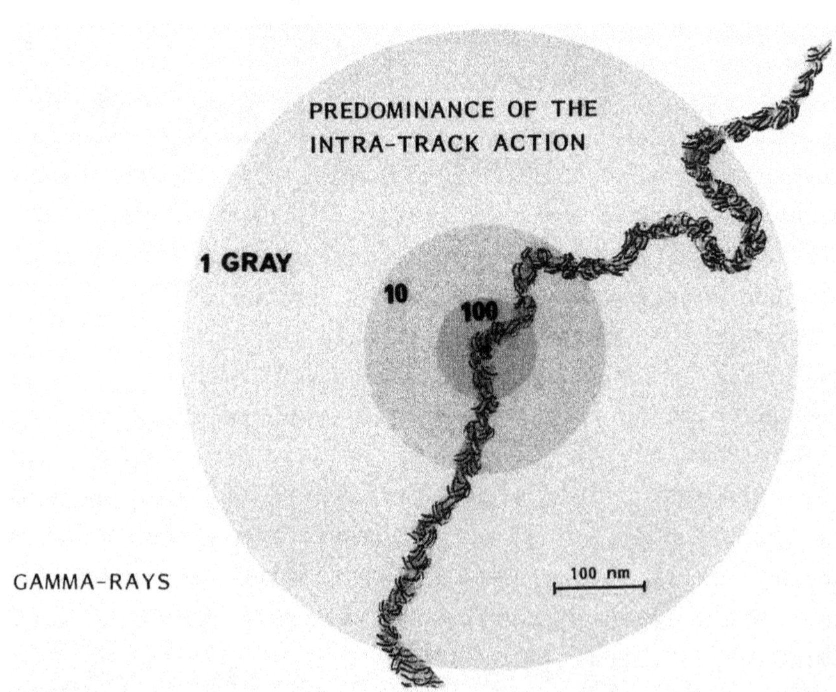

Fig.8:
Schematic representation of spherical regions of predominance of the intratrack contribution at specified doses for Co-γ-radiation superimposed on a schematic chromatin fiber.

genotoxic agents. This provides for a mechanism in radiation action that would otherwise be possible only at very high doses.

In summary, one concludes that the action of ionizing radiations - and even of sparsely ionizing radiations such as γ-rays, x-rays, or fast electrons - are governed by the high local energy concentrations that occur within individual charged particle tracks. These high local energy concentrations which can be quantified in terms of microdosimetric concepts and functions account for the specific ability of ionizing radiations to produce multiple DNA lesions in close proximity that can express themselves in socalled bulky lesions or in deletions. It has always been seen as a somewhat surprising feature of ionizing radiations that the yield of deletions relative to point mutations is higher than the corresponding ratio for most other genotoxic agents. The microdosimetric data make this understandable and they can - at least in principle - be used to quantify these relations.

Efforts to achieve a quantitative treatment would have to cover a broad range of possible mechanisms. The interaction of single-strand breaks can lead to double-strand breaks, the interaction of double-strand breaks can lead to deletions, and there is of course a broad variety of other DNA-lesions that can cause irreversible cell damage. One must also note that the interaction of energy deposits may not always be the interplay of several individual lesions, for example in misrepair. It can also be a more complex process where the repair of an individual lesion is made less efficient or is made impossible by local damage that inhibits repair processes.

References

Chmelevsky D, Kellerer AM, Terrissol M, Patau JP (1980) Proximity functions for electrons up to 10 keV. Radiat Res 84: 219-238

Cruz-Orive LM (ed) (1988) Stereology and Stochastic Geometry. Journal of Microscopy. 1st special issue, vol.151, part 3

ICRU (1983) Report 36, Microdosimetry. International Commission on Radiation Units and Measurements (ICRU), Bethesda MD

Kellerer AM (1976) A survey of approaches to radiation biophysics. In: Booz, J et al. (eds), Proc 5th Symp on Microdosim vol.1, EURATOM, Luxembourg, p 409-438

Kellerer AM (1985) Fundamentals of microdosimetry. In: Kase KR et al. (eds) The Dosimetry of Ionizing Radiation. Academic Press, Orlando FL, p 78-163

Kellerer AM (to be published) A generalized formulation of microdosimetric quantities. Radiat Prot Dosim

Kellerer AM, Chmelevsky D (1975) Concepts of microdosimetry. III. Mean values of the microdosimetric distributions. Radiat Environ Biophys 12: 321-335

Lea DE (1955) Actions of Radiations on Living Cells, 2nd edn. Cambridge Univ Press, London

Matheron G (1971) The theory of regionalized variables and its application. Les cahiers du CMM de Fontainebleau, No.5. Ecole Nationale Supérieure des Mines de Paris

Rossi HH (1960) Spatial distribution of energy deposition of ionizing radiation. Radiat Res, Suppl.2: 290-299

Rossi HH (1967) Energy distribution in the absorption of radiation. Adv Biol Med Phys 11: 27-85

Rossi HH (1968) Microscopic energy distribution in irradiated matter. In: Attix FH, Roesch WC (eds) Radiation Dosimetry, 2nd edn., vol.1. Academic Press, New York, p 43-92

Timofeeff-Ressowsky NV, Zimmer KG (1947) Das Trefferprinzip in der Biologie. Hirzel, Leipzig

Ward JF (1988) DNA damage produced by ionizing radiation in mammalian cells: identities, mechanisms, and repairability. Prog Nucleic Acid Res Mol Biol 35: 95-125

THE APPLICATION OF BIOPHYSICAL MODELS TO CELLULAR DNA DAMAGE

D. E. Charlton
Physics Department
Concordia University
1455 de Maisonneuve Blvd
Montreal H3G 1M8
Canada

Introduction

The first stages of direct damage from ionising radiation are the ionisation and excitation of atoms in the DNA molecule and the consequential production of single strand breaks (ssb). These, if on opposite strands and close together, can combine to produce double strand breaks (dsb). In turn it is likely that a small fraction of dsb combine to produce a variety of chromosome aberrations endangering cell reproduction.

In this presentation the initial energy deposition, the production of ssb and their combination into dsb will be described using electron track structure with a simple model of the DNA molecule. Results for the absolute yields of single and double strand breaks per Gy-dalton of DNA for various radiations will be presented and compared with experimental data.

Method

The calculation of the production of ssb and dsb is performed in two parts. Firstly the probability of depositing given amounts of energy E in lengths ℓ of DNA per unit dose is calculated using the techniques described by Charlton et al (1985,1989) and Nikjoo et al (1988). Secondly, the energy deposition patterns in the DNA model are converted to ssb patterns, and a general relationship is established between the energy deposited in the DNA and the probability that this energy deposition will produce a ssb or dsb.

NATO ASI Series, Vol. H 54
The Early Effects of Radiation on DNA
Edited by E. M. Fielden and P. O'Neill

The relationship between these two quantities and the yield of breaks Y(b) per Gy-dalton is

$$Y(b) = \sum_{E} \frac{p(\ell,E) * p(b,E)}{\ell * 1.94 * 10^{-3}}, \quad \ldots\ldots\ldots\ldots\ldots\ldots (1)$$

where $p(\ell,E)$ is the probability of depositing energy E in length ℓ of DNA per unit dose and p(b,E) is the probability that this energy E will produce a single or double strand break in the segment. The summation is carried out over all energy depositions and $1.94 * 10^{-3}$ converts length ℓ of DNA in nanometers to daltons. In this calculation ℓ has a value of 18.4 nm or 54 bp.

The spectrum of energy depositions in the DNA molecule; $p(\ell,E)$

Here the technique for calculating the energy deposition in the DNA per unit dose for ion tracks will be outlined. A similar method is used for x-rays and electrons.

Ions produce well defined volumes around their tracks in which the energy of the delta-ray flux is confined. This volume can be characterised by the maximum lateral displacement of their ionisations and excitations. To begin a calculation the ion track code of Wilson and Paretzke (1981) (MOCA 14) is used to generate the positions and energies of ionised and excited atoms along tracks of several micrometers in length for a chosen ion. It should be noted that for the ion calculations the ion energy is assumed to be constant over the path length whereas the energy of electrons, the delta ray flux produced by ion interactions with the medium, is followed down to 10 eV.

The tracks are scanned to find the interaction with the greatest displacement from the path of the ion and this is used to define the radius of a cylinder which contains all of the track. This virtual cylinder is used to construct random passages of chords which will define the positions of the DNA molecules.

The DNA molecule is placed at random with respect to the track spatially and in direction by generating μ-random chords through the virtual cylinder. Typically 50,000 chords are generated. The DNA molecules are to be represented by straight cylinders of diameter 2.3 nm with these chords as their axes. This arrangement is shown in figure 1. Since the frequency of hits in a well defined length of DNA is required in the calculation, a unit length is chosen (in this case 54 base pairs (bp)(54 * 0.34 nm)) and the number of these lengths in the total DNA length placed in the cylinder is calculated.

For each track, each chord is considered in turn and the co-ordinates of each ionisation or excitation lying inside the DNA cylinders is recorded as well as the energy of each ionisation or excitation.

These data are used in two ways.

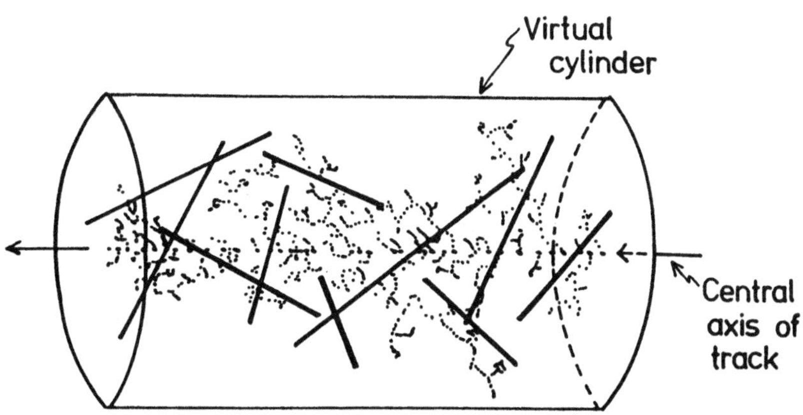

Figure 1. Showing the geometry used to score hits in 2.3 nm diameter cylinders representing the DNA molecule.

If a cylinder is "hit" (usually only one cluster of interactions is found), it is divided into sections of 54 bp beginning from its origin at the surface of the virtual cylinder . The energy in each 54 bp segment is stored and sorted into energy bins so that the frequency of hits of various sizes can be obtained. From the number of segments hit and the total number of DNA segments in the virtual cylinder, the fraction of cylinders hit gives the probability of a hit of any size. Since the energy spectrum of depositions is known the probability of depositing a given energy can also be found. These probabilities can be normalised to those per gray and so the probability of depositing energy E per 54bp segment-Gy can be evaluated. This is $p(\ell,E)$ in equation 1. Values of $p(\ell,E)$ for various ions, x-rays and electrons are given later in Table 2.

Calculation of breaks from energy deposition

Above, the probability for the total energy deposited in DNA segments for each hit was discussed. Now it is necessary to use the same data to predict the effects of these hits. From the initial data, the co-ordinates of each ionisation and excitation and their energy in each 54 bp segment are known. The segments are arranged at random in space since the co-ordinates of each interaction lie near the axes of the chords on which the DNA molecules were placed. By rotating and translating each set of co-ordinates appropriately, the interactions in each DNA molecule can be made to lie along a common axis. This makes the next step in the analysis much easier.

It is now necessary to examine where the interactions in the DNA lie with respect to the positions of the bases and the sugar-phosphate moieties. To do this the simple volume model of the DNA shown in figure 2 is used. Here the molecule is modeled as a cylinder divided into disc and arch shaped pieces of thickness 0.34 nm, where the two discs on the outside of the cylinder represent volumes occupied by the sugar-phosphate components and the central cylindrical volume that occupied by the bases. Each combination of two arches and a disc represent a base pair. The arches are rotated by 36° each base pair. When this model is superposed on the distribution of interactions in the DNA the energy deposited in the bases and sugar-phosphate groups can be found. Seven examples of the result of this superposition chosen from data for 10 MeV alpha particles are shown in table 1.

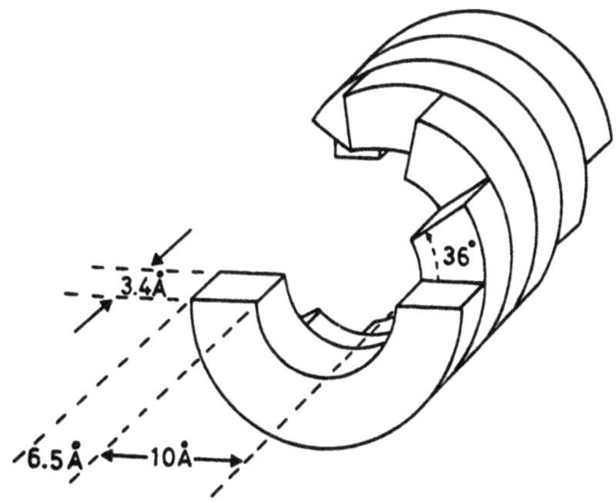

Figure 2. The volume model of the DNA.

The data in the table illustrate the energy deposition in sequential nucleotide pairs, each pair being represented by a column of three pixels. The upper and lower pixels are the depositions in sugar-phosphate volumes and the central pixel is the energy in the pair of bases.

These data show that hits in the DNA are quite limited in the length, only a few of the hits being greater than a few tens of nucleotide pairs. Also the damage (although this is not specifically shown) can be extensive in this small region.

Table 1. Examples of energy deposition in DNA from 10 MeV alpha particles.

```
..x             --...........-2-                  ....2..
3..   (68eV)   ................    (87eV)         ...xx.. (195eV)
.x.             ..........x..-.                   2x3-3x2


.-53.........x...........                      -4...x2
x2......................-   (259eV)            ....x..   (183eV)
......3.2xxx...........-.                      ..43...


.........                      3....2.......xx.
.....x623   (162eV)            2.........x2..x-   (180eV)
2.......                       ...............2
```

Key to symbols:- energy less than 10 eV;1 energy between 10 and 17.5 eV;+ energy between 17.5 eV and 20 eV;2 20-30 eV;3 30-40 eV;etc. Each column of 3 pixels is one nucleotide pair. The total energy deposited in the DNA is given in parentheses.

Data like these are now to be used to calculate the points of breakage in the DNA. In order to do this the energy (and its location) to produce a single strand break is required. This information can be obtained from an experiment by Martin and Haseltine (1981).

Energy to produce a single strand break

In one of the Martin/Haseltine experiments a 58 bp fragment of plasmid DNA was labelled at one end with 125-IUdR and the end of the same strand was labelled with 32-P. After storing the DNA to allow the 125-I to decay the double stranded DNA was denatured and the single stranded fragments analysed on sequencing gels. Using autoradiography only fragments containing the 32-P label can be detected and these have lengths equal to the distance from the 32-P atom to the furthest single strand break from the incorporated iodine position. The distribution of the distances from the 125-I end of the plasmid to the break are shown in figure 3.

The experiment was simulated using the electron track code of Paretzke (MOCA7b)and the volume model of the DNA described above as well as a code simulating the Auger

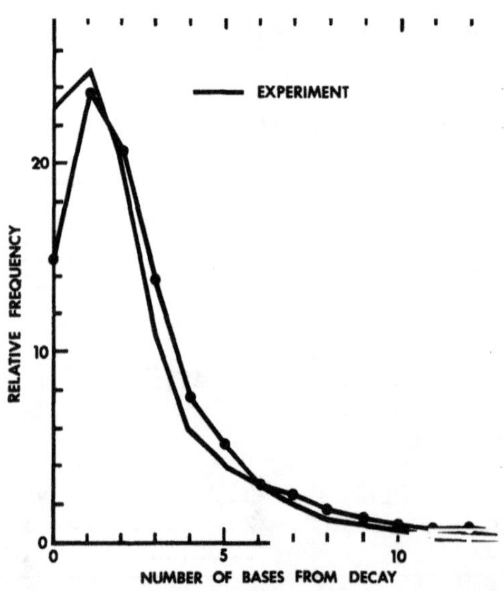

Figure 3. Frequency of distances to the furthest single strand break from a decay of 125-I. Comparison of experimental and calculated data.

electron spectrum of 125-I (Charlton and Booz 1981). Distributions of energy depositions similar to those in table 1 were produced and, assuming that single strand breaks are produced by damage to the sugar-phosphate moiety, a threshold of energy deposited in that volume was sought which produced the distribution of the measured distances (Charlton and Humm 1988). A threshold of 17.5 eV give the results shown in figure 3 - a reasonably good agreement with experiment. Other thresholds in the range 17.5 to 22.5 eV also gave reasonable agreement (Humm and Charlton 1989). The threshold value can now be used with data such as in table 1 to predict the positions of single strand breaks. Each pixel in the sugar-phosphate chain with energy greater than 17.5 eV is identified as a ssb and when these are on opposite strands and lie within 10 bp it is assumed a dsb is produced. In the table all hits shown except the one depositing 162 eV produce a dsb.

The separation between ssb on opposite strands to produce a double strand break is more restrictive than that obtained experimentally by van der Schans (1978), van Touw _et al_ (1985) and Herskind (1987) where typical values for this separation are at least 30 bp. These experiments were done in dilute solution with the breaks most likely caused by radical action whereas here only direct interactions are considered. In this case few of the hits have "lengths" as long as 30 bp.

The relationship between energy deposited in the DNA and the production of strand breaks ; p(b,E)

Raw data such as that shown in table 1 were analysed for the production of single and double strand breaks using the criteria discussed above. While a hit may contain several ssb or dsb only one dsb and a maximum of two ssb where assigned to a hit since closely spaced breaks could not be resolved experimentally. The calculations were carried out for alpha particles, protons, low energy x-rays and electrons of 20 keV and the relationship between the energy deposited in the DNA and the number of single and double strand breaks produced per hit of that size found. Typically five to ten thousand hits were analysed for each case. It was found that high LET radiations gave good statistics for the higher range of energy depositions while the low LET radiations gave better statistics for the smaller depositions. A composite curve for the probabilities obtained from all radiations is shown in figure 4 where within calculational error a single curve is adequate for the required result.

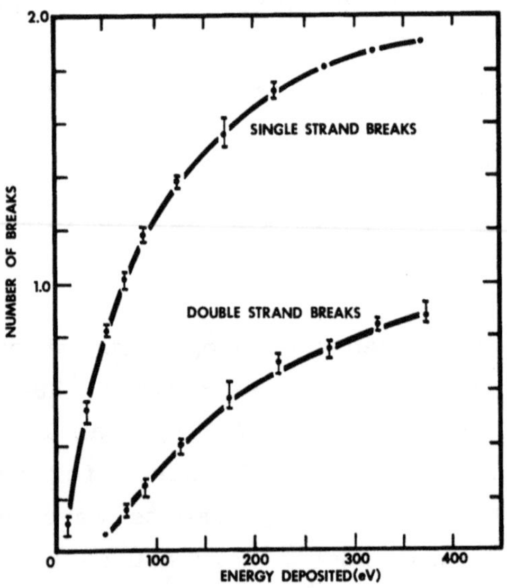

Figure 4 The relationship between energy deposited in the DNA and the probability of producing single or double strand breaks; p(b,E).

Results

The spectrum of hit sizes per gray; p(ℓ,E)

Table 2 summaries the hit size distributions per Gy in a segment of DNA 54 bp long for several radiations. At the bottom of each column the total number of hits per Gy in the segment is given. Table 2 illustrates some fundamental differences and similarities between high and low LET radiations. All radiations deposit a wide range of energies in the target, the difference between radiations being in the distributions of the hit sizes. Comparing a high LET radiation of 3 MeV α particles (LET = 125 keV/μm) with 20 keV electrons shows that in the former case 45% of hits are less than 40 eV while for the low LET radiation 78% are less than this value. According to the analysis given in figure 4, energy depositions less than 40 eV are ineffective in producing dsb.

The number of hits per unit dose decreases with increasing LET for a particular radiation but there is no consistency of this feature between radiations. For example, 49 keV/μm protons (0.38 MeV) hit the DNA at about the same rate per Gy as 76 keV/μm α particles (6 MeV).

Table 2 Hit size distributions per Gy per 54 bp segment. All values in the body of the table have been multiplied by 10^{+6}.

Energy interval	Alpha particles (MeV)					Protons (MeV)	
(eV)	20	10	6	3	1.2	2	0.38
0-20	3.795	3.025	2.103	1.491	0.749	4.960	2.239
20-40	2.613	2.188	1.444	0.954	0.475	3.346	1.501
40-60	1.569	1.313	0.970	0.597	0.294	1.513	0.994
60-80	0.935	0.963	0.687	0.432	0.211	0.809	0.799
80-100	0.642	0.526	0.548	0.337	0.176	0.483	0.619
100-150	0.678	0.858	0.804	0.639	0.311	0.638	0.837
150-200	0.198	0.308	0.374	0.390	0.136	0.136	0.385
200-250	0.071	0.137	0.188	0.247	0.137	0.057	0.163
250-300	0.035	0.061	0.084	0.153	0.106	0.018	0.065
300-350	0.013	0.033	0.053	0.094	0.115	0.009	0.042
350-400		0.016	0.020	0.051	0.096		0.030
>400			0.024	0.094	0.338		0.022
Σ	10.549	9.428	7.299	5.479	3.185	11.969	7.696

Interval	X-rays*		Electrons (keV)
(eV)	C	Al	20
0-20	2.226	2.912	5.770
20-40	1.550	2.008	2.903
40-60	1.150	1.248	1.158
60-80	0.837	0.814	0.552
80-100	0.578	0.544	0.293
100-150	0.802	0.642	0.296
150-200	0.267	0.218	0.078
200-250	0.084	0.072	0.026
250-300		0.022	0.008
300-350		0.005	
>350		0.002	
Σ	7.494	8.487	11.084

*X-rays from C and Al; 280 eV and 1.5 keV respectively.

Yields of single and double strand breaks

The sum of the product of the number of hits within a given energy interval and the probability that a single energy deposition of this size will produce a single or double strand break (see figure 4) will give the yield of breaks per Gy for a single segment. This calculation is shown in detail in table 3 for the case of 20 keV electrons since it illustrates very well the importance of the direct interaction of the secondary electron flux generated by the primary particle with the DNA molecule. Approximately 40% of hits (4.42/11.08) produce a ssb while only 4% (0.45/11.08) of hits produce a dsb. This shows that there is a large amount of direct interaction in the DNA by the ionising radiation which produces only a small but sufficient effect.

The model used to produce these data assumes that any hit within the cylinder enclosing the DNA molecule is a direct hit. It may be argued that ionisations and excitations in the major or minor groves are indirect in their action on the atoms of the DNA (Pomplun 1990). It is clear from this

Table 3 Illustrating the calculation of the yields of ssb and dsb for 20 keV electrons.These data are for a segment of DNA 54 np long per Gy.

Energy interval (eV)	1 Number[1] of hits	2 Number of ssb per hit	3 Number of dsb per hit	1*2 Number[1] of ssb	1*3 Number[1] of dsb
0-20	5.770	0.08	0.00	0.46	0.00
20-40	2.903	0.52	0.00	1.51	0.00
40-60	1.158	0.82	0.08	0.95	0.09
60-80	0.552	1.03	0.16	0.57	0.09
80-100	0.293	1.19	0.26	0.35	0.08
100-150	0.296	1.38	0.40	0.41	0.12
150-200	0.078	1.55	0.57	0.12	0.04
200-250	0.026	1.74	0.71	0.04	0.02
250-300	0.008	1.80	0.75	0.01	0.01
Σ	11.084			4.42	0.45

[1]Numbers in this column have been multiplied by 10^6.

To convert the yield of breaks to that per Gy-dalton multiply the 4.42 and 0.45 by $2.81*10^{-11}$ ($1/54*660*10^6$). Multiply by 101 ($3.6*10^{12}*10^{-6}/660*54$) to find the yield in a mammalian cell. $3.6*10^{12}$ daltons of DNA in a mammalian cell and 660 daltons per nucleotide pair are used in the calculation.

analysis however that there are sufficient numbers of events very close to the actual DNA molecule to explain the production of dsb. Other mechanisms such as long range radical diffusion appear to be unnecessary.

The ratio of ssb to dsb for 20 keV electrons from these data is 4.42:0.45 or 10:1, lying within experimental values reported by Roots et al(1985).

The data in table 2 can readily be converted to the yield of breaks per Gy-dalton of DNA (shown in figure 5) by using equation 1 or to the number of breaks in a mammalian cell per Gy (shown in table 4). The method is given in detail at the bottom of table 3.

From figure 5 the yields of single strand breaks are in excellent agreement with experiment for all comparable data. For double strand breaks the calculated yields are

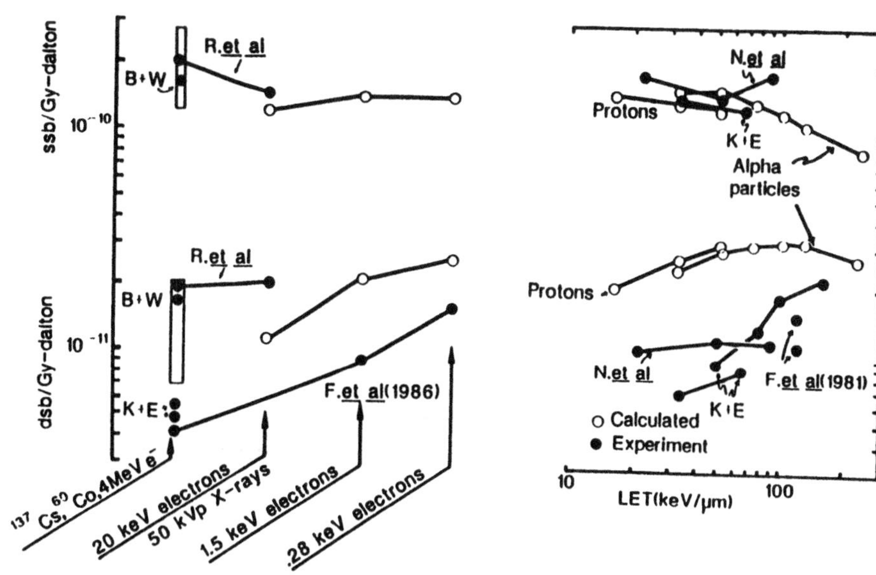

Figure 5 The yields of single and double strand breaks per Gy per dalton of DNA for various radiations:Comparison of theory and experiment.

Key to authors:B+W Baverstock and Will (1987), F et al (1981) Frankenberg et al (1981), F et al (1986) Frankenberg et al (1986), K+E Kampf and Eichhorn (1983), N et al Neary et al (1972),R et al Roots et al (1985). The blocks at the highest photon energies cover a range of data collected by Roots et al. Yields of ssb obtained by alkali treatment were reduced by 30%.

consistently higher than those measured except for the lowest LET where there is a large variation in the experimental yields.

In table 4 the data is ordered by decreasing numbers of hits in the DNA of a mammalian cell per gray of dose. While the number of hits decreases with increasing LET, the number of dsb, with the exception of the two lowest LETs, is approximately constant.

Table 4 The relationship between hits of any size and the production of ssb and dsb in a typical mammalian cell per Gy.

Particle	LET⁺ (keV/μm)	Hits per cell per Gy*	ssb per cell per Gy	dsb per cell per Gy
2 MeV p	16	1209	629	77
20 keV e⁻	(1.3)	1120	446	45
20 MeV α	33	1066	619	92
10 MeV α	51	952	624	111
Al x-ray	(10)	857	524	83
0.38 MeV p	49	777	567	118
C x-ray	(21)	757	520	92
6 MeV α	76	737	548	115
3 MeV α	125	553	463	118
1.2 MeV α	230	322	313	96

⁺For electrons the stopping powers were taken from the initial energies. A track averaged LET would give higher values.
*Using $3.6 * 10^{12}$ daltons of DNA in the genome and a mass of 660 daltons for a nucleotide pair.

Discussion

While there are many approximations in the treatment of strand breaks presented above, the results are encouraging and provide a way of simulating direct damage to the DNA. The use of a track code based on unit density water vapour and the simple volume model of the DNA limit the accuracy of the treatment. However the normalisation to the results of Martin and Haseltine (1981) appears to allow a good measure of single strand breaks to be made. It is not clear why good agreement with single strand break measurements has been obtained while the calculated yields of double strand breaks are high.

It has been suggested that at low LETs most (65%) of the damage is indirect (Roots and Okada 1975) but the data presented here are not inconsistent with experimental results at low LETs, at least down to 20 keV electrons. At the lowest LETs there is a considerable spread of experimental data (0.4 to 2.0 x 10^{-11}dsb/Gy-dalton) and it is not possible from these results to evaluate the importance of indirect damage.

One other attempt using a different method has been made to calculate absolute yields of single and double strand breaks (Holley _et al_ 1990) for various ionising radiations. Using a less-detailed ion track model and DNA in which the positions of the atoms are known these authors were able to simulate yields of ssb and dsb for very heavy fast ions. As in the treatment above breaks were assumed to originate with damage to the sugar-phosphate moiety, produced by an ionisation there rather than by using an energy deposition threshold as has been described here. The results of Holley _et al_ for the single strand yields for helium ions were below the data from Kampf and Eichhorn (1983). For double strand breaks there is reasonably good agreement for helium ions for smaller LETs which then falls below the experimental results at LETs greater than 100 keV/μm.

Most significant in the calculation described in this presentation is the nature of the direct damage and its frequency per unit dose. Data on the structure of energy deposition in the DNA as shown in table 1 clearly demonstrate the very complex and varied damage which occurs. This sort of data, in addition to predicting strand breakage, also suggests that deletions produced by direct damage should be small, less than 10 base pairs in length. In table 4 the number of hits in the DNA of a mammalian cell per Gy is surprisingly high and could account for almost all of the radiation damage.

It is likely that since energy is deposited in clusters near the ends of electron tracks that there is some production of radicals outside of the DNA molecule but sufficiently close it to add single strand breaks to those produced by direct action. These additional breaks would add to the yields of single strand breaks if no break from a direct interaction was present. In addition they could add to the number of double strand breaks if they could convert a single strand break into a double strand break.

Even though the electron track modeling and the construction of the DNA target is very elementary, this type of calculation shows how important detailed analysis of energy deposition can be and the potential contribution to fundamental radiobiology from the simulation electron tracks.

References

Baverstock KF, Will S (1987) Evidence for the dominance of the "direct" mechanism in DNA strand scission in cells. Radiation Research I Fielden EM, Fowler JF, Hendry JH Scott D (eds) Taylor and Francis London page 10

Charlton DE, Booz J, (1981) A Monte Carlo treatment of the decay of 125-I. Radiat Res 87:163-171

Charlton DE, Goodhead DT, Wilson WE, Paretzke HG (1985) Energy deposition in cylindrical volumes:(a) Protons, energy 0.3 MeV to 4.0 MeV (b) Alpha particles, energy 1.2 MeV to 20.0 MeV. MRC Radiobiolgy Unit Monograph 85/1.

Charlton DE, Humm JL (1988) A method of calculating initial DNA strand breakage following the decay of incorporated 125-I. Int J Radiat Biol 53:353-365

Charlton DE, Nikjoo H, Humm JL (1989) Calculation of initial yields of single- and double-strand breaks in cell nuclei from electrons, protons and alpha particles. Int J Radiat Biol 56:1-19

Frankenberg D, Goodhead DT, Frankenberg-Schwager M, Harbich R, Bance DA, Wilkinson RE (1986) Effectiveness of 1.5 keV aluminium K and 0.3 carbon characteristic X-rays at inducing DNA double-strand breaks in yeast cells. Int J. Radiat Biol 50:727-741

Herskind C (1987) Single strand breaks can lead to complex configurations of plasmid DNA *in vitro*. Int J Radiat Biol 52:565-575

Holley WR, Chatterjee A, Magee JL (1990) Production of DNA strand breaks by direct effects of heavy charged particles. Radiat Res 121:161-168

Humm JL, Charlton DE (1989) A new calculational method to assess the therapeutic potential of Auger electron emission. Int J Radiat Oncol Biol Phys 17:351-360

Kampf G, Eichhorn K (1983) DNA strand breakage by different radiation qualities and relations to cell killing: Further results after the influence of alpha particles and carbon ions. Studia Biophys 93:17-26

Martin RF, Haseltine WA (1981) The range of radio chemical damage to DNA with decay of iodine-125. Science 213:896-898

Neary GJ, Horgan VJ, Bance DA, Stretch A (1972) Further data on DNA strand breakage by various radiation qualities. Int J Radiat Biol 22:525-537

Nikjoo H, Goodhead DT, Charlton DE, Paretzke HG (1988) Energy deposition by C, Al, Ti and Cu X-rays in cylindrical volumes within mammalian cells. MRC Radiobiology Unit Monograph 88/1.

Pomplun E, (1990) Private communication.

Roots R and Okada S (1975) Estimation of lifetimes and diffusion distances of radicals involved in X-ray induced DNA strand breaks or killing of mammalian cells. Radiat Res 64:306-320

Roots R, Kraft G, Gosschalk E (1985) Formation of radiation induced DNA breaks:The ratio of double strand breaks to single strand breaks. Int J Radiat Oncol Biol Phys 11:259-265

van der Schans GP (1978) Gamma-ray induced double-strand breaks of DNA resulting from randomly inflicted single-strand breaks:Temporal local denaturisation, a new radiation phenomenon? Int J Radiat Res 33:105-130

van Touw JH, Verbene JB, Retel J, Loman H (1985) Radiation -induced strand breaks in ϕx174 replicative form DNA:An improved experimental and theoretical approach. Int J Radiat Biol 48:567-578

Wilson WE, Paretzke HG (1981) Calculation of distribution of energy imparted and ionisations by fast protons in nanometer sites. Radiat Res 87:521-537

Acknowledgments

The author would like to thank Concordia University, the Organising Committee of the Workshop and the Radiobiology Unit of MRC(UK) for financial assistance for this work. Also permission from the International Journal of Radiation Biology to reproduce several of the figures is acknowledged.

THE APPLICATION OF CHEMICAL MODELS TO CELLULAR DNA DAMAGE

W.R. Holley and A. Chatterjee
Cell and Molecular Biology Division
Lawrence Berkeley Laboratory
One Cyclotron Road
Berkeley, CA 94720
U.S.A.

INTRODUCTION

Understanding the biochemical mechanisms associated with the production of DNA strand breaks by ionizing radiation is a subject of great interest. Numerous attempts have been made to elucidate many aspects of these phenomena (Roots *et al.*, 1985; Pohlit and Drenkard, 1985; Ward, 1988) and indeed most of our present knowledge is due to the efforts of experimental radiation biochemists. For a better quantitative understanding these efforts need to be complemented by theoretical modeling in order to correlate energy loss events with the corresponding biochemical changes and eventually to the yields of strand breaks.

Most theoretical models on strand breaks tend to ignore the biochemical pathways. For example, Charlton and Humm (1988) have described a method for calculating DNA strand breaks induced by the decay of incorporated ^{125}I, and Goodhead and Nikjoo (1989) have analysed energy deposition of ultrasoft X-rays in target volumes with dimensions comparable to short sections of DNA. These calculations consider a detailed account of the track structure of low energy electrons but ignore the geometrical structure of DNA and the biochemical stage.

Incorporation of biochemical mechanisms in theoretical modeling is non-trivial especially in view of the variety of complex biomolecules associated with DNA in a cellular environment. To circumvent this complexity, we have considered an aqueous solution of DNA containing 0.5 M of tris. This simplification immediately raises the question of the validity of the model system compared to DNA inside the nucleus of a mammalian cell. In order to study this

NATO ASI Series, Vol. H 54
The Early Effects of Radiation on DNA
Edited by E. M. Fielden and P. O'Neill

question, we have developed a theoretical model involving an aqueous system with the inclusion of biochemical mechanisms. We have compared the normalized yields of strand breaks (per unit Gray per unit dalton) as a function of linear energy transfer (LET) based on the theory with the corresponding experimental data measured in a cellular environment but with the inhibition of enzymatic repair. This comparison provides us with information on the adequacies of the biochemical mechanisms considered and the simplified aqueous-DNA system.

Based on experimental data, the most important biochemical mechanisms associated with the production of strand breaks are: (1) water radical abstraction of H from the sugar rings (indirect mechanism) and (2) ionization of the sugar-phosphate backbone followed by deprotonation (direct mechanism). It has been reported in the literature (Ward, 1988) that the average ·OH migration distance in a cellular environment is about 3.5 nanometers. The presence of 0.5 M tris (an efficient ·OH scavenger) mimics, approximately, this migration distance in the aqueous system. However, this system will not necessarily reproduce effects of secondary radicals that may contribute to the overall yield of strand breaks.

The theoretical model discussed here has the following features: (a) representation of a linear DNA molecule (in the B form) in a three-dimensional configuration based on X-ray diffraction data (Arnott and Hukins, 1972), (b) simulation of the diffusive motion of water radicals in three dimensions, (c) reactions between sibling water radicals, (d) reactions of water radicals with scavengers (tris), (e) reactions of water radicals with sugars and bases resulting in strand breaks at sites of sugar damage only, and (f) ionization events on the DNA backbone and corresponding strand breaks. The details of these features have already been published (Chatterjee *et al.*, 1986; Chatterjee, 1989; Holley *et al.*, 1990) and, hence, in this report we have only provided a brief review of each of them.

Consideration of track structure has played a significant role in the modeling effort, and has enabled us to calculate the yields of strand breaks for different qualities of radiation: photons, electrons, and heavy charged particles. High-LET radiation is an excellent probe for studying the different biochemical mechanisms in a systematic manner using selected track-segments.

One of the important results of the theoretical calculations is the variation of the yields of single and double strand breaks (breaks produced on opposite strands with a maximum separation of ten base pairs) as a function of LET. These results can be compared with many experimental measurements available in the literature, thus providing a way to evaluate the utility of the approach we have taken.

METHOD OF CALCULATION

The overall method of calculation has been developed in various stages which we now describe very briefly.

Track Structure

Consideration of the track structure of energetic electrons is unavoidable under any irradiation condition with ionizing radiation. They are involved either as primary or secondary charged particles. Very low energy (~ 100 eV) as well as very high energy electrons (~ few MeV) should be included in any theoretical computation. In the present work, the concepts of track entities such as spurs, blobs, short tracks, and branch tracks have been used since there is a long history of their usefulness in radiation chemistry. These concepts and their applications in the Fricke dosimeter system to evaluate the ferric yields have been reported elsewhere (Magee and Chatterjee, 1978). Since the chemical mechanisms for DNA damage in aqueous solutions are qualitatively similar (·OH reactions with sugar sites), no change in the method of calculation seems necessary as far as track structure is concerned.

For heavy charged particles, the microscopic pattern of energy distribution is taken into account through the concept of "core" and "penumbra". Again, these concepts have previously been applied to the Fricke dosimeter system, and, for consistency, the procedure has remained the same in the DNA aqueous solution situation. The radius, r_c, of a core is defined as the radius within which a major fraction of the glancing collisional losses are contained. In addition, the core contains the energy deposited by those low energy knock-on electrons which are stopped completely in this region. The radius of the core is given by $r_c = \hbar v/2\varepsilon_1$, where v is the velocity of the charged particle, ε_1 is the lowest electronic transition energy and $\hbar = h/2\pi$ is the modified Planck's constant. This radius is independent of the charge on the projectile but the energy density is not.

For a 600 MeV/n particle, $r_c = 82$ Å, and when the energy decreases to 10 MeV/n this radius is 15 Å. The variation of its size as a function of energy has been published elsewhere (Magee and Chatterjee, 1980). The region of energy deposition outside the core is called the penumbra. The average penumbra radius is given by the empirical relation, $r_p = 396\ v^{2.7}$. Here v is in units of 10^9 $cm.s^{-1}$ and the radius is in angstroms. In the present work, each electron track in the penumbra has been considered independently and the concepts of spurs, blobs, and short tracks have been utilized for the evaluation of the radical mechanism.

Structure of DNA

In order to do the proper bookkeeping as to whether a given water radical has reacted with a base or a sugar, consideration of the DNA structure in three dimensions is essential. Only then can the accessibility of a given site (sugar or base) by a water radical be properly represented. It is known that the rate constants for ·OH reactions with isolated sugar or base molecules are about ten times greater than the corresponding rate constants in a structured DNA molecule. Hence, in the present calculations, it is not satisfactory to represent DNA by isolated sugar and base molecules even if present in the proper concentrations. The constituent molecules of the DNA polymer are positioned in Cartesian coordinates based on X-ray diffraction data and in this manner the locations of the sugar or base molecules which have undergone reactions can be accurately stored in the computer memory. From this information, one can easily determine whether a given break is an isolated one or part of a double strand break based on the definition that two breaks on opposite strands within ten base pairs of each other create a double strand break.

Indirect Mechanism

The production of damaged sites by water radicals is called the indirect mechanism. From the radiolysis of water, four major chemical species, ·H, e_{aq}^-, ·OH, H_3O^+, are produced in a picosecond or less with yields of 0.88, 5.00, 5.88, and 5.00 respectively. These species undergo diffusive motion in the system and their individual fates are determined by: (1) reactions with each other to produce molecular products, (2) scavenging reactions with the tris molecules, and (3) reactions with sugars or bases. Of the radicals which interact with the DNA, most of the ·H react with the bases, all of e_{aq}^- react with the bases, 20% of

·OH react with the sugars and the rest (80%) with the bases. H_3O^+ does not react with DNA but it can react with other water radicals. Hence, as far as strand breaks are concerned, our focus has been on ·OH. Our procedure has been to determine the decay curves for ·OH as a function of time using a Monte Carlo approach to simulate the diffusion phenomenon. These decay curves have been obtained in pure water and in systems which include the scavenging effects of tris. For electron tracks as well as heavy particle tracks, these curves have been determined before proceeding to calculate DNA strand breaks. In order to calculate the production of strand breaks, each sugar and each base molecule have been positioned in three-dimensional space as mentioned earlier.

For each OH radical whose diffusive motion is to be followed, the initial position coordinates distributed uniformly in space around the DNA are determined using random numbers. Each radical is followed in space and time by simulating diffusive motion. Corresponding to an interval Δt of "jump" time, a reaction probability P for an OH radical under consideration is determined from the appropriate previously obtained decay curve. Next, a random number between 0 and 1 is selected and if it happens to be smaller than or equal to P, the ·OH is removed from further consideration. If the number is greater than P, the next jump for ·OH is taken. In this manner a surviving OH radical is followed until it is found within the reaction radius of a sugar or a base. The respective reaction radii of the various sites have been calculated using Smoluchowski's theory and the rate constants for isolated molecules. The respective radii are: ·OH + sugar, 1.0 Å; ·OH+A, 3.6 Å; ·OH + T, 3.6 Å; ·OH + G, 5.2 Å; and ·OH + C, 3.6 Å. Although the rate constants used for determining the reaction radii are for isolated molecules, in a separate calculation it has been verified that, when the respective sites are present in the structural configuration, there is a reduction in the resultant rate constant per molecule compared to the isolated cases by a factor of about ten. Furthermore, it has also been verified that with the above-mentioned reaction radii, about 20% of all the ·OH that react with DNA are absorbed at the sugar sites and the rest (80%) with the bases. This is an experimental fact that should be satisfied by any reasonable theory.

We keep a record of: (a) how many OH radicals are lost from the initial number on their way to a DNA molecule, (b) how many have reacted with the sugar moiety, and (c) how many have reacted with the bases. The total number of trials we take for hydroxyl radicals is typically about 200,000 which yields results reproducible within ± 2%. The use of precalculated decay curves saves an enormous amount of computer time. Based on these procedures, the D_{37} (indirect effect) values for the induction of single strand breaks have been calculated. The D_{37} dose is defined as the dose required to reduce the fraction of the number of undamaged DNA molecules to 37% of the initial (unirradiated) value. Separate calculations have been done for the core and the penumbra, using appropriate decay curves. From the D_{37} values we have computed the indirect effect cross sections for single strand breaks for high-LET particles according to the formula

$$\sigma_{SSB} = KL\left(\frac{f_{core}}{D_{37}^{core}} + \frac{1-f_{core}}{D_{37}^{pen}}\right), \tag{1}$$

where K = 16.02, σ_{SSB} is in Å^2, L is the LET in eV.Å^{-1}, D_{37} values are in krad and f_{core} is the fraction of energy deposited in the core.

Indirect production of double strand breaks (breaks on opposite strands within 10 base pairs) have been computed by assuming that 2.0% of the single strand breaks (experimental result) also result in breaks on opposite strands. This is somewhat unsatisfactory. However, this procedure is adopted in the absence of a mechanism for the production of double strand breaks due to indirect effects. Two hydroxyl radicals attacking opposite strands in our calculations do not produce enough breaks to be in agreement with the experimental data. This mechanism yields smaller values by about two orders of magnitude.

Direct Mechanism

The main features of our theoretical model of the direct effect are summarized in the following. Track structure and stopping power theory in combination with the detailed geometry of DNA molecules are used to calculate the average energy deposition by fast charged particles on DNA. Since the actual physical energy deposition processes are stochastic in nature, we use Poisson statistics to provide a detailed description of the distribution of the resulting states of excitation and ionization of the molecule. The average energy of an excitation (or ionization) on the DNA has been calculated from existing oscillator strength data for biological molecules to be approximately $\langle E \rangle = 30$ eV. Ionization of an atom on the DNA sugar-phosphate backbone is assumed to be followed by one of two different types of biochemical change, deprotonation or direct dissociation, both of which then lead to DNA strand breaks. Base damage has been assumed not to lead to the formation of strand breaks.

For convenience and consistency, we have used the same three-dimensional geometric model of DNA for direct effect calculations as used in the indirect effect calculations. The spatial orientation of a given track can be represented by its impact parameter or distance from the DNA central axis, the angle between the track and the DNA axis, and the core radius r_c. The coordinates of a particle track and the DNA model can be correlated to determine which sugar-phosphate molecular groups lie within the core. Restricting ourselves to glancing collisions only and using the Bragg rule in combination with the stopping-power formula, we can calculate the average energy E_g lost to a collection of atoms (sugar-phosphate) inside the track core as discussed in detail in Holley *et al.* (1990). Since the energy density in the core is a sum of glancing collisions and low-energy-knock-on collisions, we have modified E_g to take into account the δ-rays by writing

$$E_d(r \leq r_c) = E_g \left[1 + \frac{1}{1+2\ln(r_p/r_c)} \right] \tag{2}$$

Similarly, for the penumbra region (again to account for the δ-rays) we write

$$E_d(r_c < r < r_p) = \frac{E_g r_c^2}{r^2[1 + 2\ln(r_p/r_c)]} \tag{3}$$

The average number of ionizations (and or excitations) on a sugar-phosphate molecule lying a distance r from the track trajectory is

$$n = E_d(r)/\langle E \rangle \ . \tag{4}$$

Using Poisson statistics, the respective probabilities of no ionization, one ionization, two ionizations, etc., are

$$P_i = \frac{n^i e^{-n}}{i!}, \qquad i = 0, 1, 2, ... \tag{5}$$

The probability of one or more ionizations is given by

$$P_{\geq 1} = 1 - P_o = 1 - e^{-n} \ . \tag{6}$$

In our calculations, for each incident particle track, we determine the probability of ionization, $P_{\geq 1}$, at each reaction site within the track penumbra radius r_p. We generate an "event" by choosing a random number for each site. If $r\# < P_{\geq 1}$ for a particular site, we consider an ionizing reaction to have occurred with the production of a DNA strand break, either by direct dissociation or deprotonation. Breaks occurring close enough to each other on opposite strands (i.e., within 10 base pairs) lead to double strand breaks. An event is then classified into one of the following categories:

(1) no breaks (NB),

(2) one or more single strand breaks (SSB), and
(3) one or more double strand breaks (DSB).

If we consider N incident particle tracks uniformly distributed over an area A, yielding N_{SSB} single strand breaks and N_{DSB} double strand breaks on DNA of molecular weight M, the corresponding normalized cross sections are given by

$$\sigma_{SSB} = N_{SSB}/(MN/A), \tag{7}$$

$$\sigma_{DSB} = N_{DSB}/(MN/A). \tag{8}$$

It is frequently more convenient to discuss strand break production in terms of yields of breaks per unit radiation dose rather than cross sections. The yields, Y_x, corresponding to the normalized cross sections in eqs. (10) and (11) are given by the following expressions:

$$Y_{SSB} = \sigma_{SSB}\rho/LET, \tag{9}$$

$$Y_{DSB} = \sigma_{DSB}\rho/LET, \tag{10}$$

where ρ is the mass density of the medium and LET is the total linear energy transfer of the incident particle. In terms of conventional units of $(\text{Å})^2$ for σ, keV/μm for LET and rad for dose, the expression for the strand break yield per rad per dalton becomes

$$Y = 6.2 \times 10^{-10}\,\sigma/LET. \tag{11}$$

In the expression for Y, it should be kept in mind that σ depends on the track structure.

RESULTS

Cross-Sections

Results of our calculated strand-break cross-sections for representative particles, He and Ne, are shown in Figure 1. Single strand break cross-sections vary approximately linearly with LET and the corresponding dependence for double strand scissions is LET^2. For single strand breaks, it appears that both the indirect mechanism and the direct mechanism have roughly equal cross-sections which introduces a possible new phenomenon. Namely, one strand can be broken by the indirect mechanism and the other by direct mechanism. This effect has not been included in the present model. Below 30 keV/μm, radical mechanisms seem to create double strand breaks with a greater efficiency than the deprotonation events. Above this LET value, the reverse is true.

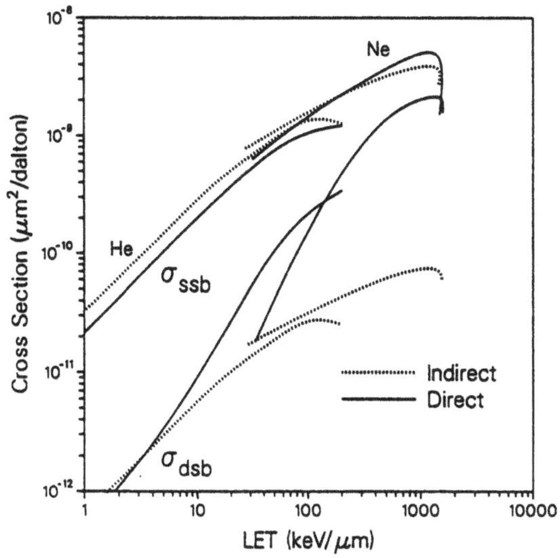

Figure 1. Calculated cross-sections for strand break formation by direct and indirect effects for He and Ne as functions of LET. The total cross-section is the sum of the two contributions. At the lowest LET's, the indirect effect cross-sections are larger and at the highest LET's, the direct effect dominates.

Negligible Effects of Tris Radicals

In Figure 2, we have plotted the variation of the D_{37} values as a function of tris concentration for ^{60}Co-γ radiation. When the molarity is low, the average ·OH migration distance is very large which results in dominance of the indirect action. The dotted line in Figure 2 is the result of indirect action only and the solid line represents the sum of indirect and direct actions according to the relationship:

$$1/D_{37}^{total} = 1/D_{37}^{indirect} + 1/D_{37}^{direct} \qquad (12)$$

For low concentration of tris, there is no difference between $1/D_{37}^{total}$ and $1/D_{37}^{indirect}$ indicating that the contribution due to direct action is negligible. This feature is expected because ·OH can migrate from large distances and direct excitation and ionization of DNA has a much smaller probability than reactions with water molecules which are present at 55 M. However, as the tris concentration becomes larger than 0.04 M, the $1/D_{37}^{direct}$ values start influencing the $1/D_{37}^{total}$ values, i.e., the direct action becomes increasing important. These results are supported by the experimental data (Chatterjee and Holley, 1990) as indicated by the solid circles. Since our calculation does not take into account effects of tris-radicals (mainly formed from the reactions of tris with ·OH), the agreement between theory and experiment suggests that these species do not cause appreciable strand breaks.

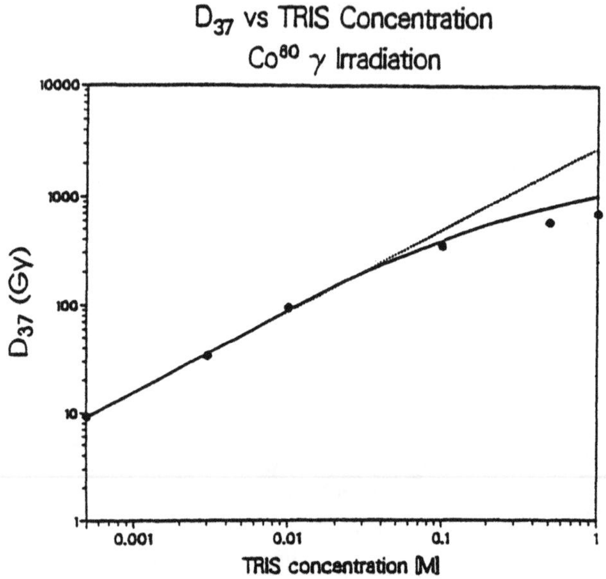

Figure 2. Variation of D37 for single strand breaks as a function of tris-HCl concentration. The dotted line represents calculated results for indirect effects only, while the solid curve includes direct and indirect effects. Solid circles are data points from experiments of Ruth Roots, Ernst Henle, and Conrad Trumbore (private communications).

It is important to note that at 0.5 M tris, the direct effect is appreciable. At this molarity, the average ·OH migration distance is about 3 nm, similar to what it is believed to be in a cellular complex. The calculated D_{37} value for single strand breaks at this concentration is 749 Gray and a value of 700 Gray has been measured experimentally.

One of the aims of this work is to compare our calculations with strand break data when cells are irradiated with various qualities of radiation. When the LET is high, the contribution to strand breaks from the indirect mechanism can be less than 20% (Roots *et al.*, 1985). In Figure 3, such a comparison has been made by using data from reported measurements of several investigators. In these measurements, the experimental conditions were manipulated so that no (or minimal) enzymatic repair of strand breaks was allowed. It can be seen from the comparison that the theoretical and experimental results are qualitatively similar, providing some confidence in the calculation procedure described in this paper. Even the quantitative agreement at high LET values seems quite reasonable.

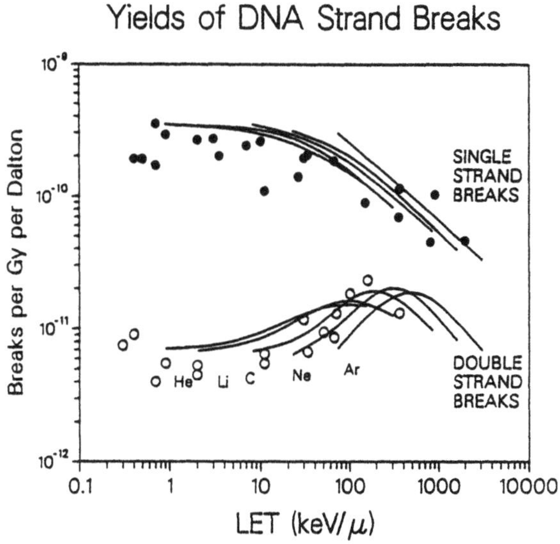

Figure 3. The smooth curves are calculated yields (including both direct and indirect effects) of single and double strand breaks plotted *vs*. LET. For comparison, a selection of experimental measurements from the literature of radiation-induced initial (enzymatic repair inhibited) strand break yields for a variety of mammalian cell types is also plotted.

Clusters of DNA Damage

We have used our present model to estimate quantitatively the extent of sugar and base damage at or near the sites of radiation-induced single and double strand breaks. Figures 4a, 4b, and 4c represent schematically the distribution of DNA damage for typical events. Sugar damage resulting in a strand break is indicted by a missing "S" and base damage is indicated by "B*". Figure 4a shows typical low LET events, characterized by isolated single strand breaks and damaged bases with an occasional double strand break. Figures 4b and 4c show typical double strand break events for 10 MeV/n neon (450 keV/μm). These events are characterized by a substantial number of damaged sites (sugars and bases) spread over a region of 10-15 base pairs.

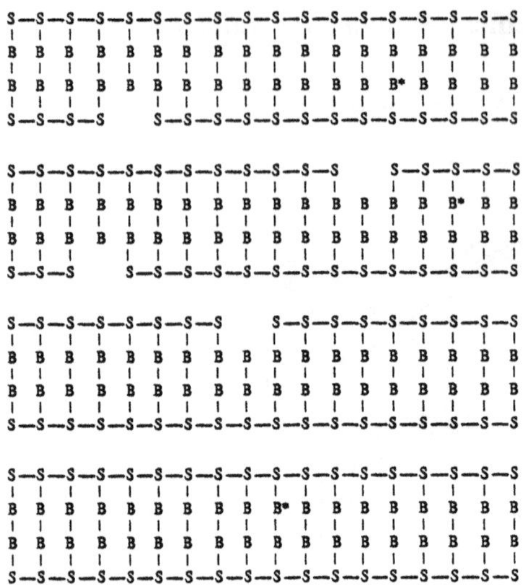

Figure 4a. Diagrams of typical low LET strand break events. Second event from top is a double strand break event.

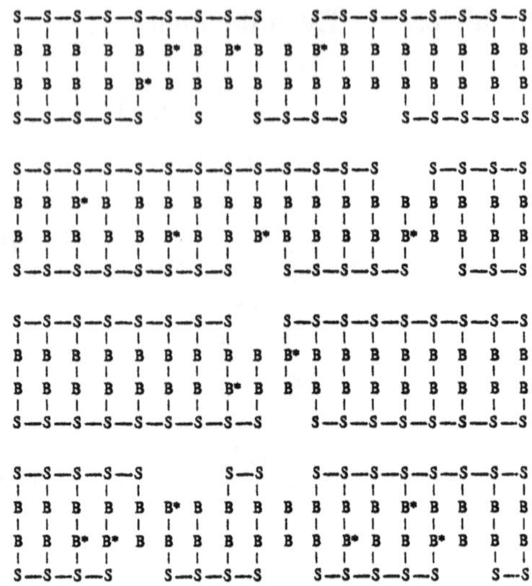

Figure 4b. Diagrams of typical high LET (450 keV/μm) double strand break events. Strand scissions (damaged sugars) are indicated by a missing "S" and damaged bases bases by "B*".

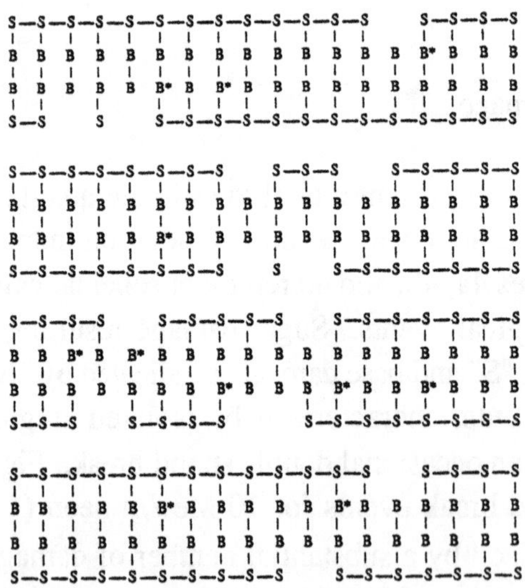

Figure 4c. Diagrams of additional high LET (450 keV/μm) double strand break events.

The results of these calculations indicate that there is a qualitative difference in the nature of double strand breaks induced by low-LET and by high-LET radiation. Most lesions produced by low LET radiation are simple, isolated, single (and sometimes double) strand breaks. On the other hand, double strand breaks produced by high LET radiation almost always are associated with multiple lesions either on adjacent sugars or bases or frequently both.

DISCUSSION

A theoretical model has been presented in this paper which correlates the yields of single and double strand breaks in a DNA molecule produced by different qualities of radiation. The model correlates track structure of energy deposition with chemical and biochemical changes which ultimately result in the formation of observable strand breaks. Although there are parameters involved in the method outlined above, they have not been adjusted to match the experimental data. For example, 17 eV has been taken as the energy needed to create a radical pair such as ·H and·OH. This number is not known very well for liquid water and hence has to be considered only as a parameter without a strong fundamental basis. However, this same value has been used in our previous models in the analysis of the Fricke dosimeter system (an aqueous solution of ferrous sulphate) and we believe that this number should not be arbitrarily varied to fit experimental data for another aqueous system. Since in the present context we have been dealing with DNA in aqueous solution, we have used 17 eV as the energy required to create a radical pair. Other parameters used are well known diffusion constants, rate constants, base pair separation distance, etc.

In order to avoid the complexities of a cellular media, a very simple system consisting of linear DNA molecules in water has been used for modeling purposes. This is an oversimplification of the actual biological case. However, when compared with experimental data from exposures of mammalian cells, the results from the proposed theoretical model seem to be in fairly good agreement. Hence, it appears that the model has some validity with respect to the production of initial damage.

In spite of this success, we believe that there are several drawbacks to the model in its present form which need improvement. For example, energy migration along the DNA chain is a well known phenomenon which suggests that the site of damage may be quite different from the location where energy was deposited. What effect this energy migration has on strand break formation is not clear at the present time. Similarly, no attempt has been made in the model discussed here to understand the effects of structured water associated with a DNA molecule. It is possible that structured water behaves differently than bulk water. It is, therefore, important to understand how a hydroxyl radical propagates through the bound water before it gets to a particular DNA site (sugar or base).

Another weakness in the present model is the lack of a suitable mechanism for the production of double strand breaks by hydroxyl radicals. As discussed previously, reactions by two or more hydroxyl radicals from the same incident track do not yield an appreciable number of double strand breaks. It is possible that a single ·OH may be responsible for the production of these types of alterations in a DNA molecule. If this is so, one needs to understand the mechanisms involved before they can be incorporated in a model.

In the present model no account has been taken of the creation of double strand breaks by cooperative phenomena in which one strand is broken by the indirect effect and the other strand by the direct effect. We do not expect this mechanism to introduce a large contribution to the overall yield of double strand breaks; nevertheless, inclusion of this effect should improve the accuracy of the results presented in this report.

From these and other considerations, it is clear that a mechanistic model such as the one presented here needs further improvements. It is hoped that other investigators will interest themselves in this problem and provide added insight. At present the authors are considering some of the deficiencies of the model presented here, including the effects of higher order DNA structure on the yields of strand breaks.

ACKNOWLEDGMENT

This work was supported by the Office of Health and Environmental Research, Office of Energy Research, U.S. Department of Energy, under Contract No. DE-AC03-76SF00098.

REFERENCES

Arnott S and Hukins DWL (1972) Optimized parameters for A-DNA and B-DNA. Biochem Biophys Res Comm 47:1504-1509

Charlton DE and Humm JL (1988) A method of calculating initial DNA strand breakage following the decay of incorporated ^{125}I. Int J Radiat Biol 53:353-365

Chatterjee A, Koehl P, and Magee JL (1986) Theoretical consideration of the chemical pathways for radiation-induced strand breaks. Adv Space Res 6:97-105

Chatterjee A (1989) Radiobiological effects of high-LET particles: DNA strand breaks. Nucl Inst & Methods in Physics Res A280:439-448

Chatter A and Holley WR (1990) Direct and Indirect Effects in DNA Aqueous Solution. J Cell Biochem Suppl 14A

Goodhead DT and Nikjoo H (1989) Track structure analysis of ultrasoft X-rays compared to high- and low-LET radiations. Int J Radiat Biol 55:513-529

Holley WR, Chatterjee A, Magee JL (1990) Production of DNA strand breaks by direct effects of heavy charged particles. Rad Res 121:161-168

Magee JL and Chatterjee A (1978) Theory of the chemical effects of high-energy electrons. J Phys Chem 82:2219-2226

Magee JL and Chatterjee A (1980) Radiation chemistry of heavy particle tracks. 1. General considerations. J Phys Chem 84:3529-3536

Pohlit W and Drenkard S (1985) Quantitative determination of the contribution of indirect and direct radiation action to the production of lethal lesions in mammalian cells. Rad Prot Dosimetry 13:195-198

Roots R, Chatterjee A, Chang P, Lommel L, and Blakely EA (1985) Characterization of hydroxyl radical induced damage after sparsely and densely ionizing radiation. Int J Radiat Biol 47:157-166

Ward JF (1988) DNA Damage Produced by Ionizing Radiation in Mammalian Cells: Identities, Mechanisms of Formation, and Repairability. Prog Nuclei Acid Res & Mol Biol 35:95-125

What basis for the development of radiation-induced DNA damage?

Hooshang Nikjoo,
MRC Radiobiology Unit, Chilton, Oxon OX11 ORD

To account for the observed biological effects a multitude of different physical and chemical processes have been put forward todate. In terms of time domains molecular effects are referred to as 'direct-effect' (D) and 'indirect-effect' (ID). There exist a number of definitions for these terms: (D) effect due to ionisation or excitation of the DNA molecule, (D) effects not modified by radical scavenging species, (D) thermal excitation of DNA by internal conversion, (ID) damage caused by reactive species and (ID) damage due to finite migration of free radicals. These definitions collectively describe a set of very sophisticated modes of energy transfer by ionising radiation to a DNA molecule either directly or by secondary chemical interactions which raise a number of questions. What is the nature of the ionisation of the DNA molecule? What is the relevance of aqueous solution chemistry to the condensed phase of the cell? Does the energy migrate along the chain of the DNA? Normally, there are four methods of assessing the contributions from direct and indirect effect, namely, addition of a radioprotector to the system, freezing, dry DNA method and estimation by computational modeling. These data are grouped according to the cell type used. Examination of the published data of the contributions to the biological effects from direct and indirect effects shows a broad range without offering conclusive evidence as to the nature of the radiation action. The general belief is that OH radical scavengers are a major source of lethal damage and removal of OH leads to protection. It is believed that over 50% contribution to the total effect is due to such effects at low-LET. Is this contribution a pure indirect-effect from radical scavengers or a combination of direct and indirect contribution to the total biological response. The data from most experiments involving radioprotectors, to some degree, are dependent on experimental conditions and parameters which makes it difficult to infer conclusive remarks. Therefore, some degree of bench-marking is required. A useful comparison is the efficiency of production

NATO ASI Series, Vol. H 54
The Early Effects of Radiation on DNA
Edited by E. M. Fielden and P. O'Neill

of ssb and dsb in DNA molecules by radiations of different qualities. Yields of ssb from experiments under dry conditions (Baverstock & Will 1989), to maximise the direct effect contribution, are in good agreement with those under cellular conditions . Estimation of the initial biophysical damage in terms of the yields of sb in DNA have been obtained from frequency of the energy deposition events. Table(1) shows a comparison of calculations with experimental data. Method (1) is based on deposition of 100 eV energy in a 2x2 nm cylinder as a gross DNA volume (Goodhead and Nikjoo 1989, Nikjoo et al 1989). Method (2) is based on a more detailed analysis of the patterns of energy deposition events as described by Charlton (this volume) and Charlton et al(1989). The data presented here, including the recent calculations by Holley et al (1990) which give a maximum of 20% contributions for the indirect effect at high-LET, show a dichotomy as to the relative contribution by direct and indirect effects and the role of OH radical in producing lethal damage. Although most experimental data give 50% or more for indirect effect contributions to the biological effect but the initial biophysical calculations of single and double strand breaks show similar yields and ratios. Model calculations by Ito (1987) give the contributions from purely indirect OH radicals at 1% while the direct effect at 46% from low-LET radiations. Because of the complexity of the cell environment we may need to consider mechanisms of energy deposition at the atomic level taking into account the non-homogeneity of the biological systems and determine whether large molecules require different mechanisms. It may be useful to avoid the use of the term indirect-effect in attempting to model the mechanisms of energy deposition in DNA.

Table 1 number of (dsb/ Gy/ dalton)* 10^{-11}

Radiation	mesaured	expected (1)	expected (2)
C X-ray	1.6	1.9	1.7
Al X-ray	0.9	1.3	1.4
Ti X-ray	-	0.8	-
100 keV e	0.7	0.6	0.7

STOCHASTIC METHODS IN RADIATION CHEMICAL KINETICS

N.J.B. Green
Physical Chemistry Laboratory
South Parks Road
Oxford OX1 3QZ
United Kingdom

In radiation tracks reactive particles are formed with clustered distributions. Reaction rates and yields depend on the initial distributions: at short times the short range structure is most important, whereas later on, larger scale structure is important.

Classical treatments of reaction kinetics (Burns et al) use diffusion–reaction equations to describe the concentration profiles in the system. Continuum equations are useful for macroscopic systems, but break down for small clusters because the particles should be treated as discrete, and rate constants measured in steady–state homogeneous conditions are not appropriate in transient nonhomogeneous systems such as tracks (Clifford et al 1986). It is not possible to solve multi–body diffusion encounter problems exactly, except in special cases, so it is hard to analyse the errors incurred by these approximations. However, it is possible to simulate multi–body systems. Here we describe how this may be done. We also describe a stochastic theory, which has been applied to nonhomogeneous systems with some success.

The Monte Carlo simulation method is based on the mathematics of diffusion processes (Karlin and Taylor). A diffusion process obeys a stochastic differential equation (*sde*), and can be simulated by time–discretisation: for a particle at **R** with a deterministic force **F**, and diffusion coefficient D the discretised sde is (Green, 1989)

$$\delta\mathbf{R} \approx (\mathbf{F}/kT)\, D\, \delta t + \sqrt{(D\, \delta t)}\, \mathbf{N}_3(0,I)$$

$\mathbf{N}_3(0,I)$ is a sample from a spherical normal distribution of unit variance. There are discretisation errors if **F** depends on **R**; δt must be sufficiently small that **F** does not change significantly during δt. Another problem arises when particles are close to encounter: if the simulation only reacts particles in reactive configurations at the timesteps it misses trajectories which pass through a reactive zone during δt, but are unreactive at the sampled times. This problem can be fixed by reducing δt when particles are close, but simulations become expensive, and still underestimate the encounter probability except in the limit. A better solution is to estimate the probability of entering a reactive zone during δt, which can be done using a bridging diffusion process (Green, 1989). A single pair with relative diffusion coefficient D' whose

NATO ASI Series, Vol. H 54
The Early Effects of Radiation on DNA
Edited by E. M. Fielden and P. O'Neill

separation goes from x to y in δt, encounters with probability exp $[-(x-a)(y-a)/D'\delta t]$. If the chance of meeting more than one partner in δt is negligible, this probability can be applied pairwise. The method was tested by Clifford et al (1986) and Green (1989).

The standard theory of diffusion–controlled reaction kinetics is based on an independent pairs (IP) approximation (Noyes 1961). We have developed a theory of kinetics in microscopic systems based on the same approximation (Clifford et al 1986, Green et al 1989a). There is an analytic form of the theory, which uses coupled differential equations to describe the distribution of the number of each species in a cluster, and is most useful for small systems with simple chemistry. There is also a simulation method (Independent Reaction Times) which uses the IP approximation to simulate kinetics without following the particle trajectories, and can be used for more complex reaction schemes and extended systems. The method has been tested for clusters of uncharged particles (Clifford et al 1986), and ions (Clifford et al 1987, Green et al 1989b); in each case the method has been in excellent agreement with MC results. The technique has also been applied to tracks (Zaider and Brenner, Green et al 1990), but has not yet been tested under such conditions. Some of our recent results (unpublished) imply that corrections will have to be made when clusters overlap.

Burns WG, Sims HE, Goodall JAB (1984) Radiation chemical diffusion kinetic calculations with prescribed and non–prescribed diffusion. *Radiat Phys Chem* 23:143–180

Clifford P, Green NJB, Oldfield MJ, Pilling MJ, Pimblott SM (1986) Stochastic models of multi–species kinetics in radiation–induced spurs. *J Chem Soc Faraday Trans I* 82:2673–2689

Clifford P, Green NJB, Pilling MJ, Pimblott SM (1987) Stochastic models of diffusion–controlled ionic reactions in radiation–induced spurs – 1 high permittivity solvents. *J Phys Chem* 91:4417–4422.

Green NJB (1989) On the simulation of diffusion processes close to boundaries. *Molec Phys* 65:1399–1408

Green NJB, Pilling MJ, Pimblott SM (1989a) Stochastic models of short–time kinetics in irradiated liquids. *Radiat Phys Chem* 34:105–114

Green NJB, Pilling MJ, Pimblott SM, Clifford P (1989b) Stochastic models of diffusion–controlled ionic reactions in radiation–induced spurs – 2 low permittivity solvents. *J Phys Chem* 93:8025–8031

Green NJB, Pilling MJ, Pimblott SM, Clifford P (1990) Stochastic modelling of fast kinetics in a radiation track. *J Phys Chem* 94:251–258

Karlin S, Taylor HM (1981) A second course in stochastic processes. Academic Press, New York London Toronto Sydney San Francisco

Noyes RM (1961) Effects of diffusion rates on chemical kinetics. *Prog React Kinet* 1:129–160.

Zaider M, Brenner DJ (1986) On the stochastic treatment of fast chemical reactions. *Radiat Res* 100:245–256

This work has been supported by the Office of Basic Energy Sciences, US Department of Energy. This is Document No NDRL 3293 from the Notre Dame Radiation Laboratory.

THE EFFECT OF ENVIRONMENT UPON DNA FREE RADICALS

E. Sagstuen[1], E.O. Hole[1], W.H. Nelson[2], and D.M. Close[3]
[1]Dept. Physics, Univ. Oslo, P.O. Box 1048 Blindern, 0316 Oslo 3
[2]Dept Phys. and Astron., Georgia State Univ., Atlanta, GA 30303
[3]Dept. Physics, East Tenn. State Univ., Johnson City, TN 37614

INTRODUCTION

Since 1985 a continued effort has been made to understand the primary damage caused by direct radiation effects to the constituents of the DNA molecule. The main motivation for initiating these studies was the report from a Swedish group (Gräslund et al. 1971). Using ESR spectroscopy it was proposed that in oriented fibres of DNA mainly two radical species are formed and trapped at low temperatures. These were the reduced thymine base, $\mathbf{T^{\bullet -}}$, and the oxidized guanine base, $\mathbf{G^{\bullet +}}$. More recently, Symons and co-workers have shown that these primary damage centers are potential precursors for strand break (Boon et al. 1984).

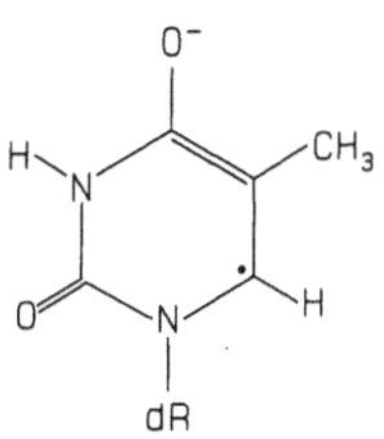

$\mathbf{T^{\bullet -}}$

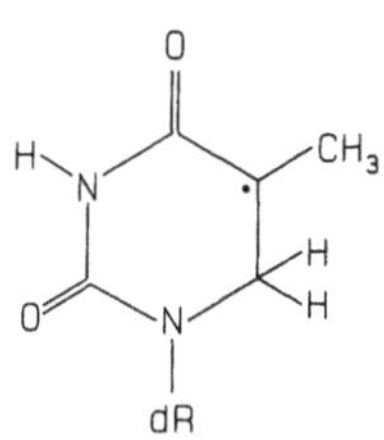

•TH

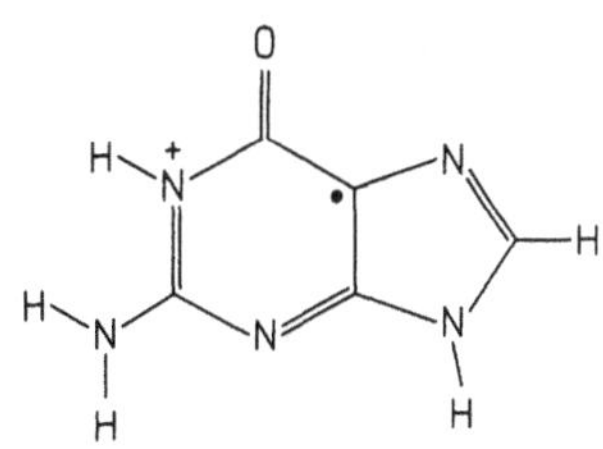

$\mathbf{G^{\bullet +}}$

Detailed studies of oriented DNA and frozen aqueous glasses of DNA provided results which apparently corroborated the $T^{\bullet -}$ and $G^{\bullet +}$ model of primary direct radiation damage to DNA (Hüttermann et al. 1984, Cullis and Symons 1986, Gregoli et al. 1982). It is noticable that in these studies only primary base damage seemed to be involved. In particular, indirect effects due to ionization of surrounding water molecules reportedly were of minor importance. This should indicate that the manifold of secondary products

NATO ASI Series, Vol. H 54
The Early Effects of Radiation on DNA
Edited by E. M. Fielden and P. O'Neill

observed, including strand breaks, mainly are due to these two ionized base constituents.

The ESR-characterization of the radiation-induced radical products of the purine constituents of DNA has developed more slowly than that of the pyrimidine constituents. Whereas the magnetic parameters of thymine base radicals were thoroughly studied at the time of the oriented DNA-studies (Bernhard 1981), the ionic purine base radicals had not been detected in the solid state. An extensive set of studies of ionic guanine products in a variety of surroundings was therefore undertaken using ESR and ENDOR spectroscopy of single crystals x-irradiated at low temperatures. These studies revealed amounts of new information of the primary radical formation in the solid state. This prompted an extension to other purines as well, of which the adenine derivatives constituted the largest group. Some of these results will be reviewed in the present paper.

Recently, evidence has accumulated indicating that the two-component model for the primary direct radiation damage to DNA must be modified (Close 1984, Hüttermann 1988, Zell et al. 1988). It seems necessary to take other sites for trapping of the primary oxidation and reduction products into account (Bernhard 1989). Other structures of the electron-loss and electron-gain products as well as alternative mechanisms for their secondary reactions must be taken into account. One key-piece of evidence for suggesting that the thymine anion was trapped at low temperatures in oriented DNA was the observation of the famous thymine octet resonance growing in upon annealing above 200 K. It is well known that this resonance is due to the thymine C6 H-addition radical (the 5-yl radical **•TH**) (Ehrenberg et al. 1963, Pershan et al. 1964). Radiation-chemical experiments have shown that the 5-yl radical is formed in aqueous solution by protonation at C6 of a parent thymine anion (Ormerod 1965). However, it has **not** been shown that this process also occurs in crystalline matrices. Based on the knowledge gained from the radiation response of the purines, it seemed worthwhile to re-investigate the structure and reactions of the ionic pyrimidine base radicals. In this report some results from recent studies of several thymine derivatives will also be reviewed.

EXPERIMENTAL

The data have mainly been recorded in the ESR laboratory in Atlanta. The experimental procedures including a description of the spectrometer system, x-irradiation and measurements at temperatures close to that of liquid helium (lowest available temperature is about 8 K) have previously been described (Nelson _et_ _al_. 1988). Recently the spectrometer was improved by including a 200 W ENI rf-amplifier for the ENDOR experiment and a Hughes 24 dB low-noise μwave preamplifier. This permits the operation at very low power levels (in the nanowatt range) preventing excess power saturation of the ESR at low temperatures. A new computerized data aquisition system has been implemented.

Details from studies of several purine and pyrimidine derivatives have previously been published. The guanine base derivatives include: guanine hydrochloride monohydrate (**GmHCl**) (Close _et_ _al_. 1985, 1987, 1988), guanine hydrochloride dihydrate (**GdHCl**) (Strand _et_ _al_. 1987, Nelson _et_ _al_. 1988), guanine hydrobromide monohydrate (**GmHBr**) (Hole _et_ _al_. 1989, 1990), guanosine 5'-monophosphate (**5'-GMP**) (Sagstuen _et_ _al_. 1988), 2'-deoxyguanosine 5'-monophosphate (**5'-dGMP**) (Hole and Sagstuen 1987, Hole _et_ _al_. 1987,1989), guanosine:dimethylformamide (**GR:DMF**) (Sagstuen _et_ _al_. 1989) as well as unpublished results for cyclic guanosine monophosphate (**cGMP**). The adenine base derivatives include adenosine (**AR**) (Close and Nelson 1989), deoxyadenosine monohydrate(**AdRm**) and adenosine hydrochloride (**ARHCl**) (Close _et_ _al_. 1987b, Nelson _et_ _al_. 1989) together with unpublished results for adenine hydrochloride hemihydrate (**AhHCl**), anhydrous deoxyadenosine (**AdRa**) and 9-methyladenine (**9MA**). The related purine derivative xanthosine (**XR**) (Nelson and Close 1983) has also been studied. The thymine base derivatives include thymidine (**TdR**) (Sagstuen _et_ _al_. 1989b), 1-methylthymine (**1MT**) (Close _et_ _al_. 1989, Hole _et_ _al_. 1990b) and unpublished results for thymine monohydrate (**Tm**) and the co-crystal of 1-methylthymine and 9-methyladenine (**1MT:9MA**).

In these systems the DNA constituent is a part of a hydrogen-bonded network including between zero and five water molecules and in some cases also small polar molecules. For 5'-dGMP sodium acted as counter ions for the negatively charged phosphate group.

RESULTS AND DISCUSSION

In this section an attempt will be made to outline some of the most important results pertaining the structure, charge state and reactions of the primary DNA base radicals stabilized at low temperatures.

In the guanine model systems studied, the base occurs in two different charge states depending upon the protonation state at N7. The base is protonated in **GmHCl**, **GmHBr**, **GdHCl**, and **GMP**, whereas it is neutral in **5'-dGMP**, **GR:DMF**, and **cGMP**. In the following, these two groups will be discussed separately.

<u>N7-protonated guanine</u>

In Table 1 the main primary radicals trapped in the four crystals at 10 K are summarized. The structures referred to with roman numerals are shown below.

Table 1. Primary Radicals in N7-protonated guanine bases

Crystal	I	II	III	OH-add	Other
GmHCl	*	*	*[1]		IV
GmHBr	*	*	*[1]	V	IV
GdHCl		*	*	VI	
GMP		*	*		sugar

[1] Minor amounts

In all four systems, charge neutralization following primary ionization takes place immediately. Thus, the reduction product **II** invariably was protonated at O6. MO calculations show that in the anion, O6 is the site for largest negative charge, followed by N7. O6 usually participate in hydrogen bonding. Thus, proton transfer across the hydrogen bond, forced by coulombic attraction between $H^+ \cdot\cdot O6^-$ is a straightforward mechanism for charge neutralization. The protonated anion **II** is fairly stable. In GMP and GmHBr it decayed about 250 K without any successor radical, whereas in GmHCl and GdHCl it decayed near room temperature with the concomitant formation of an N9 H-abstraction radical.

Similarly, the primary electron-loss center rapidly re-establishes charge balance with the environment. The dominant

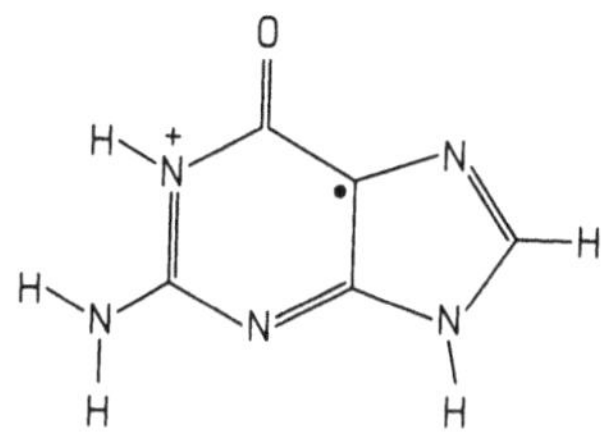

I

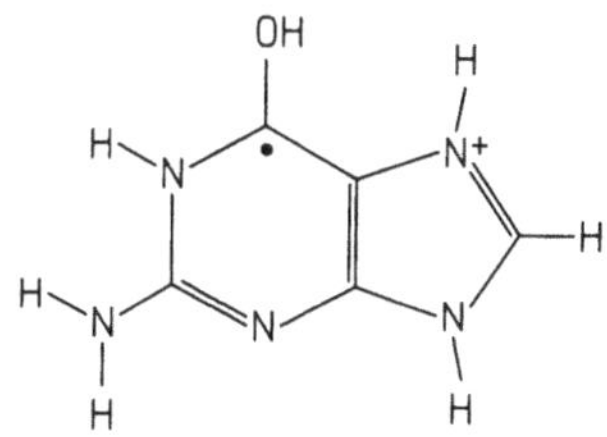

II

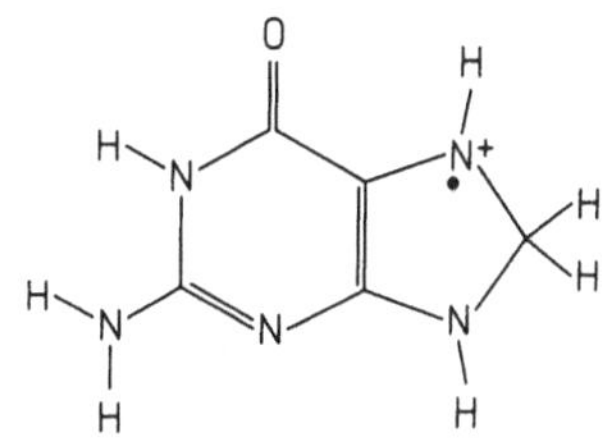

III

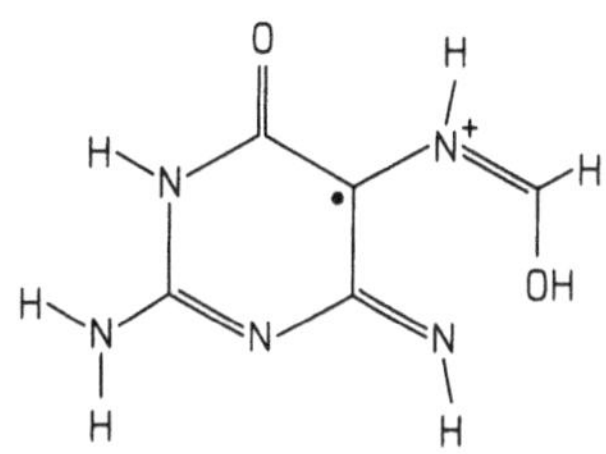

IV

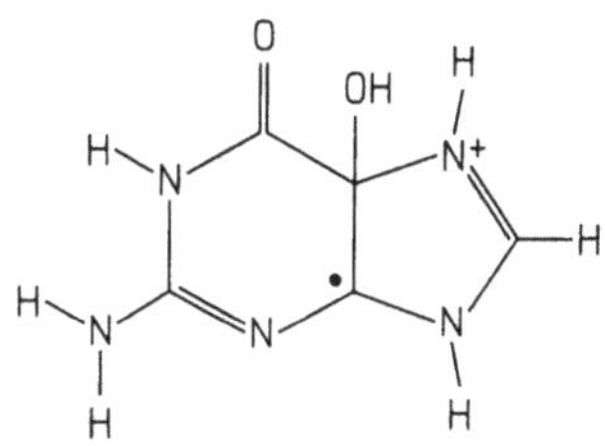

V

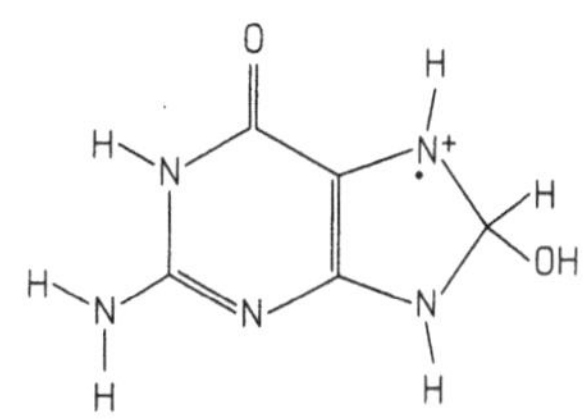

VI

mode is deprotonation of the most acidic proton (HN7) yielding Radical **I**, but OH^- addition may also occur. Thus, in GmHBr Radical **V** is formed by OH^- addition concurrent to deprotonation at N7. In GmHCl and GmHBr Radical **I** is stable until about 60 K where it decays without detectible successor radicals. In GdHCl and GMP, however, it is suggested that Radical **I** is transient. The difference is due to the nature of the acceptor group for HN7. In GmHCl and GmHBr this is a second guanine base which means that the excess positive charge easily is delocalized over the lattice. In GdHCl and GMP the acceptor is a water molecule. H_3O^+ is an effective trap for secondary electrons and the dissociation reaction

$$H_3O^+ + e^- \longrightarrow \cdot H + H_2O$$

produces H-atoms which either may return to Radical **I** restoring the virgin molecule, or may add to C8 of a neighboring base.

In all four systems radicals are formed which may be due to •OH addition to the base. •OH may be formed by primary ionization of crystalline water. The site of its attack on the base, and the stability of the resulting product, is critically dependent upon the localization of the water molecule. In GdHCl the C8 OH-adduct, **VI**, is observed. In GmHCl, GmHBr and GMP it is suggested that •OH attacks at C8 immediately followed by fragmentation of the guanine ring, yielding Radical **IV**. In GmHCl Radical **IV** transforms into Radical **III** at 150 K. No successors from Radical **IV** were observed the case of GmHBr and GMP.

The diversity of end products in these four model systems clearly demonstrates how critical the microscopical environment is with respect to reaction mechanisms for the primary electron-loss and electron gain centers.

<u>The neutral guanine base</u>

In this group of model systems the guanine base is not protonated at N7 and thus exhibits the same structure as that of guanine in DNA. The data analysis is not completed for these systems, but from the available results some conclusions may be made. In Table 2 the major primary radicals trapped in the three crystals at 10 K are summarized. The structures referred to by roman numerals are shown below.

Table 2. Primary radicals in neutral guanine bases

Crystal	VII	VIII	IX	X	XI	Other
5'-dGMP	*	*			*[2]	XII, [1]
GR:DMF	*		*		*	[1]
cGMP	*			*	*[2]	[1]

[1] An unidentified base radical
[2] Minor amounts

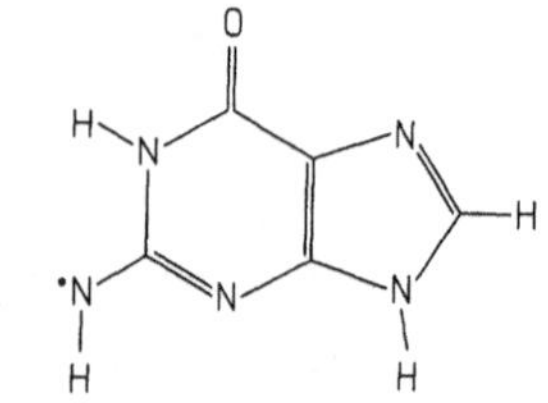

VII

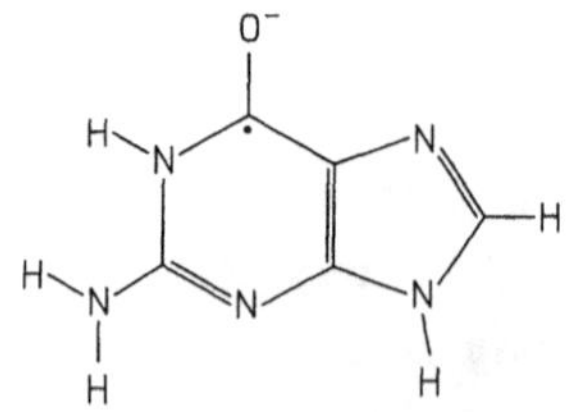

VIII

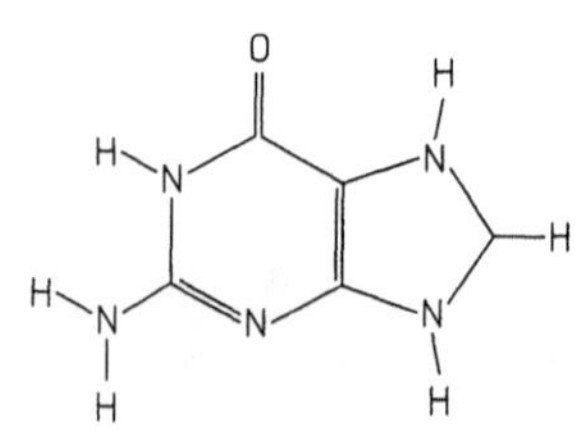

IX

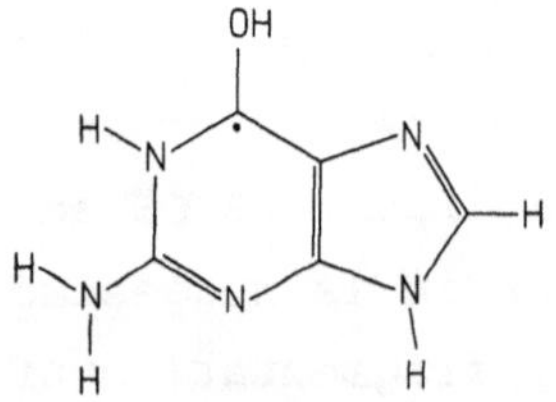

X

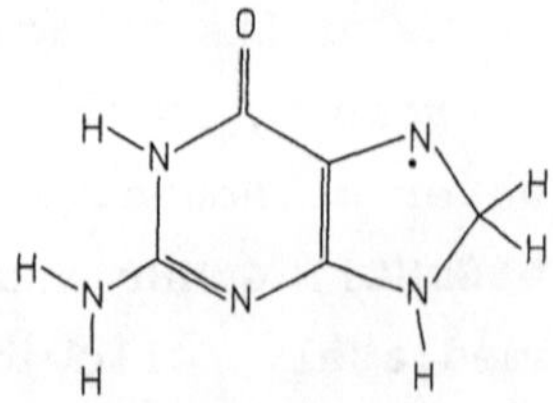

XI

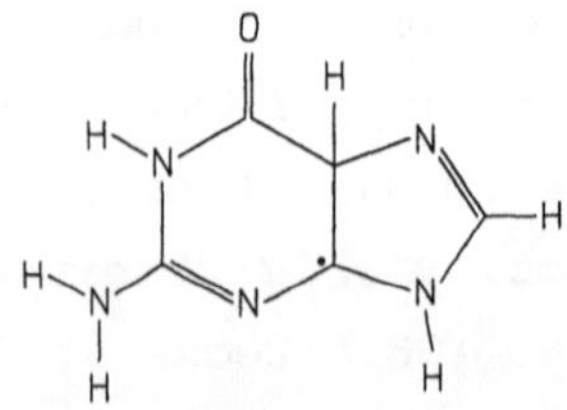

XII

The major result from these studies is that charge neutralization follows immediately after the initial ionization event by deprotonation (oxidation products) or protonation (reduction products). For the oxidation products, the picture is uniform. In all systems deprotonation takes place at the amino group (bonded to C2) resulting in Radical **VII**. A large coupling from the remaining hydrogen of the NH_2-group serves as fingerprint for this species (the N10 spin density is about 0.33). Furthermore, 16-18% of the spin reside at C8 and the spin density at N3 is estimated to 0.31. In Fig. 1 is shown an experimental spectrum recorded from a cGMP crystal after irradiation at 10 K. This spectrum is compared with a simulated spectrum obtained using the above spin densities calculated from ESR and ENDOR-derived hyperfine coupling tensors. The fit is excellent, and noting that an INDO calculation for Radical **VII** closely reproduces this spin density distribution (Hole et al. 1987) the assignment is unambigous. No successor could be detected upon the decay of this radical in the temperature range 100 - 200 K.

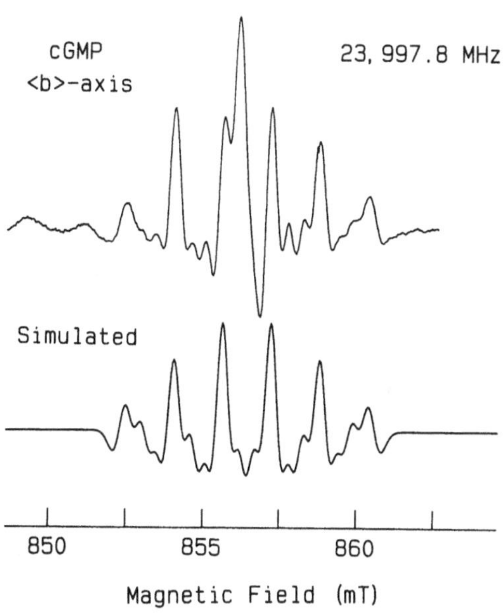

Fig. 1. The upper trace is the experimental spectrum from a single crystal of cyclic 3',5'-guanosine monophosphate after irradiation at 10 K. The magnetic field is along <b>, almost perpendicular to the guanine base. The lower spectrum is calculated using spectral parameters as described in the text.

For the reduction products, the picture is not as simple as for the oxidation products. Generally, the primary electron-gain center is protonated at 10 K. The site of protonation is determined by the availability of protons in the hydrogen bonding network. A preliminary analysis indicates that in GR:DMF the protonation takes place at N7 (Radical **IX**), whereas in cGMP it takes place at O6 (Radical **X**). In both cases the added proton was detected by ENDOR. The O6-protonated anion was also observed in the related compound xanthosine (XR). In 5'-dGMP the charged anion (Radical **VIII**) was observed at 10 K. It is not clear why this species is not protonated, but the sodium counter ion is localized in the O6-N7 region. This may prevent the protonation reaction by coulombic interaction. Radical **VIII** is unstable and decay upon exposure to white light at 10 K or by thermal annealing at 30 K without any detectible successor.

In 5'-dGMP the C8 H-adduct, Radical **XI**, grew in strongly at about 200 K upon the decay of a sugar-centered radical. A similar process was observed in XR. In GR:DMF and cGMP the protonated anion decayed by warming into new (as yet unidentified) species. In particular, none of those seem to be associated with the C8 H-addition radical **XI**.

The results presented for 5'-dGMP differ from those previously published by other authors (Rakvin et al. 1987). Specifically, evidence for N1 H-abstraction radicals were not found in the present work. Kim et al. (1989) have studied crystals of cGMP after irradiation at 4.2 K. These authors suggest that ionic species are stabilized at this temperature.

Also for the adenine derivatives included in this study, the base occurs in two different charge states depending upon whether it is protonated at N1 or not. The base is protonated in **AhHCl** and **ARHCl** whereas it is neutral in **AR**, **AdRa**, **AdRm** and **9MA**. The data analysis for this group of DNA constituents is as yet not finished, but some general considerations may be made.

N1-protonated adenine

In both ARHCl and AhHCl the primary electron-loss center rapidly deprotonates. In ARHCl deprotonation takes place at the amino group, forming Radical **XIII**, whereas in AhHCl the leaving proton tentatively is suggested to be the one bonded to N1. In ARHCl the deprotonated cation decays above 100 K with no successor radical. The primary electron-gain radical protonates at N3 in AhHCl (Radical **XIV**) and decays without any successor at about 110 K. In ARHCl the anion or its protonated successor may well be present, but has not yet been identified.

In ARHCl sugar-centered radicals dominate the spectra at 10 K, but both C2- and C8 H-addition radicals (**XV** and **XVI**) are clearly present. Warming above 250 K leads to a decrease in the yield of sugar-centered radicals and a concomitant increase in the yield of H-addition radicals. In AhHCl H-adducts are weakly present at 10 K, and grow further in upon warming above 200 K.

<u>The neutral adenine base</u>

In the group consisting of AR, AdRm, AdRa and 9MA the base is not protonated at N1. In Table 3 the main primary radicals trapped at 10 K are summarized. Their structures are shown below.

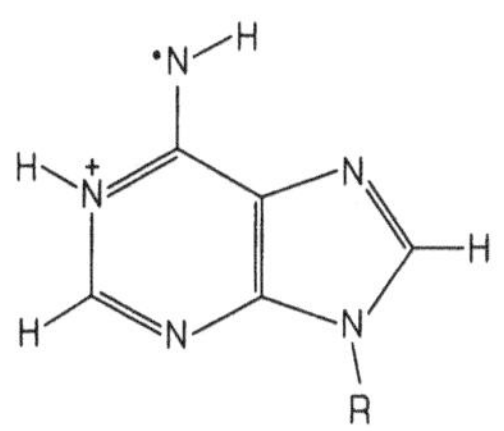

XIII

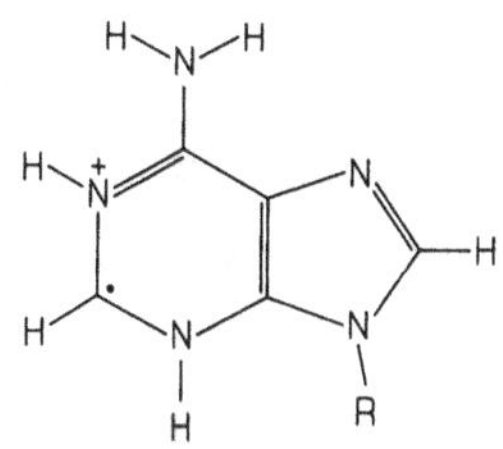

XIV

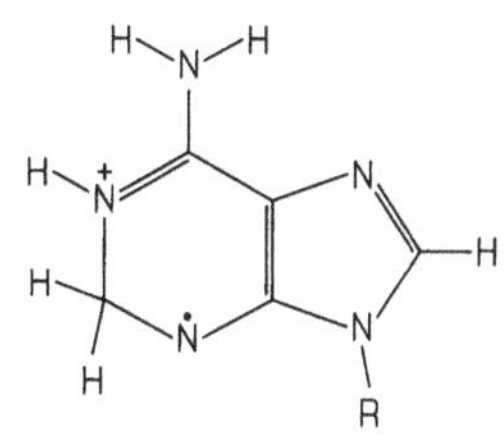

XV

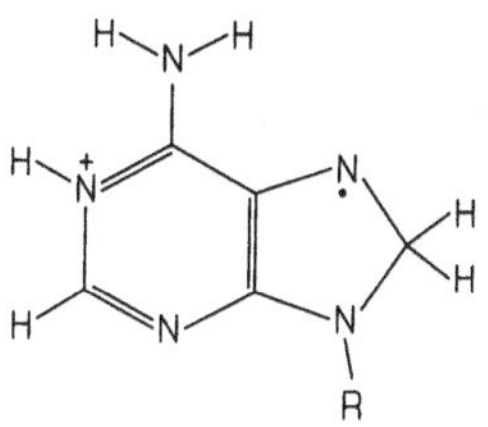

XVI

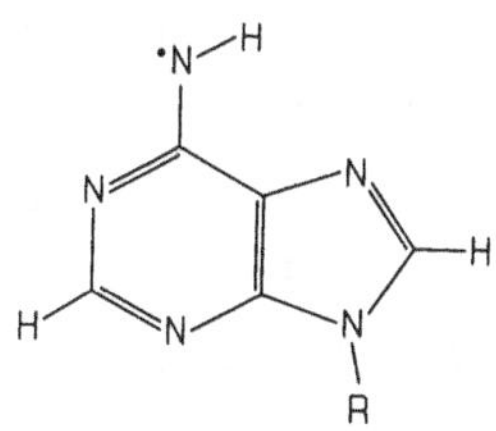

XVII

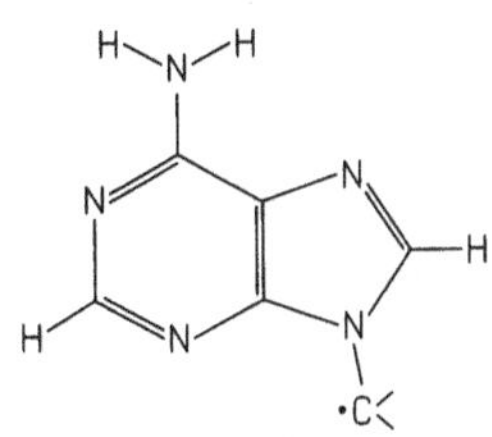

XVIII

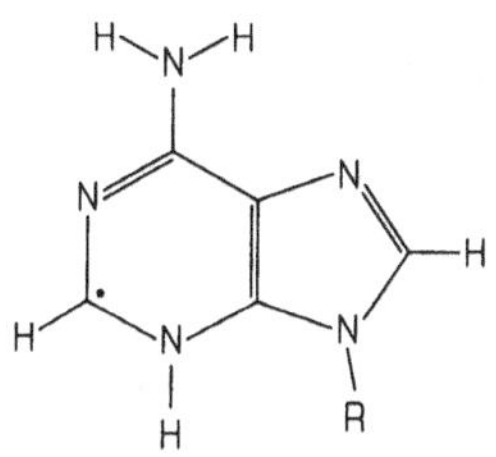

XIX

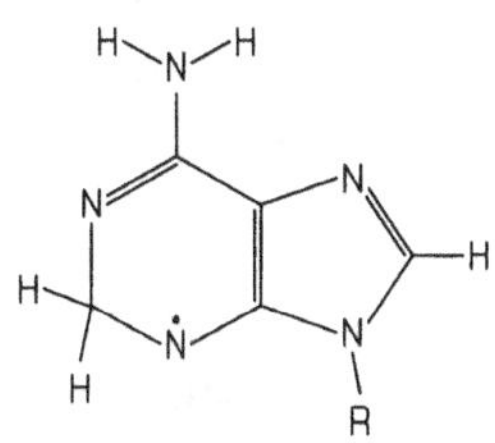

XX

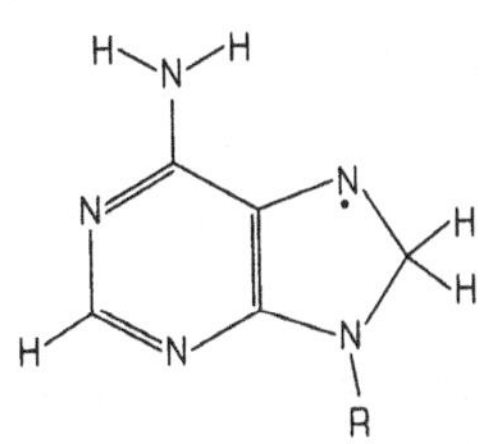

XXI

Table 3. Primary radicals in neutral adenine bases

Crystal	XVII	XVIII	XIX	XX	XXI
AR	*		*		
AdRa	*		*		
AdRm		*	*	*	*
9MA		*	*	*	*

[1] Other base radicals not yet identified

In the case of oxidation, there seems to be a scattering of results. In AR and AdRa the electron-loss center deprotonates at the amino group (Radical **XVII**). This product is not observed in AdRm and 9MA. However, a species formed by net hydrogen loss from C1' (the methyl group in the case of 9MA), Radical **XVIII**, is observed. This product may represent the deprotonated cation. The site of deprotonation may possibly depend upon the degree of delocalization of the cationic electronic wave function to the group bonded to N9. In 9MA delocalization by hyperconjugation is clearly feasible. Furthermore, in AdRm the torsion angle C8-N9-C1'-O5' is close to 0° which also enhances delocalization. This torsion angle deviates significantly from 0° in the other adenosine derivatives. Although still a hypothesis the implications of such a model for DNA are interesting.

In AR, AdRm and 9MA the primary electron-gain center have become protonated at N3 (Radical **XIX**). This seems to be the case also in AdRa. In all of these crystals the protonated anion decays in the range 40-100 K without any successor radical.

In AR both types of hydrogen addition radicals **XX** and **XXI** grow in at about 200 K without any connection to the decay of the primary radicals above. In AdR sugar-centered radicals dominate the resonance, but both types of H-addition radicals **XX** and **XXI** are present at low temperature. Upon warming to temperatures above 150 K the H-addition radicals grow in further concomitantly with the decay of sugar radicals. In 9MA both types of hydrogen addition radicals are present at 10 K, and seem to grow in further at about 200 K.

The thymine derivatives constitute so far the last member in this set of studies on primary direct radiation damage in DNA components. Only two model systems have been thoroughly investigated, **TdR** and **1MT**, but some very preliminary data from **Tm** and the co-crystal **1MT:9MA** is also available.

In Table 4 are included the main base-centered radicals trapped in thymine derivatives at 10 K. The structures in this Table, referred to by roman numerals, are shown below.

Table 4. Primary radicals in thymine bases

Crystal	XXII	XXIII	XXIV	XXV
TdR	*	*	*	*[1]
1MT	*	*		*
Tm	*	*		
1MT:9MA	*	*	*[1]	

[1] Minor amounts

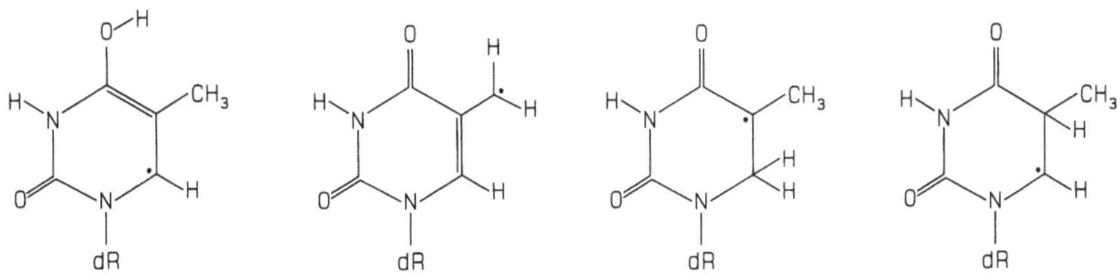

XXII **XXIII** **XXIV** **XXV**

In TdR the O4-protonated anion, Radical **XXII**, is unambigously identified at 10 K, and a comparison with a previous report (Box and Budzinski 1975) suggests that protonation has taken place already at 4.2 K. The main reason for suggesting this structure was simply the experimental observation of the proton bonded to O4. This coupling was not detected by Box and Budzinski due to their use of partially deuterated crystals only.

In TdR a C5-CH_3 coupling was observed, characterized by a small negative spin density at C5. In addition, the dipolar part of the coupling tensor reflected large positive spin density on both C4 and C6 (about 45 %). The sensitivity of this coupling tensor to the spin density distribution suggested that it can serve as indicator for the protonation state of the thymine anion. In 1MT the geometry of the bases prohibited the direct observation of an eventual proton at O4 for the thymine anion.

However, a C5-CH_3 interaction identical to that in TdR strongly suggests that the anion is protonated in 1MT, thus exhibiting the structure **XXII**. In Tm previously published data (Flossmann et al. 1980) together with our preliminary data suggest a similar structure of the anion. Also in the co-crystal 1MT:9MA this structure of the thymine anion seems most probable. If the spin density at C6 should be used as an indicator upon the structure of the anion radical, Bernhard (1984) has argued that the dipolar part of the hyperfine coupling tensor is the most reliable reporter of the spin density at α-carbons. Using his procedure, a number varying between 0.51 and 0.53 is arrived at for all four tensors. Thus, this procedure indicates all four anions to have the same structure.

In TdR and 1MT the protonated anion decayed at 40 K with no apparent successor radical. In particular no connection to the formation of C6 H-adduct radicals , **XXIV**, was observed. In 1MT the C6 H-adduct could not be observed until prolonged annealing above 250 K (in deuterated crystals mainly H**H**-adducts). In TdR, Radical **XXIV** was present even at 10 K, in deuterated crystals mainly as H**D**-adducts.

The C7 H-abstraction radical **XXIII** was present as a dominant radical in all four systems at 10 K and remained stable upon annealing above 300 K.

The C5 H-addition radical **XXV** was present as a minor constituent of the TdR resonance at 10 K and a major constituent in 1MT. Flossman et al. (1980) reported that this radical converts into the 5-yl radical **XXIV** upon annealing at high temperatures. This agree with our observation that the 5-yl radical only grows in upon prolonged annealing above 250 K in 1MT.

In 1MT the 7-yl radical **XXIII** is the most probable candidate for the oxidation product. The primary electron-loss center in thymine exhibits a large spin denty at C5 (Rhodes et al. 1988) making deprotonation at the β-methyl group feasible. Subsequent addition of the ejected proton to the C5-position of a neighboring molecule followed by electron capture provide the most probable mechanism for the formation of Radical **XXV**. This also agrees with the non-exchangeable nature of the added proton determined experimentally for Radicals **XIV** and **XXV** in 1MT.

In TdR other sites for deprotonation of a pristine electron-loss center is possible. Alkoxy radicals are observed as well as C1' H-abstraction radicals. The 5-yl radicals trapped at 10 K are almost exclusively HD-adducts thus precluding the 7-yl radicals from being an immediate precursor for **XXIV**.

In Tm the 5-yl radical **XXIV** seems to grow in upon the decay of the anion resonance above 120 K. Further study of this transformation is in progress.

The general notion of the thymine anions being protonated at O4 in the solid state agrees with results from solution studies. Deeble <u>et</u> <u>al</u>. (1985) and Novais and Steenken (1986) have shown that for the pristine π^* anion protonation at O4 is a fast and reversible process, whereas protonation at C6 is slow and irreversible. In solution ultimately all the protonated anions are C6-protonated. In the solid state (and in DNA) a ready supply of protons at O4 is generally present and the O4 protonated product is expected to be stabilized in the rigid frame of the hydrogen-bonded network. At C6, however, no similar source of protons is available (in contrast to the situation in aqueous solution). Even if protons were available near C6, the pre-existing O4 protonation would reduce the coulombic force driving an eventual C6-protonation. In agreement with this, Herak <u>et</u> <u>al</u>. (1982) have shown that in TdR crystals the 5-yl radical is <u>not</u> protonated at O4.

CONCLUSIONS

The most important conclusions from these studies may be summarized as follows:

1) The DNA base anion radicals protonate rapidly following ionization. The site of protonation is generally the exocyclic oxygen atom O6 and O4 in guanine and thymine, respectively, and N3 in adenine-type derivatives. Variations to this scheme occasionally may occur depending on the environment of a given molecule.
2) The DNA base cation radicals rapidly deprotonate following ionization. The site of deprotonation is generally N7 in N7-protonated guanine bases, and the NH_2 group in neutral gua-

nine and in adenine derivatives although some variations to this have been observed. In the thymine base, the C5-methyl group is the most probable candidate for deprotonation, while in thymine nucleosides the reaction of the base oxidation product is not established.

3) The purine and pyrimidine hydrogen addition radicals are generally formed, either at low temperatures, or as a result of secondary reactions. With one possible exception (Tm) no connection between the formation of H-adducts and the decay of (protonated) anions could be established.

4) The secondary reactions of the charge-neutralized primary radicals are critically dependent upon the microscopic environment of the base.

5) In hydrated crystals, addition of OH^- to the charged cation as well as •OH radical formation from the hydration water with subsequent addition to the DNA base was observed.

ACKNOWLEDGEMENTS

This research has been supported by NIH Grant CA 36810 and by NATO Travel Grant RG 0426/88. Discussions with Drs. W.A. Bernhard and S. Steenken are acknowledged.

REFERENCES

Bernhard WA (1981) Solid-state radiation chemistry of DNA: The bases. Adv Radiat Biol 9:199-280

Bernhard WA (1984) The use of alpha hyperfine coupling tensors as a measure of unpaired spin density and free radical geometry. J Chem Phys 81:5928-5936

Bernhard WA (1989) Sites of electron trapping in DNA as determined by ESR of one-electron reduced oligonucleotides. J Phys Chem 93:2187-2189

Boon PJ, Cullis PM, Symons MCR, Wren BW (1984) Effects of ionizing radiation on deoxyribonucleic acid and related systems. Part I. The rôle of oxygen. J Chem Soc Perk Trans II:1393-1399

Box HC, Budzinski EE (1975) Primary radiation damage in thymidine. J Chem Phys 62:197-199

Close DM (1984) General discussion. Faraday Discuss Chem Soc 78:227-228

Close DM, Sagstuen E, Nelson WH (1985) ESR study of the guanine cation. J Chem Phys 82:4386-4388

Close DM, Nelson WH, Sagstuen E (1987) Radical formation in x-irradiated single crystals of guanine hydrochloride monohydrate. II. ESR and ENDOR in the range 10-77 K. Radiat Res 112:283-301

Close DM, Nelson WH, Sagstuen E (1987b) EPR and ENDOR study of x-irradiated single crystals of purines at 4.2 K. In: Weil JA (ed) Electron magnetic resonance of the solid state. The Canadian Soc Chem, Ottawa, p 237

Close DM, Sagstuen E, Nelson WH (1988) Radical formation in x-irradiated single crystals of guanine hydrochloride monohydrate. III. Secondary radicals and reaction mechaninsms. Radiat Res 116:379-392

Close DM, Nelson WH (1989) ESR and ENDOR study of adenosine single crystals x-irradiated at 10 K. Radiat Res 117:367-378

Close DM, Nelson WH, Sagstuen E, Hole EO (1989) ESR and ENDOR studies of x-irradiated single crystals of guanine derivatives. Free Rad Res Comms 6:83-85

Cullis PM, Symons MCR (1986) Electron Spin Resonance Studies of the Mechanisms of Radiation Damage to DNA. In:Simic MG, Grossmann L, Upton AC (eds) Mechanism of DNA damage and repair. Plenum, New York London, p 29

Deeble DJ, Das S, von Sonntag C (1985) Uracil derivatives: Sites and kinetics of protonation of the radical anions and the uv spectra of the C(5) and C(6) H-atom adducts. J Phys Chem 89:5784-5788

Ehrenberg A, Ehrenberg L, Löfroth G (1963) Thymidine-like electron spin resonance spectra in gamma-irradiated deoxyribonucleic acid. Nature 200:376-377

Flossmann W, Zehner H, Müller A (1980) Radicals produced in single crystals of thymine derivatives by uv and x-rays. Z Naturforsch 35c:20-29

Gräslund A, Ehrenberg A, Rupprect A, Ström G (1971) Ionic base radicals in gamma-irradiated DNA. Biochem Biophys Acta 254:172-186

Gregoli S, Olast M, Bertinchamps A (1982) Radiolytic pathways in gamma-irradiated DNA: Influence of chemical and conformational factors. Radiat Res 89:238-254

Herak JN, Velenik A, McDowell CA (1982) An ENDOR study of long range spin interactions in molecular crystals. II. The 5-thymyl radical in single crystals of thymidine. J Magn Reson 49:64-72

Hole EO, Sagstuen E (1987) Free radical formation in crystals of 2'-deoxyguanosine 5'-monophosphate irradiated at 15 K: An ESR study. Radiat Res 109:190-205

Hole EO, Nelson WH, Close DM, Sagstuen E (1987) ESR and ENDOR study of the guanine cation: Secondary product in 5'-dGMP. J Chem Phys 86:5218-5220

Hole EO, Sagstuen E, Nelson WH, Close DM (1989) Free radical formation in single crystals of 2'deoxyguanosine monophosphate and guanine hydrobromide monohydrate after x-irradiation at 10 and 65 K: An ESR, ENDOR, and FSE study. Free Rad Res Comms 6:87-90

Hole EO, Sagstuen E, Nelson WH, Close DM (1990) Radiation damage to guanine hydrobromide monohydrate. Solid state ESR and ENDOR between 8 and 295 K. Radiat Res (to be published)

Hole EO, Sagstuen E, Nelson WH, Close DM (1990b) Primary reduction and oxidation of thymine derivatives: ESR/ENDOR of thymidine and 1-methylthymine x-irradiated at 10 K. J Phys Chem (to be published)

Hüttermann J, Voit K, Oloff H, Köhnlein W, Gräslund A, Rupprecht A (1984) Specific formation of electron gain and loss centres in x-irradiated oriented fibres of DNA at low temperatures. Faraday Discuss Chem Soc 78:135-149

Hüttermann J (1989) Free radicals from solid oriented DNA-fibers: Facts and fancy. Free Rad Res Comms 6:103-104

Kim H, Budzinski EE, Box HC (1989) The radiation induced oxidation and reduction of guanine: ESR/ENDOR studies of irradiated guanosine cyclic monophosphate. J Chem Phys 90:1448-1451

Nelson WH, Close DM (1983) Radiation chemistry of purines in the solid state: ESR and ENDOR studies of x-irradiated xanthosine dihydrate single crystals. J Chem Phys 79:3240-3250

Nelson WH, Hole EO, Close DM, Sagstuen E (1988) ESR/ENDOR study of guanine•HCl•H_2O x-irradiated at 20 K. Int J Radiat Biol 54:963-986

Nelson WH, Close DM, Sagstuen E, Hole EO (1989) Radiation chemistry of adenine derivatives following direct ionization in solids: ESR and ENDOR investigations. Free Rad Res Comms 6:81-82

Novais HM, Steenken S (1986) ESR studies of electron and hydrogen adducts of thymine and uracil and their derivatives and of 4,6-dihydroxypyrimidines in aqueous solution Comparison with data from solid state. J Amer Chem Soc 108:1-6

Ormerod MG (1965) Free-radical formation in irradiated deoxyribonucleic acid. Int J Radiat Biol 9:291-300

Pershan PS, Schulman BJ, Wyluda BJ, Eisinger J (1964) Electron spin resonance of irradiated DNA. Physics 1:163-182

Rakvin B, Herak JN, Voit K, Hüttermann J (1987) Free radicals from single crystals of deoxyguanosine 5'-monophosphate (Na salt) irradiated at low temperatures. Radiat Environ Biophys 26:1-12

Rhodes CJ, Podmore ID, Symons MCR (1988) Radical cations of N,N-dimethyluracil and N,N-dimethylthymine. J Chem Res (S) 120-121

Sagstuen E, Hole EO, Nelson WH, Close DM (1988) ESR/ENDOR study of guanosine 5'-monophosphate (free acid) single crystals x-irradiated at 10 K. Radiat Res 116:196-209

Sagstuen E, Hole EO, Nelson WH, Close DM (1989) Free radical formation in nucleosides and nucleotides of guanine: ESR and ENDOR of guanosine 5'-monophosphate and guanosine:dimethylformamide x-irradiated at 10 K. Free Rad Res Comms 6:91-92

Sagstuen E, Hole EO, Nelson WH, Close DM (1989b) Structure of the primary reduction product of thymidine after x-irradiation at 10 K. J Phys Chem 93:5974-5977

Strand P, Sagstuen E, Lehner T, Hüttermann J (1987) Free radical formation in crystals of guanine hydrochloride dihydrate: an ESR and ENDOR study. Int J Radiat Biol 51:303-318

Zell I, Hüttermann J, Gräslund A, Rupprecht A, Köhnlein W (1989) Free radicals in irradiated oriented DNA fibers: Results from B-form DNA and from deuterated DNA samples. Free Rad Res Comms 6:105-106

CHEMICAL REACTIVITY OF DNA RADICALS - A REFLECTION OF THEIR REDOX PROPERTIES

E.M. Fielden, P. O'Neill and S. Steenken*
MRC Radiobiology Unit
Chilton
Didcot
Oxon OX11 ORD
U.K.

The lethal and mutagenic effects of ionising radiation in biological systems are thought to involve chemical changes induced within essential biomolecules following the initial energy deposition events (Painter, 1979). There is much evidence that, within cells, DNA is a critical target which is sensitive to ionising radiation, damage to which results in various biological effects. These effects have been shown to be critically dependent upon the cellular environment whereby various agents are known to modify the biological consequences. For instance, oxygen and radiosensitisers, on the one hand, enhance cell killing by radiation when compared to the effect on cells under hypoxia, whereas thiols (reductants) protect cells, to varying degrees, from the deleterious effects of ionising radiation. Since these agents have to be present during irradiation to exert their influence, it is thought that fast free radical processes are involved (Adams et al., 1975) as distinct from slower, biochemical processes. Considerable evidence has accumulated over the years that radical scavengers, such as DMSO (Chapman et al., 1975; Roots et al., 1982; Millar et al., 1987), afford considerable protection to biological systems against ionising radiation. Indeed, it is thought that this "scavengable" component of the radiation effect is due to water radicals and, in particular, hydroxyl radicals which are produced in the water associated with or near to the DNA (Ward, these proceedings). Since energy is deposited in discrete packages of varying amounts, the radiation damage induced within DNA will result from direct energy deposition

* Max Planck Institut für Strahlenchemie, Stiftstr. 34-36, D4330 Mülheim/Rühr, F.R. Germany

NATO ASI Series, Vol. H 54
The Early Effects of Radiation on DNA
Edited by E. M. Fielden and P. O'Neill

within the DNA together with that deposited in associated water to give potentially scavengable OH radicals.

Various techniques such as esr (Bernhard, 1989; Symons,1987; Hüttermann et al., 1984; Boon et al., 1984; Close et al., 1985), diffuse reflectance (O'Neill et al., 1989) and radiation-induced luminescence (Al-Kazwini et al., 1988, 1990) have been utilised to explore the chemical species and events which occur following the initial energy deposition. To characterise the various chemical events, added solutes, known to be dose modifiers, are frequently employed. Indeed, a vast amount of information has been accumulated on the interactions of water radicals with DNA and model compounds (O'Neill & Fielden, 1986; von Sonntag, 1987; Steenken, 1989). Since the hydroxyl radical is implicated in cellular damage at the DNA level, it is important to understand in detail the properties of the various DNA radicals produced on its interaction with DNA. Over the past few years not only the distribution of sites of attack of OH on the base moieties has been characterised, but it has also been recognised that the resulting DNA radicals have different redox properties with respect to their subsequent interactions with oxidants and reductants (Fujita & Steenken, 1981; O'Neill & Fielden, 1986; Steenken,1989). In this paper, we will discuss the properties of the hydroxyl radical adducts of the DNA bases and highlight:

(1) their interactions with dose modifiers, and
(2) the similarities of such radicals with those induced upon one-electron oxidation of the base, e.g. electron loss centres resulting from ionisation as a consequence of direct energy deposition within the DNA.

Sites of Addition of OH radicals to DNA bases

Several studies over recent years have quantified the sites of addition of hydroxyl radicals to the purine and pyrimidine bases (Fujita & Steenken, 1987; Hazra & Steenken, 1983; O'Neill, 1983; O'Neill et al., 1985; Vieira & Steenken, 1990). The interaction of hydroxyl radicals with the sugar phosphate backbone has been studied in detail as discussed by

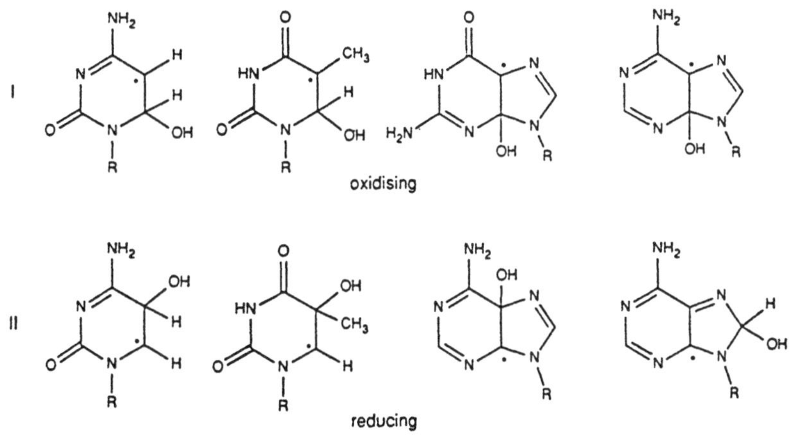

Figure 1: Hydroxyl radical adducts - redox properties

Sonntag (1987). Several of the base radicals produced by OH addition are shown in Figure 1. With the pyrimidines, the major site of addition is at C(5) to produce the 6-yl radical, which is reducing due to the adjacent nitrogen (at N(1)). A minority of OH radicals add to C(6) yielding the 5-yl radical which is weakly oxidising due to the adjacent carbonyl oxygen at C(4) or the delocalisation of spin to N(3) or to O(2). With the purines, however, a greater percentage of radicals which may act as oxidants are produced. These radicals have been identified through a redox titration method developed by Fujita & Steenken (1981) using either a strong oxidant, e.g. tetranitromethane, or a strong reductant, e.g. ascorbate, tetramethylphenylenediamine (see Steenken, 1989) to differentiate between the oxidising and reducing radicals. From this titration method, the redox properties of the OH-adducts have been shown to be critically dependent upon the site of addition of the OH radical to the base moiety. Not surprisingly, the redox property of the respective base-OH adducts may exert a major influence upon their subsequent chemical reactivity with dose modifiers. Furthermore, the induction of strand breaks (ssb) by base radicals produced by the hydroxyl radical through their abstraction of a H-atom from the sugar phosphate backbone has also been suggested to be critically dependent upon the redox property of the base-OH adduct (see Jones and Bothe, these proceedings). Both types of radicals involved may interact with dose modifiers and this has important mechanistic implications when considering the possible increase in multiplicity of localised radical sites leading to the ultimate biological end points. This has been discussed in some detail by O'Neill et al. (these proceedings).

The sites and yields of addition of OH to the DNA bases have generally been determined for monobase systems or for single stranded polymers (Deeble et al., 1986) such as poly U. It is as yet not known whether the distribution of attack of OH radicals at the base moieties of DNA is different from the model systems due to the double stranded nature of DNA. The double-stranded structure may restrict the accessibility to the base sites of the OH ($H_2O^{\cdot+}$) radical produced in the water associated with the DNA and the subsequent interactions of base radicals with dose modifiers (see later).

In the case of purine bases, some of the initially produced hydroxyl radical adducts have been shown to undergo transformations with, in some cases, inversion of the redox properties of the radical (O'Neill et al., 1985; Vieira & Steenken, 1987). With adenine, for example, the OH-adducts may undergo both dehydration (O'Neill et al., 1985; Vieira & Steenken, 1987, 1990) and ring opening reactions (van Hemmen, 1975; Vieira & Steenken, 1987, 1990) as shown in Scheme (1). In order to identify these conversions, it was necessary to

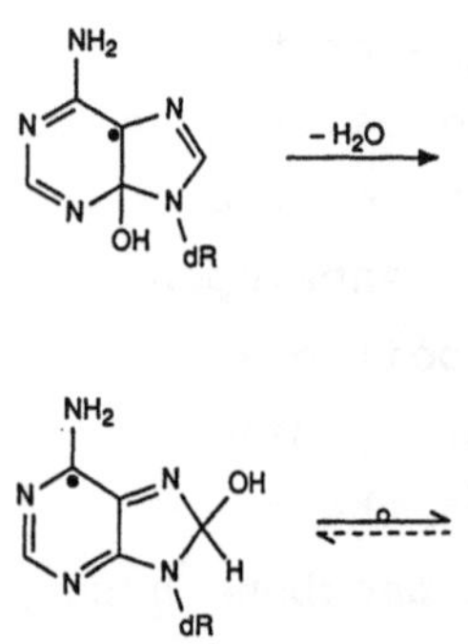

Scheme 1

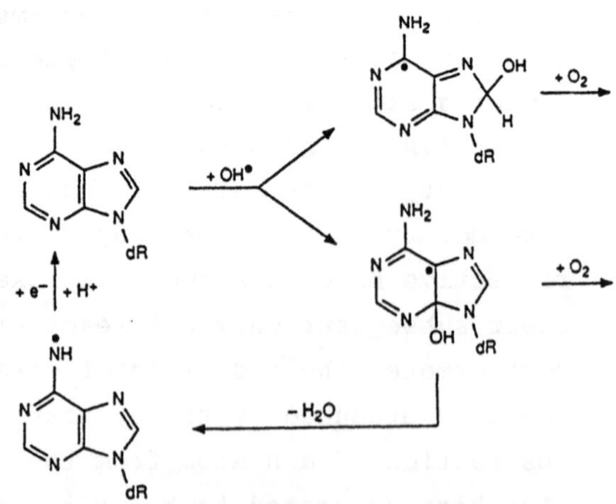

Scheme 2

investigate, in detail, the temperature dependences of these subsequent reactions. It was established that the activation parameters of the dehydration and ring-opening reactions are different. In the case of the dehydration reaction, the radical produced initially is converted into an even stronger oxidising radical. Recent studies with guanine infer that similar reactions may occur, i.e. dehydration and imidazole-ring opening (Candeais, these proceedings). With adenine, it has been shown (Vieira & Steenken, 1990) that addition of the OH radical to C(8) yields a reducing radical which may undergo a ring-opening of the imidazole-ring to yield the reducing radical (I). In model systems the half life for such conversions of the adenine moiety is of the order of 50 µs at room temperature. Similar reactions are observed following interaction of OH with poly A. The rate constant for the dehydration reaction is $2.6 \times 10^4 s^{-1}$. If such rearrangements occur with the OH-induced base radicals produced within double-stranded DNA, then their subsequent interactions with dose modifiers (e.g. thiols, oxidants) will be dependent upon their concentration. Further, for those radicals which undergo redox inversion, their interaction with dose modifiers will be determined by the competition reactions as shown in Scheme (2). For instance, at low oxygen tension, the radical produced initially may undergo redox inversion prior to its interaction with oxygen. This competition will lead to a radical which subsequently may be repaired chemically in contrast to the fixation by O_2. This suggests a role for both thiols (reductants) and oxygen/sensitiser (oxidants) by modifying radicals which originate from a common precursor although the two types of dose modifiers are not in direct competition for the same radical. This highlights the importance of redox inversion reactions when considering the concentration dependence of dose modifiers in their biological action.

As will be discussed below, a knowledge of the properties of the radicals derived from OH attack is essential for understanding the role of dose modifiers. It is thought that these particular radicals are considered to contribute significantly to the "scavengeable" effect in radiobiology, which is significantly greater under aerobic compared with anaerobic conditions.

One-electron oxidation of DNA bases to mimic electron loss centres

Since direct energy deposition within DNA has been shown to result in electron loss centres and in particular guanine radical cations, several studies have recently been undertaken (Willson *et al.*, 1975; Schuchmann *et al.*, 1987; O´Neill & Davies, 1987; Fujita *et al.*, 1988; Deeble *et al.*,1990) in an attempt to produce the same electron loss centres in an aqueous environment so that their chemical properties may be characterised in the presence of water. One method of producing these electron loss centres has utilised a series of one electron oxidising agents such as $Br_2^{\bullet -}$ (E = 1.66 V) and $SO_4^{\bullet -}$ (E ≈ 2.43 V) (Wardman, 1989) produced as shown in reactions (1) and (2).

$$^{\bullet}OH + Br^- \longrightarrow [BrOH]^{\bullet -} \xrightarrow{+Br^-} {}^-OH + Br_2^{\bullet -} \qquad (1)$$

$$e^-_{aq} + S_2O_8^{2-} \longrightarrow SO_4^{\bullet -} + SO_4^{2-} \qquad (2)$$

The sulphate radical anion in particular is a powerful one-electron oxidising agent and may be used to abstract an electron from the purines and pyrimidines.

The rate constant for interaction of $SO_4^{\bullet -}$ with deoxynucleosides were determined directly from the rate of loss of $SO_4^{\bullet -}$ and are in the range 2.1-4.1 x 10^9 dm^3 mol^{-1} s^{-1}. With the possible exception of deoxyguanosine, these directly determined values are greater than those previously derived (O´Neill & Davies, 1987) using a competition method. It was subsequently established that the indirect competition method was complicated by interactions of the redox standard compared with the one-electron oxidised base radicals. This factor, with the exception of deoxyguanosine, resulted in rate constants which are considerably lower than those in Table 1. One-electron oxidation of the purines, deoxyguanosine and deoxyadenosine, results in the formation of radical species (O'Neill & Chapman, 1985; O'Neill & Davies, 1987; Candeias & Steenken, 1989) as shown in Scheme (3). The pKa of the radical cations of deoxyguanosine and deoxyadenosine have been determined to be 3.9 (Asmus et al., 1978) and <1 (Steenken, 1989) and this value for deoxyguanosine has recently been confirmed independently (Candeias & Steenken, 1989). Therefore at pH ~7, the one-electron loss centres of the purine bases in aqueous solution are neutral and not present as radical cations. These radical assignments have recently been supported by conductometric methods and photochemical studies (Candeias & Steenken, 1989; Steenken, 1989).

The one-electron oxidised radical from deoxyadenosine has previously been shown to have oxidising properties and is identical to that produced upon dehydration of the C(4) hydroxyl radical adduct of deoxyadenosine (O'Neill & Davies,1987). Similarly with deoxyguanosine, the radical produced upon one-electron oxidation is similar to that produced following interaction with the OH radical and subsequent dehydration. That the purine radicals produced by the loss of one electron have oxidising properties was also confirmed by using tetranitromethane as the indicator which failed to detect any reducing species. It is evident that a proportion of the purine radicals produced by OH are identical to those expected to be produced upon ionisation of the purine moiety. Therefore, in the case of purines, it is impossible to distinguish whether particular oxidising radicals result from direct ionisation of the base or from energy deposition within the water associated with the DNA. This lack of distinction emphasises the futility of trying to categorise the initial energy deposition events into "direct" and "indirect" effects, a sentiment often expressed by numerous workers in the field.

In contrast, one electron oxidation of the pyrimidine, deoxycytidine, results in radicals that are different from those produced following interaction with ·OH. One-electron oxidation of deoxycytidine results in the formation of a radical, the pKa of which is <4 and which probably has

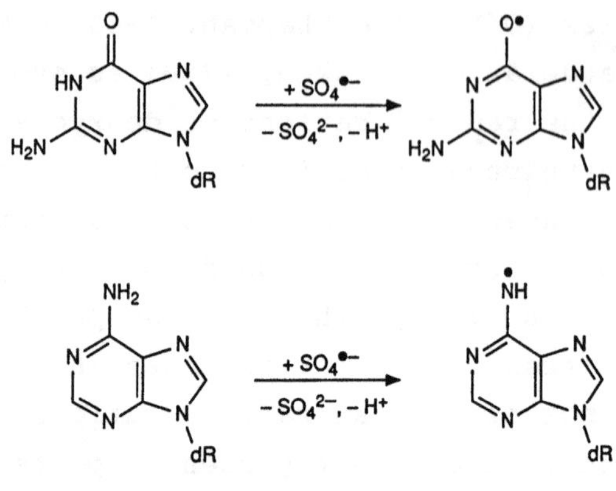

Scheme 3

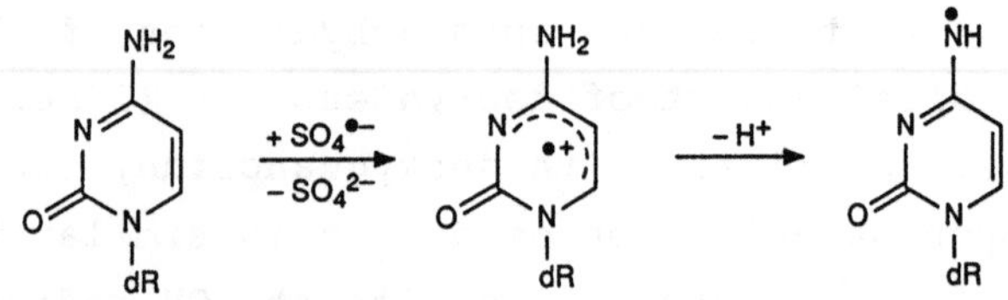

Scheme 4

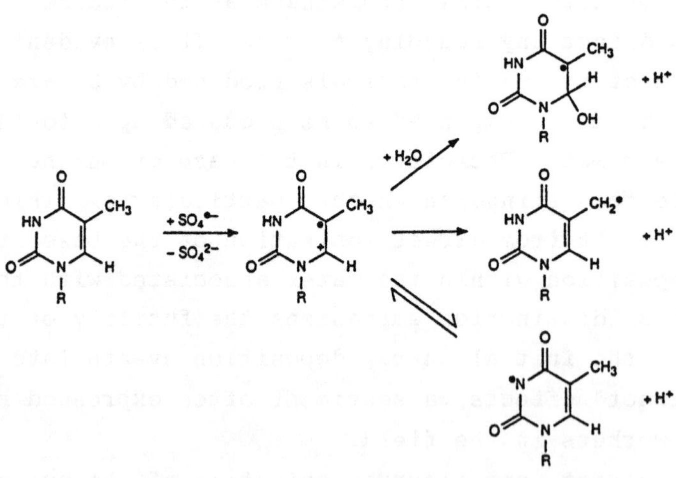

Scheme 5

oxidising properties, inferred from the fact that it does not possess reducing properties (O'Neill & Davies, 1987; Steenken, 1989). This may be interpreted in terms of deprotonation from the radical cation to give the N(4)-centred neutral anilino-type radical or, alternatively, by reaction of the radical cation with water to yield the 5- or 6-hydroxy-5,6-dihydrodeoxycytidine-5- or 6-yl radicals (not shown). If these latter radicals are produced, they would be expected to interact with O_2, due to their considerable spin density on carbon (C(6) or C(5)). However, the radical formed by reaction of $SO_4^{\cdot-}$ with deoxycytidine was found not to interact at an appreciable rate with O_2 (O'Neill & Davies, 1987), excluding formation of 5- or 6-yl radicals. The Scheme (4) is proposed to explain the experimental observations. Such a radical is expected to have oxidising properties and, indeed, has been shown not to interact with TNM (O'Neill & Davies, 1987); therefore indirectly supporting the suggestion of an oxidising radical.

One-electron oxidation of thymidine by $SO_4^{\cdot-}$ may be described by the following Scheme (5). The species produced initially is the radical cation of thymidine with a pKa of ~ 3.6 (Deeble *et al.*, 1990). At pH 7 the radical cation deprotonates in competition with its reaction with water to yield N(3) centred radicals. At pH <7 the water reaction yielding the thymidine-6-OH-5yl radical assumes increasing importance. The identity of the radical resulting from the addition of water was verified by both spectral data and the absence of reducing radicals using TNM (O'Neill & Davies, 1987) as the indicator. Interestingly, the N(3) centred radical interacts with oxygen with a rate constant of ~3 x 10^7 dm^3 mol^{-1} s^{-1} (Deeble *et al.*, 1990). Of the one-electron oxidised radicals produced from the DNA-bases, only the thymidine radical reacts with oxygen at a significant rate (O'Neill & Davies, 1987). However, the reaction with oxygen would not be expected to compete with positive charge migration occurring at times <1 ns as inferred from our data on DNA. The importance of the reaction with O_2, is its subsequent role in modifying the product distribution especially when considering DNA ionisation events. The question arises as to whether hole migration to guanine in DNA competes effectively with proton transfer to H_2O or to its complementary base as previously discussed?

Based upon these findings the electron loss centres produced by ionisation of the DNA base are expected to have oxidising properties and would be expected to interact with reductants such as thiols in competition with alternative modes of decay.

Interactions involving dose modifiers

(a) Oxidants

Many oxidants, e.g. oxygen and radiosensitisers, enhance the biological effects of ionising radiation. Their ability to influence these biological responses is generally related to interference with free radical reactions at times short compared to conventional biochemical processes. It has previously been proposed and evidence presented (O´Neill, *et al*., these proceedings) that oxidants interfere with the initial ion recombination processes by acting as electron sinks. However, a further mechanism should be considered when considering radiosensitisation of cellular systems in the light that certain of the hydroxyl-radical adducts of the bases possess reducing properties (see above).

In order to assess the ability of the OH-adducts of the purines and pyrimidines to act as reducing agents, the rate constants were determined for their interaction with a series of oxidants of differing one-electron reduction potentials. The oxidants used were either quinones or viologens. The dependence of the rate constants for interaction of the OH-adducts of the DNA bases upon the redox potentials of the oxidants is shown in Figure 2. With the pyrimidines, the reactions are essentially diffusion controlled for the more powerful oxidants (E^1_7-values > -200mV) whereas they become activation controlled and dependent on the redox potential of the oxidant for E^1_7-values < -200mV. The purines appear to be much less reactive with oxidants. From these observations the ability of the OH-adducts to act as reducing

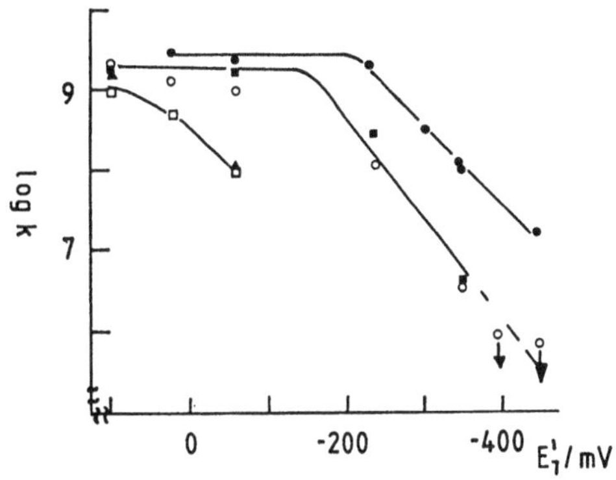

Figure 2: The dependence of the rate constant for interaction of oxidants with OH-adducts of cytosine (●), uracil (■), thymine (o), dG (□) and dA (Δ) at pH 7 on the E^1_7 value of the oxidant

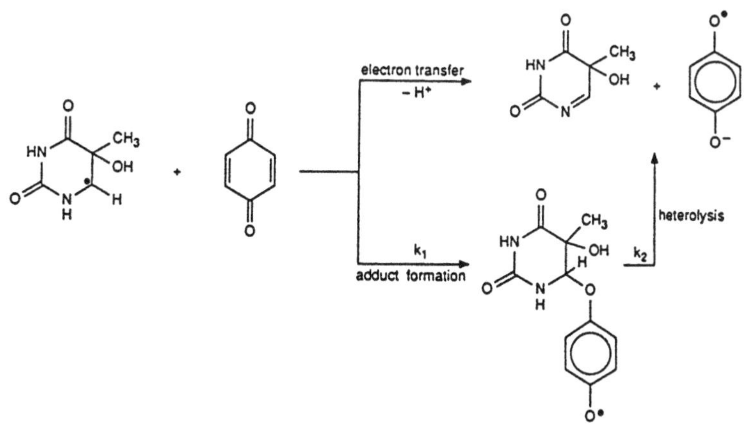

Scheme 6

agents is in the order cytosine > thymine ≈ uracil >> deoxyguanosine ≈ deoxyadenosine, where the radical from cytosine is the most powerful reducing agent. The yields of reducing radicals based upon their interaction with napthaquinone-2-sulphonate are consistent with those determined using TNM (Fujita & Steenken, 1981; Hazra & Steenken, 1983; O´Neill, 1983), a more powerful oxidant.

Scheme (6) is proposed to account for these observations using thymine as an example.

In all cases, the one electron reduced radical of the oxidant is observed following interaction of the oxidant with the OH-adducts of the bases. From these observations at the concentrations of oxidant used, it is inferred that if an adduct is initially formed between the oxidant and the base radical then $k_2 > k_1$ [oxidant]. From this data, it is not possible to differentiate between outer and inner sphere mechanisms of electron transfer. As shown in Figure 2, the purine and pyrimidine base undergoes a kinetically controlled reaction, the rate of which is governed by the redox potential of the oxidant. Since the established relationship for hypoxic cell radiosensitisation exhibits a dependence upon the redox potential of the radiosensitiser (Adams *et al.*, 1979), in the first instance, it is tempting to propose that the interaction of the radiosensitiser with hydroxyl radical adducts of the DNA bases involves a kinetically controlled, redox process in the mechanism of action of hypoxic cell radiosensitisation. Indeed, the various dependences determined from Figure 2 are very similar with a slope of ~ 10-12 V^{-1} and not very different from the dependence for cellular response of 9.5 V^{-1}. In further support of the involvement of diffusible radicals (e.g. OH radicals) which induce DNA damage which subsequently interact with the oxidants, is the fact that radical scavengers, e.g. DMSO, considerably protect irradiated cells against the effects of radiosensitisers and oxygen (Chapman *et al.*, 1975; Roots *et al.*, 1982; Millar *et al.*, 1987). As shown below, the interaction of a base radical with an oxidant could ultimately yield a carbocation which may subsequently interact with water or other nucleophiles.

In the case of oxygen, it is generally assumed that reaction with the DNA base radicals yields an addition product, namely a peroxyl radical (von Sonntag, 1987). In polynucleotides, it has been shown that the peroxyl radical induces strand breakage via an H-atom abstraction from the sugar moiety (Bothe & Schulte-Frohlinde, 1982; Jones & O'Neill, 1990; Bothe *et al.*, 1986). In the cellular environment, the oxidants may also interact by addition which, in the case of a nitro aromatic compound, would yield a nitroxyl free radical that could either undergo heterolysis to the nitro radical anion or H-atom abstraction from the sugar moiety to give a strand break. Indeed, nitrobenzenes have been shown (Jagannadham & Steenken, 1984) to interact with pyrimidines and mononucleotides via an inner sphere electron transfer reaction with the adducts having lifetimes of several milliseconds. The possible H-atom abstraction reaction from the sugar phosphate backbone of polynucleotides by such adducts is presently under investigation.

(b) Reductants

The generally accepted mechanisms by which thiols protect cellular systems against ionising radiation is their ability to donate an H-atom to carbon centred radicals such as are produced on the deoxyribose moiety. Over the last few years it has become apparent that thiols may also interact with base radicals with oxidising properties via electron transfer processes (O'Neill, 1983; O'Neill & Chapman, 1985). The same is probably true of other reductants such as ascorbate and dihydroxybenzenes. Interestingly, the base radicals with oxidising properties do not appear to react with oxygen or other oxidants. There is evidence that ascorbate may act as a weak radioprotector in mammalian cells.

If, as proposed from esr studies (Gräslund *et al*., 1975; Boon *et al*., 1984; Gregoli *et al*., 1982; Hüttermann *et al*., 1984), the majority of the electron loss centres ultimately migrate to guanine, it is interesting to speculate that the resulting guanine radical would be expected to interact with reductants. Such an interaction would lead to the restitution of the purine base, a radioprotective interaction. Does a role exist for the guanine electron loss centres to be involved in radiosensitisation as well as radioprotection? The studies of Boon *et al*. (1984) would imply that the electron loss centres ultimately yield strand breaks; a possible competing pathway for its interaction with reductants. In discussing the interaction of reductants with DNA base radicals, we have not eluded to the possible role of thiyl/thiyl peroxyl radicals in their interactions with DNA bases.

Conclusions

We have attempted to show in this short review that the scavengeable component of radiation action generally related to the action of hydroxyl radical adducts of DNA bases may show very different facets when considering their interactions with radiation dose modifiers. Such differences are related to the different redox properties of the hydroxyl radical adducts, a property which is governed by the particular site of addition of the OH radical. Whether this distribution is similar in the cellular environment is, as yet, not known since the water radicals will be formed near to or within water closely associated with DNA. The subsequent rearrangement of certain radicals with possible redox inversion has been highlighted. The role of dose modifiers has been presented with respect to their interaction with the base radicals. Indeed, dose modifiers may not always compete for the same radical sites as has been assumed on the basis of the behaviour of the sugar/ phosphate backbone. An additional mechanism for radiosensitisation has also been proposed, taking into account the well established redox relationship observed with cellular systems.

Finally, it has been demonstrated that radicals produced upon ionisation of DNA bases, particularly for purine electron loss centres, may be the same as those produced via hydroxyl radical attack and will therefore possess the same chemical reactivity. It is hoped that this simplified chemical view of cellular effects may stimulate biophysicists to advance their models to incorporate the effects of dose modifiers and assist the biologists to better understand the primary effects of ionising radiation.

References

Adams GE, Michael BD, Asquith JC, Shenoy MA, Watts ME, Whillans DW (1975) Rapid-mixing studies on the time-scale of radiation damage in cells. Radiat.Res. 478-492.

Adams GE, Clarke ED, Flockhart IR, Jacobs RS, Sehmi DS, Stratford IJ, Wardman P, Watts ME, Parrick J, Wallace RG, Smithen CE (1979) Structure-activity relationships in the development of hypoxic cell radiosensitizers. I. Sensitization efficiency. Int.J.Radiat. Biol. 35: 133-150.

Al-Kazwini AT, O'Neill P, Fielden EM, Adams GE (1988) Radiation-induced luminescence from dry and hydrated DNA and related macromolecules. Radiat.Phys.Chem. 32: 385-389.

Al-Kazwini AT, O'Neill P, Adams GE, Fielden EM (1990) Radiation-induced energy migration within solid DNA: the role of misonidazole as an electron trap. Radiat. Res. 121: 149-153.

Asmus K-D, Deeble DJ, Garner A, Ali Idriss KM, Scholes G (1978) Chemical aspects of radiosensitization. Reactions of sensitizers with radicals produced in the radiolysis of aqueous solutions of nucleic acid components. Br.J. Cancer 37 (Suppl. III): 46-49.

Bernhard WA (1989) Sites of electron trapping in DNA. J.Phys. Chem. 93: 2187-2194.

Boon PJ, Cullis PM, Symons MCR, Wren BW (1984) Effects of ionizing radiation on deoxyribonucleic acid and related systems. Part I. The role of oxygen. J.Chem.Soc. Perkin Trans. II: 1393-1399.

Bothe E, Schulte-Frohlinde D (1982) Release of K^+ and H^+ from poly U in aqueous solution upon γ and electron irradiation. Rate of strand break formation in poly U. Zeitschr. für Naturforsch. 37c: 1191-1204.

Bothe E, Behrens G, Böhm E, Sethuram B, Schulte-Frohlinde D (1986) Hydroxyl radical-induced strand break formation of poly (U) in the presence of oxygen: comparison of the rates as determined by conductivity, esr and rapid-mix experiments with a thiol. Int.J.Radiat.Biol. 49: 57-66.

Candeias LP, Steenken S (1989) Structure and acid-base properties of one-electron-oxidized deoxyguanosine, guanosine and 1-methylguanosine. J.Amer.Chem.Soc. 111: 1094-1099.

Chapman JD, Gillespie CJ, Reuvers AP, Dugle DL (1975) The inactivation of Chinese hamster cells by X-rays: The effects of chemical modifiers on single and double events. Radiat.Res. 64: 365-375.

Close DM, Sagstuen E, Nelson WH (1985) ESR study of the guanine cation. J.Chem.Phys. 82: 4386-4388.

Deeble DJ, Schulz D, von Sonntag C (1986) Reactions of OH radicals with poly (U) in deoxygenated solutions: sites of OH radical attack and the kinetics of base release. Int.J.Radiat.Biol. 49: 915-926.

Deeble DJ, Schuchmann MN, Steenken S, von Sonntag C (1990) Direct evidence for the formation of thymine radical cations from the reaction of SO_4^- with thymine derivatives: a pulse radiolysis study with optical and conductance detection. J.Phys.Chem. (in press).

Fujita S, Steenken S (1981) Pattern of OH radical addition to uracil and methyl- and carboxyl-substituted uracils. Electron transfer of OH adducts with N,N,N´,N´-tetramethyl-p-phenylenediamine and tetranitromethane. J.Am. Chem.Soc. 103: 2540-2545.

Fujita S, Nagata Y, Dohmaru T (1988) Radicals produced by the reactions of SO_4^- with uridine and its derivative. Studies by pulse radiolysis and γ-radiolysis. Int.J. Radiat.Biol. 54: 417-427.

Gräslund A, Ehrenberg A, Rupprecht A, Strom G, Crespi H (1975) Ionic base radicals in γ-irradiated oriented non-deuterated and fully deuterated DNA. Int.J.Radiat. Biol. 28: 313-323.

Gregoli S, Olast M, Bertinchamps A (1982) Radiolytic pathways in γ-irradiated DNA: influence of chemical and conformational factors. Radiat.Res. 89: 238-254.

Hüttermann J, Voit K, Oloff H, Köhnlein W, Gräslund A, Rupprecht A (1984) Specific formation of electron gain and loss centres in X-irradiated oriented fibres of DNA at low temperature. Faraday Disc. of the Chemical Society, 78: 135-149.

Jagannadham V, Steenken S (1988) One-electron reduction of nitrobenzenes by OH and H radical adducts to 6-methyl-uracil and 6-methylisocytosine via electron transfer and addition/elimination. Effects of substituents on rates and activation parameters for formation and heterolysis of nitroxyl-type tetrahedral intermediates. J.Phys. Chem. 92: 111-118.

Jones GDD, O´Neill P (1990) The kinetics of radiation-induced strand breakage in polynucleotides in the presence of oxygen: a time-resolved light-scattering study. Int.J. Radiat.Biol. 57: 1123-1139.

Millar BC, Sapora O, Fielden EM, Loverock PS (1981) The application of rapid-lysis techniques in radiobiology. IV. The effect of glycerol and DMSO on Chinese hamster cell survival and DNA single-strand break production. Radiat.Res. 86: 506-514.

O´Neill P (1983) Pulse radiolytic study of the interaction of thiols and ascorbate with OH-adducts of dGMP and dG: Implications for DNA repair processes. Radiat.Res. 96: 198-210.

O´Neill P, Chapman PW (1985) Potential repair of free radical adducts of dGMP and dG by a series of reductants. A pulse radiolytic study. Int.J.Radiat.Biol. 47: 71-80.

O´Neill P, Chapman PW, Papworth DG (1985) Repair of hydroxyl radical damge of dA by antioxidants. Life Chemistry Reports 3: 62-69.

O´Neill P, Fielden EM (1986) Molecular aspects of DNA damage and its modification. In: Burns FJ, Upton AC, Silini G (eds) Radiation Carcinogenesis and DNA Alterations. New York, Plenum, pp. 425-438.

O'Neill P, Davies SE (1987) Pulse radiolytic study of the interaction of SO_4^- with deoxynucleosides. Possible implications for direct energy deposition. Int.J. Radiat.Biol. 52: 577-587.

O'Neill P, Al-Kazwini AT, Land EJ, Fielden EM (1989) Diffuse reflectance pulse radiolysis of solid DNA: the effect of hydration. Int.J.Radiat.Biol. 55: 531-537.

Painter RB (1979) The role of DNA damage and repair in cell killing induced by ionizing radiation. In: Meyn RE and Withers HR (eds) Radiation biology in cancer research. Raven Press, New York, pp. 59-68.

Roots R, Chatterjee A, Blakely E, Chang P, Smith K. Tobias C (1982) Radiation response in air- nitrous oxide- and nitrogen-saturated mammalian cells. Radiat.Res. 92: 245-254.

Schuchmann PH, Deeble DJ, Olbrich G, von Sonntag C (1987) The SO_4^- induced chain reaction of 1,3-dimethyluracil with peroxodisulphate. Int.J.Radiat.Biol. 57: 441-453.

Steenken S (1989) Purine bases, nucleosides and nucleotides: aqueous solution redox chemistry and transformation reactions of their radical cations and e^- and OH adducts. Chem.Rev. 89: 503-509.

Symons MCR (1987) Application of esr spectroscopy to the study of the effects of ionising radiation on DNA and DNA complexes. J.Chem.Soc. Faraday Trans. 1, 83: 1-11.

van Hemmen JJV (1975) Some reactions of the hydroxyl adduct of adenine. Int.J.Radiat.Biol. 27: 403-407.

Vieira AJSC, Steenken S (1987) Pattern of OH radical reactions with 6- and 9-substituted purines. Effect of substituents on the rates and activation parameters of the unimolecular transformation reactions of two isomeric OH adducts. J.Phys.Chem. 91: 4138-4144.

Vieira AJSC, Steenken S (1990) Pattern of OH radical reaction with adenine and its nucleosides and nucleotides. Characterization of two types of isomeric OH adduct and their unimolecular transformation reactions. J.Amer. Chem.Soc. (in press).

von Sonntag C (1987) The chemical basis of radiation biology. Taylor & Francis, London.

Wardman P (1989) Reduction potentials of one-electron complex involving free radicals in aqueous solution. J.Phys. Chem. Refer. Data 18: 1637-1755.

Willson RL, Wardman P, Asmus K-D (1974) Interaction of dGMP radical with cysteamine and promethazine as possible model of DNA repair. Nature 252: 323-324.

Chemical properties of 'radiation modifiers' of DNA damage and their radiobiological effects

Peter Wardman

Cancer Research Campaign
Gray Laboratory, PO Box 100,
Mount Vernon Hospital
Northwood, Middx. HA6 2JR
England

Introduction

Interest in radiation damage to DNA is easily justified by the recognition that DNA is known to be a target of major importance in radiobiology, i.e. in radiation-induced cellular lethality. Interest in the early effects of radiation in DNA reflects the importance of fast, free-radical damage in the processes which lead eventually to cell death. Much of our information concerning the molecular mechanisms which damage DNA, such as strand breaks, has been obtained by the use of chemical additives which react with these transient radical intermediates to modify the damage.

These chemical 'radiation modifiers' usually express their effects in terms of shifting the curve of biological response *vs.* dose, such that they appear as dose-modifiers for practical purposes: radiosensitizers or radioprotectors. The simplest examples would include oxygen, the presence of which is easily controlled in many model systems, and cellular constituents such as thiols. Other oxidants or reductants can be added or removed from the test system and the dose-modifying effects observed. Mechanistic interpretation of such experiments relies on an appreciation of the chemical properties of the modifiers. However, particularly in model systems involving intact cells and certainly with *in vivo* models, chemical reactivity is only one aspect which has to be considered. Problems with drug distribution and biotransformation may eliminate effects easily seen in simpler models. Hence evaluation of the effects of chemical radiation modifiers in radiobiology must encompass concepts of medicinal chemistry broader than merely free-radical reactivity.

NATO ASI Series, Vol. H 54
The Early Effects of Radiation on DNA
Edited by E. M. Fielden and P. O'Neill

Most papers in science seek to present advances in knowledge; too infrequently we take stock of what we do *not* know. In this short outline, some paradoxes and uncertainties will be emphasized, as well as presenting recent progress in our understanding of the biodistribution of some radiation modifiers.

'Chemical modifiers': main features

Damage to DNA can be enhanced or diminished by chemical substances: radiation sensitization or protection. These phenomena include several effects not necessarily related to early events in DNA damage when higher-order models such as whole cells or tissue are considered, but for simplicity we consider here only chemical effects involving the fast, free-radical stage of radiation damage.

Chemically, the main feature of radiosensitizers thought to interact with DNA radicals is oxidizing ability or electrophilicity; oxygen is representative and nitroaryl compounds have been most extensively investigated. Quinones, nitroxyls and viologens are other examples. The main feature of chemical radioprotectors is the converse: reducing ability or nucleophilicity. Thiols are representative but other good reductants such as ascorbate protect in some systems (see below) and *it is not clear* whether a labile hydrogen atom is an essential feature of radioprotectors. What is clear is that both these general groups of chemical modifiers must act by reaction with short-lived species, presumably mainly DNA-derived and presumably free-radicals. Addition of oxygen or nitroaryl radiosensitizers to bacterial or mammalian cells a few milliseconds after irradiation has no effect on survival, whilst pre-irradiation contact only requires sufficient time for diffusion to the nucleus (see e.g. Howard-Flanders and Moore 1958, Adams and Jameson 1980, and the presentation by Michael in this volume). A similar situation, although less clear-cut, is seen with radioprotection by thiols (Hodgkiss *et al.* 1987, 1989).

Both sensitizers and protectors are radiation 'dose-modifying' in that they exert their effects upon the exponential portion of the cellular-survival curve. The *ca.* three-fold maximal effect (in such dose terms) of oxygen appears to be extended to nitroaryl radiosensitizers, but the breadth of the approximate 'plateau' established in the response *vs.* concentration curve for oxygen *has not been established* with these other radiosensitizers. Factors such as solubility, hydroxyl radical scavenging and hypoxic cytotoxicity or other effects of metabolic activation preclude this desirable comparison. On the other hand, the variation in the oxygen effect with cell cycle is mirrored by the typical nitro radiosensitizer, misonidazole (Asquith *et al.* 1974).

For many years the oxygen effect has been numerically modelled in terms of simple competition between damage 'fixation' by oxygen and 'repair' or even 'restitution' by protective thiols, either endogenous or exogenous. Thiol depletion or addition enhances or diminishes, respectively, the radiosensitizing effect of nitroimidazoles (e.g. Koch and Howell 1981, Hodgkiss and Middleton 1983). Such phenomena raise the question, *not yet answered*, as to whether these competing agents are indeed competing at the molecular level for the same lesion.

The electron-affinic nature of radiosensitizers prompted the development *ca.* 20 years ago of quantitative structure-activity relationships (QSARs) relating sensitization efficiency to redox properties such as reduction potential or Hammett sigma substituent 'constants'. The development of pulse radiolysis methods to measure reduction potentials involving free-radical couples (reviewed by Wardman 1989a) facilitated extension of such studies to different groups of nitroaryl compounds (Adams *et al.* 1979). Oxygen and a quinone apparently followed the same redox relationship as the nitro compounds, but whether this was a fortuitous coincidence *is still not clear*, and the relevance of these measures of redox properties is considered further below. What *is* established is that radiosensitization by nitroaryl compounds ($ArNO_2$) varies in efficacy according to the magnitude of the mid-point reduction potential at pH 7 of the one-electron couple: $E_{m7}(ArNO_2/ArNO_2^{\bullet-})$ with $d(\log C)/dE \approx 8$ V^{-1}, where C is the concentration required for a constant effect. This result in cellular systems provides a crucial test for any model of radiosensitization.

Redox properties of radiation modifiers

Leaving aside the question of drug distribution (see below), reduction potential is the property dominating radiosensitization efficiency. There seems little doubt that the appropriate potential here is that for adding a single electron: the illogical 'mixing' of one- and two-electron potentials in this context has been discussed (Wardman 1989b). Of course, it is not the only important property. For example, if outer-sphere electron-transfer is involved in the critical or rate-limiting step, then the Marcus theory of such reactions indicates that the rates of the 'zero-energy' self-exchange reactions of the electron acceptors are reflected in the kinetics of electron transfer. This may explain why some typical electron-transfer reactions are more than an order of magnitude slower for nitroaryl compounds than quinones, for example (Meisel 1975, Wardman and Clarke 1985), although inner-sphere electron transfer may also be the explanation. Much more dramatic differences are reflected in the behaviour of other redox pairs. Thus there are several orders of magnitude difference in the rates of oxidation of methyl viologen radical cation by oxygen and by hydrogen peroxide. Finkelstein and Glatstein (1988) discussed hydrogen peroxide as an example to demonstrate the breakdown of the 'electron affinity relationship', but judging from these known rate data for its radical reactivity, its inefficient radiosensitization is entirely in accord with chemical experience.

Many electron-transfer reactions proceed *via* intermediate adducts (Steenken 1989) and there are obvious similarities in the reaction pathways of carbon-centred radicals from DNA base or sugar components, or models thereof, reacting with oxygen and nitroaryl radiosensitizers. (See Steenken (1989), Wardman (1989b) for summaries and von Sonntag (1987) for further discussion.) The redox-controlled step may not necessarily be that of the initial bimolecular reaction, as in the scheme below, based on the work of Steenken and Jagannadham (1985) and Jagannadham and Steenken (1988). They showed the effectively unimolecular heterolysis of the nitroxide radical intermediate in the reaction between 5-hydroxy-pyrimidin-6-yl radicals and nitrobenzenes had a redox relationship with the (Hammett-type) plot of log *k* *vs*. E having a slope of *ca.* 6-7 V^{-1}, quite similar to the observed redox dependence of radiosensitization.

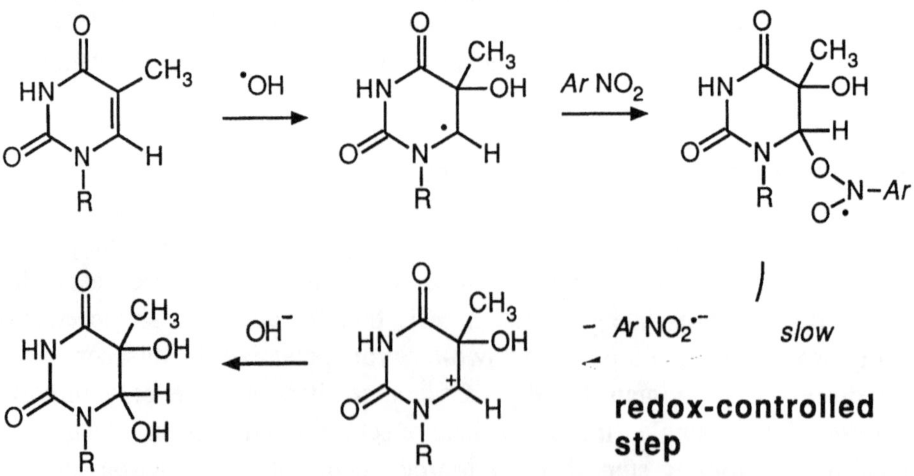

Redox properties of radioprotectors are less well defined, quantitatively at least, compared to radiosensitizers. However, the most important values such as $E_{m7}(RS^{\bullet}, H^+/RSH)$ and $E_{m7}(AH^{\bullet}/AH^-)$ for some aliphatic thiols (RSH) and ascorbate (AH^-) are known (see Wardman 1989a); the former couple is *ca.* 1.0 V more positive than the latter. In accordance with prediction (Wardman and Clarke 1985), the redox properties of different common thiols do not appear to vary greatly at pH 7.

Critical DNA damage: oxidative or reductive?

By 'critical' we mean, in the present context, that damage which reacts with chemical modifiers. Some of the articles in this volume reflect interest in the site of localization and chemical interception of electrons or electron excess centres in irradiated DNA; others describe electron loss or oxidation sites. Some methods involve studying solid DNA or solid solutions; others dilute aqueous solutions, fluid or solid. Yet further systems include bound histones. It is not possible to state with confidence which of these diverse matrices are the more relevant models in the present context. Apparently at odds with this diversity of chemical approaches, conventional wisdom in radiobiology is that *ca.* two-thirds of the damage *leading to lethality* arises from hydroxyl radicals produced in the immediate vicinity of DNA. There is general acceptance that the most important lesions are double-strand breaks (or locally multiply-damaged sites, see Ward 1981, 1988). Some of the evidence for hydroxyl radical importance is itself not without alternative interpretation; as pointed out by Wardman (1984), short-lived oxidizing intermediates such as the guanine radical-cation would have reactivity towards alcohols, DMSO etc. paralleling hydroxyl radical reactivity. However, the work of Milligan *et al.* presented in this volume provides further support for the importance of hydroxyl radicals. There is an apparent contradiction with results from experiments with solid solutions of DNA in water at low temperatures, where electron spin resonance detection fails to indicate reactions of $^{\bullet}OH$ in the glassy region close to DNA (Symons 1987). Whilst a simple explanation might be that the bulk of the early damage to DNA does not lead to cellular lethality, more work in this area is clearly needed.

The products of oxidative attack can be themselves either oxidative or reductive. Thus $^{\bullet}OH$ addition to the pyrimidine 5,6-double bond yields oxidizing (6-OH-5-yl) and reducing (5-OH-6-yl) radicals. The former are expected (and found) to be more reactive towards reductants, whilst the latter react with oxidants (summarized by von Sonntag 1987).

Oxidative targets: sugar or base?

As discussed above, nitro compounds - and similarly oxygen - can sensitize the formation of pyrimidine glycols via redox-controlled, inner-sphere electron transfer oxidation of 5-hydroxy-pyrimidin-6-yl radicals. Experimental evidence (Nishimoto *et al.* 1983) that the glycol yield is dependent on reduction potential can not at present be analysed in the same way as the QSARs for cellular radiosensitization. A more serious question as to the relevance of this model is the reported absence of sensitization of thymine glycol formation in mammalian cells by misonidazole (Remsen 1985). Although briefer studies of *base* product sensitization were reported much earlier (e.g. Varghese 1975), redox-controlled sensitization of *sugar* damage was postulated as a plausible mechanism of radiosensitization as soon as the importance of reduction potential was recognised (Raleigh *et al.* 1974).

The initial proposal of electron-transfer oxidation of sugar radicals to liberate phosphate (in simple aliphatic phosphates) or yield a strand break (in polynucleotides) can thus be modified to include the probable formation of radical-adducts as intermediates with both oxygen and nitro compounds, and the probable generation of sugar radicals by intramolecular abstraction by the C-6-yl radicals of pyrimidines (e.g. Lemair *et al.* 1987). For a more extensive discussion, see von Sonntag (1987) or Wardman (1987). Whilst the abstraction of hydrogen at a sugar site (resulting in a strand break) by the C-6-yl peroxyl radical is now well documented - see the paper by Schulte-Frohlinde in this volume - the analogous reactivity of the nitroxyl radicals from e.g. nitrobenzenes or nitroimidazoles and pyrimidin-6-yl radicals *is a completely open question.* In fact, any such abstraction would compete with sensitization of pyrimidine glycols (as in the reaction scheme shown above) and could therefore conceivably have a negative redox dependence for strand break sensitization. In this connexion it may be noted that redox-controlled *protection* by nitrobenzenes of phosphate release from $^{\bullet}$OH attack on guanosine-3'-phosphate is already documented (Raleigh *et al.* 1974), although the reverse scenario was found for the 5'-phosphate.

Radiation sensitizers: no nitro, no use?

The dominance of the requirement for improved therapeutic index (efficacy: toxicity) of radiosensitizers has tended to delay desirable experimental evaluation of types of molecules other than those which showed the most clinical promise: nitroaryl compounds. Yet other types of oxidant can point to the mechanisms which may be involved. We have already noted that whilst reduction potential may be a dominant property in a single class of compounds, in a wider context other factors such as electron-exchange rates must be considered. This points to the need for a wider vista in exploring model systems. Thus the work by Fielden *et al.* (reported in this volume) concerning the oxidation of pyrimidinyl radicals by bipyridinium compounds and quinones is timely. We are ourselves comparing the effects of bipyridinium, quinone, and nitro oxidants in sensitizing oxidation of pyrimidines in steady-state systems. Such comparisons reveal the need to answer more basic questions such as the possibility of inner-sphere electron-transfer reactions between these non-nitro oxidants and even the simplest of electron donors, reactions which seem to have been little investigated. It would be of obvious mechanistic interest to investigate the various radiosensitization phenomena with a type of oxidant which, unlike oxygen or nitroaryl compounds, can *not* form radical-adducts as reaction intermediates with free radicals.

Radiation protection: no hydrogen, no hindrance?

For forty years, radiobiologists have, in the main, appeared content to accept as a model for radioprotection hydrogen-atom donation:

$$-\overset{|}{\underset{|}{C}}\cdot \quad + \quad RSH \longrightarrow -\overset{|}{\underset{|}{C}}-H \quad + \quad RS^{\bullet}$$

despite questions over the sterochemical identify of the 'repaired' carbon centre and the reactivity of the thiyl radical formed (see e.g. von Sonntag 1987, Akhlaq *et al.* 1987, Schöneich *et al.* 1989). Experimental evidence for other reductants, such as ascorbate, reacting much more rapidly than thiols with e.g. radical products of purine oxidation, and the increasing reactivity of thiols with pH in such model systems, has established beyond doubt that *electron* rather than *hydrogen atom* donation can be important in some systems. An example is reaction between ascorbate (AH^-) or thiolate (RS^-) ions and a radical obtained on reaction of $^{\bullet}OH$ with guanine derivatives (O'Neill 1983):

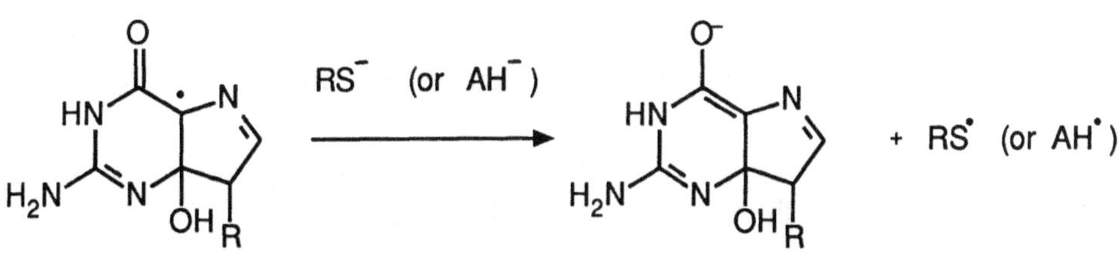

Such observations seem bound to be valuable. On the one hand, if ascorbate or other electron donors do not radioprotect in higher models such as mammalian cells *in vitro*, then such 'repair' reactions may have no importance in radiation lethality. On the other hand, if such electron- rather than hydrogen-donors *do* radioprotect, then the pulse radiolysis measurements may point the way to novel classes of radioprotector.

In the simplest of chemical experiments in dilute aqueous solutions, there are generally no doubts as to the availability of reactant at the reaction centre. As discussed below, in more complex models such as with intact cells and certainly *in vivo*, but also even merely with polynucleotides in solution, major questions arise about the distribution of potential reactants in the vicinity of the short-lived target radicals. A good illustration, relevant to the discussion above, is the availability of ascorbate in the commonest of *in vitro* mammalian cell models. Whilst tumours contain roughly half as much ascorbate as glutathione, V79 hamster fibroblast-like cells as commonly cultured contain negligible concentrations of ascorbate. More importantly, incubation of such cells with ascorbate does not result in significant uptake until several hours' exposure at 37°C. This could have led to an erroneous conclusion - that ascorbate does not influence the radiosensitization efficacy of misonidazole - had not Stratford and Hodgkiss (1986) measured intracellular concentrations. Earlier work (Koch *et al.* 1979) had involved exposure of cells to ascorbate at 4°C to minimise toxicity. The absence of an effect in these early experiments was hardly surprising in the light of the subsequent uptake measurements.

Hodgkiss and Stratford (1988) showed that an intracellular concentration of ascorbate not dissimilar to those reported in tumours was sufficient to modify significantly the radiosensitivity of cells in the presence of misonidazole, but the effect could only be demonstrated if the intracellular thiol levels were depleted by inhibiting glutathione biosynthesis with buthionine sulphoximine. Of course, these phenomena do not define the molecular mechanisms involved. Aside from electron donation from ascorbate to base radicals such as that illustrated above for a guanine radical, it is likely that both glutathione and ascorbate would be reactive towards radical/sensitizer adducts such as the nitroxyl radicals of the type shown in the reaction scheme above involving a thymine radical. The analogous reactions of peroxyl radicals have been already reported (e.g. Schulte-Frohlinde *et al.* 1986).

If electron donation plays a role, then a critical equilibrium could be the reaction where the thiyl radical - including those from reactions of thiols with the diverse radicals which do not relate to critical 'targets' - is conjugated with the thiolate anion:

$$RS^{\bullet} + RS^{-} \rightleftharpoons (RSSR)^{\bullet -}$$

The three-electron-bonded disulphide radical anion is a powerful reductant whereas the thiyl radical precursor is a fairly good oxidant (see Wardman 1989a). Prütz (1989) has suggested that $RSSR^{\bullet -}$ could serve as an electron donor to positive 'holes' and that this reaction could be an alternative basis for radioprotection. Simple neutralization of radical-cations such as that from guanine:

$$(RSSR)^{\bullet -} + G^{\bullet +} \rightarrow RSSR + G$$

would be in competition both with extremely rapid deprotonation of $G^{\bullet +}$ and (in oxygenated systems) electron transfer to oxygen:

$$(RSSR)^{\bullet -} + O_2 \rightarrow RSSR + O_2^{\bullet -}$$

Thus, quite apart from the heterogenous kinetics of cellular systems, the kinetic behaviour would be complex. The deprotonation of $G^{\bullet +}$ seems likely to be much faster than formation and reaction of $(RSSR)^{\bullet -}$, but alternative sites for reactivity of the latter could be e.g. pyrimidin-6-yl peroxyl radicals which are known strand break precursors.

Arguably the single most important reaction of thiols in biology is the ionization:

$$RSH \rightleftharpoons RS^{-} + H^{+}$$

Not only is thiolate ($RS^{\bullet -}$) a much better electron donor or nucleophile than thiol (RSH), this equilibrium controls the formation of ($RSSR^{\bullet -}$) *via* the conjugation of $RS^{\bullet}$ with RS^{-}. The positions of both these equilibria will differ considerably for thiols such as glutathione or polyaminothiols such as WR-1065: the author is currently measuring these equilibrium constants for WR-1065, etc. and this should help to evaluate the importance in radioprotection of both these key equilibria. Whether the hydrogen-*abstracting* properties of thiyl radicals (the reverse of the 'repair' reaction, see Akhlaq *et al.* 1987, Schöneich *et al.* 1989) differ significantly for differing thiols is also currently under investigation by the author. Other effects of molecular structure, such as ion condensation, may also be a factor (see below), but it remains something of a paradox that despite considerable research, in broad terms cysteamine is as good a radioprotector as any.

The need for better chemical characterization of the experimental system

Other papers in this volume attest to the need for better characterization of the DNA target, such as the amount, site and nature of water in its immediate vicinity. When working with cellular systems with added 'radiomodifiers' it is important to recognize the possibility not only of exclusion from the cell or intracellular concentration gradients (see below), but also of undesired changes in the chemical environment of the target. Thus Milligan *et al.* (this volume) are considering whether artefacts from displacement of hydrating water near DNA by high concentrations of $^{\bullet}OH$ scavengers could be a factor. Bioreduction of nitroaryl compounds by hypoxic cells will inevitably involve (rapid) reaction of nitrosoaryl reduction intermediates with thiol groups and probably explains the well-known 'pre-incubation' sensitizing effects of such compounds.

A common procedure in testing numerical models such as those based on the simple competition kinetics of Alper and Howard-Flanders is to add varying concentrations of exogenous thiols. Often neglected, though long recognised, is the possibility of thiol-disulphide exchange between cellular free or protein-bound thiols and the added modifier:

$$R^1S^- + R^2SSR^3 \rightleftharpoons R^1SSR^3 + R^2S^-$$

Dennis *et al.* (1989) showed low concentrations of dithiothreitol could reduce cystine in cell culture medium and thus produce cysteine, which accumulated in V79 cells to a level about as high as the endogenous glutathione. Intracellular protein thiols may also be reduced by this mechanism. Thus attempts to fit competition kinetics to the results of such experiments without an attempt at characterization of the relevant intracellular chemical environment seems at best naïve and possibly even quite valueless.

Accessibility of radiomodifers to the cellular target

QSARs between chemical properties and radiobiological response can include terms which may reflect the probability of diffusion to, or association with, the presumed target(s). A superior approach is to measure the effective concentrations of radiomodifiers e.g. in the vicinity of nuclear DNA. Some information can be gained from measurements of *average* intracellular concentrations (as in the work with ascorbate described above), especially where molecular properties define extracellular/intracellular gross concentration gradients (as in the effects of prototropic equilibria on uptake, Dennis *et al.* 1985). However, *subcellular* concentration gradients prove more elusive.

One approach to probe concentration gradients around DNA has shown promise (Wardman *et al.* 1989). The fluorophore, acridine orange (AO) binds strongly to DNA; a small fraction of the emission (which is actually enhanced on binding) arises from triplet precursors (^{3}AO*) to the emitting singlet excited states (^{1}AO*). Because the 'natural' lifetime of the triplet is a few milliseconds, it is quenched by additives (Q) at much lower concentrations than singlet states with lifetimes τ_f typically of a few nanoseconds. Further, a commercial luminometer operating in the phosphorescence mode can 'gate out' the unwanted prompt fluorescence and monitor only the delayed fluorescence produced *via* the long-lived triplet precursor. (Fluorescence *via* ^{3}AO* is delayed because τ_{isc} >> τ_f, see the scheme below.) Hence monitoring the *delayed* fluorescence from cells stained with AO provides in a single experiment information about the extracellular/intracellular uptake of radiomodifiers and the effective concentration in the immediate vicinity of the nuclear DNA.

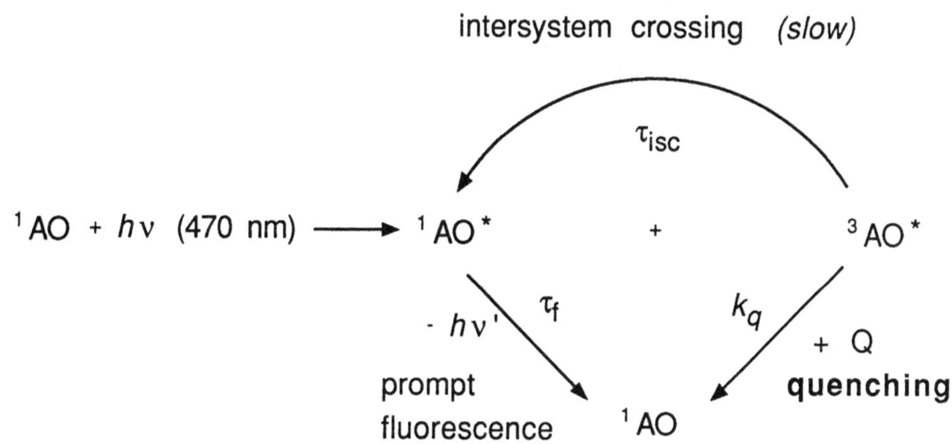

This work immediately resolved an earlier 'discrepancy' between the efficient radiosensitization observed with a hydrophilic analogue of pimonidazole, Ro 31-0368 and its low *average* intracellular uptake (see the structures below). In spite of marked differences in average uptake (Stratford *et al.* 1988, 1989) the delayed fluorescence from AO-stained V79 cells was quenched equally efficiently by pimonidazole and Ro 31-0368 (Wardman *et al.* 1989). In the same study highly-efficient quenching of AO bound to 'naked' DNA by ascorbate was found, but not when cell suspensions were exposed to ascorbate (or at least not until after prolonged exposure at 37°C), in accord with what Stratford and Hodgkiss (1986) had found earlier. Whilst the ascorbate result with AO was just that expected, it did help establish confidence in the technique.

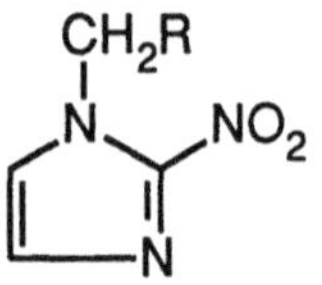

R =		
	$CH(OH)CH_2OCH_3$	(misonidazole)
	$CH(OH)CH_2N$ 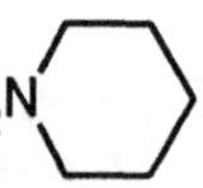	(pimonidazole)
	$CH(OH)CH_2NH_2$	(Ro 31-0368)
	CO_2H	(Ro 31-0258)

A potential problem with this approach was the possibility that quenching could occur *via* the small fraction of the probe which is free at any instant and has diffused some distance from the DNA. This possibility was negated by experiments by Mr. John White with AO/DNA in solutions of varying ionic strengths (μ). The triplet state of AO, $^3AO^*$ has similar prototropic properties to the ground state; thus $^3AO^*$ has a single positive charge at pH 7.4. According to the Debye-Hückel-Brønsted-Bjerrum treatment of the primary kinetic salt effect, increasing ionic strength should *increase* the rate of quenching by largely protonated bases (at pH 7.4) such as pimonidazole, or *decrease* the rate of quenching by largely dissociated acids such as Ro 31-0258 (see formulae above). If the rate data are fitted to the extended Debye-Hückel-Onsager-Davies expression:

$$f(\mu) = \sqrt{\mu}(1 + \sqrt{\mu})^{-1} - 0.2\ \mu$$

then we expect the rate dependence to be of the form:

$$\log k_{obs} \approx \log k_o + z_a z_b f(\mu)$$

where z_i is the net charge on a reactant. In fact, although the slope of the plots of log k_{obs} *vs.* $f(\mu)$ was zero for misonidazole ($z_i = 0$), it was *negative* for pimonidazole ($z_i = +1$) and *positive* for Ro 31-0258 ($z_i = -1$). The charged compounds showed ionic strength dependencies which were opposite in sign to that expected if quenching occurred *via* triplets free in solution; further, the magnitude of the kinetic salt effect was several times that expected. The only reasonable explanation for this behaviour is that quenching occurs with $^3AO^*$ bound to a polynucleotide having multiple negative charges (although the theoretical treatment above is only applicable quantitatively to reactions between point charges).

These results undoubtedly reflect the ion condensation/exclusion phenomena which have also been seen with radioprotectors (e.g. Smoluk *et al.* 1986, 1988). Of major interest in radioprotection is the effective concentration of thiols near nuclear DNA - either of exogenous thiols or the endogenous levels, e.g. after treatment with buthionine sulphoximine (BSO). Measurements with isolated nuclei are subject to the uncertainty associated with possible changes resulting from the extraction process. Experiments by Mr. John White and Mrs. Madeleine Dennis using the AO technique described above yielded the unexpected result that the delayed fluorescence from V79 cells decreased rather than increased after depletion of the endogenous quencher, glutathione, using buthionine sulphoximine. The explanation seems to be that the uptake of the dye was decreased by the treatment, since the prompt fluorescence was reduced. A cytoplasmic dye also showed reduced uptake after similar BSO treatment; such effects have not been previously reported, and illustrate the pitfalls of assuming only glutathione levels are affected. Measurements of emission lifetimes (in progress) may shed some light on these problems.

Conclusion

It is a convenient, but surely not a complete, explanation to generalize damage to DNA in terms of products of hydroxyl radical reactions repairable by hydrogen donation from a thiol. It is conceivable, but it may be merely coincidental, that radiosensitization by both oxygen and nitroaryl compounds involves radical-adduct formation, i.e. peroxyl and nitroxyl radicals, as intermediates. Most will concur that reduction potential is the property dominating radiosensitization efficiency by some simple oxidants, but the constraints of electron transfer theory include kinetic parameters, not directly related to redox factors, which may ultimately control reactivity. Leaving aside quantum-mechanical tunnelling, contiguity or proximity of reactants is a necessary condition for chemical radiomodifiers, and concentration at the target site is a parameter essential to the reliable interpretation of most relevant experiments. Whilst some progress is being made in this latter area, there are still many basic questions which remain unanswered.

Acknowledgement

This work is supported by the Cancer Research Campaign. The author gratefully acknowledges the assistance of Mr. John White and Mrs. Madeleine Dennis in the work outlined above.

References

Adams GE, Jameson DG (1980) Time effects in molecular radiation biology. *Radiat Environ Biophys* **17:** 95-113

Adams GE, Clarke ED, Flockhart IR, Jacobs RS, Sehmi DS, Stratford IJ, Wardman P, Watts ME, Parrick J, Wallace RG, Smithen CE (1979) Structure-activity relationships in the development of hypoxic cell radiosensitizers: I. Sensitization efficiency. *Int J Radiat Biol* **35:** 133-150

Akhlaq MS, Schuchmann H-P, von Sonntag C (1987) The reverse of the 'repair' reaction of thiols: H-abstraction at carbon by thiyl radicals. *Int J Radiat Biol* *51:* 91-102

Asquith JC, Watts ME, Patel K, Smithen CE, Adams GE (1974) Electron affinic sensitization. V. Radiosensitization of hypoxic bacteria and mammalian cells *in vitro* by some nitroimidazoles and nitropyrazoles. *Radiat Res* **60**: 108-118

Dennis MF, Stratford MRL, Wardman P, Watts ME (1985) Cellular uptake of misonidazole and analogues with acidic or basic functions. *Int J Radiat Biol* **47:** 629-643

Dennis MF, Stratford MRL, Wardman P, Watfa RR (1989) Increase in intracellular cysteine after exposure to dithiothreitol: implications in radiobiology. *Int J Radiat Biol* **56**: 877-883

Finkelstein E, Glatstein E (1988) Seduced by oxygen. *Int J Radiat Oncol Biol Phys* **14:** 205-207; *ibid* (1989) **16:** 287-288

Hodgkiss RJ, Middleton RW (1983) Enhancement of misonidazole radiosensitization by an inhibitor of glutathione biosynthesis. *Int J Radiat Biol* **43**: 179-183

Hodgkiss RJ, Stratford MRL (1988) Competitive dose-modification between ascorbate and misonidazole in human and hamster cells: effects of glutathione depletion. *Int J Radiat Biol* **54:** 601-610

Hodgkiss RJ, Roberts IJ, Watts ME, Woodcock M (1987) Rapid-mixing studies of radiosensitivity with thiol-depleted mammalian cells. *Int J Radiat Biol* **52:** 735-744

Hodgkiss RJ, Vojnovic B, Woodcock M, Michael BD (1989) A single-shot rapid-mixing device for radiobiological studies with mammalian cells. *Int J Radiat Biol* **55:** 705-715

Howard-Flanders P, Moore D (1958) The time interval after pulsed irradiation within which injury to bacteria can be modified by dissolved oxygen. I. A search for an effect of oxygen 0.02 second after pulsed irradiation. *Radiat Res* **9:** 422-437

Jagannadham V, Steenken S (1988) One-electron reduction of nitrobenzenes by OH and H radical adducts to 6-methyluracil and 6-methylisocytosine via electron transfer and addition/elimination. Effect of substituents on rates and activation parameters for formation and heterolysis of nitroxyl-type tetrahedral intermediates. *J Phys Chem* **92:** 111-118

Koch CJ, Howell RL (1981) Combined radiation-protective and radiation-sensitizing agents. II. Radiosensitivity of hypoxic or aerobic Chinese hamster fibroblasts in the presence of cysteamine and misonidazole: implications for the 'Oxygen Effect' (with appendix on calculation of dose-modifying factors). *Radiat Res* **87:** 265-283

Koch CJ, Howell RL, Biaglow JE (1979) Ascorbate anion potentiates cytotoxicity of nitro-aromatic compounds under hypoxic and anoxic conditions. *Brit J Cancer* **39**: 321-329

Lemaire DGE, Bothe E, Schulte-Frohlinde D (1987) Hydroxyl radical-induced strand break formation of poly(U) in anoxic solution. Effect of dithiothreitol and tetranitromethane. *Int J Radiat Biol* **51:** 319-330

Meisel D (1975) Free energy correlation of rate constants for electron transfer between organic systems in aqueous solutions. *Chem Phys Lett* **34:** 263-266

Nishimoto S, Ide H, Wada T, Kagiya T (1983) Radiation-induced hydroxylation of thymine promoted by electron-affinic compounds. *Int J Radiat Biol* **44:** 485

O'Neill P (1983) Pulse radiolytic study of the interaction of thiols and ascorbate with OH adducts of dGMP and dG: implications for DNA repair processes. *Radiat Res* **96:** 198-210

Prütz WA (1989) 'Chemical repair' in irradiated DNA solutions containing thiols and/or disulphides. Further evidence for disulphide radical anions acting as electron donors. *Int J Radiat Biol* **56:** 21-33

Raleigh JH, Whitehouse R, Kremers W (1974) Effect of oxygen and nitroaromatic radiosensitizers on radiation-induced phosphate release from 3'- and 5'-nucleotides: a model for nucleic acids. *Radiat Res* **59:** 453

Remsen JF (1985) Effect of misonidazole on formation of thymine damage by gamma rays. *Radiat Res* **101:** 306-311

Schöneich C, Bonifačić M, Asmus K-D (1989) Reversible H-atom abstraction from alcohols by thiyl radicals: determination of absolute rate constants by pulse radiolysis. *Free Rad Res Comms* **6:** 393-405

Schulte-Frohlinde D, Behrens G, Önal A (1986) Lifetime of peroxyl radicals of poly(U), poly(A) and single- and double-stranded DNA and the rate of their reaction with thiols. *Int J Radiat Biol* **50:** 103-110

Smoluk GD, Fahey RC, Ward JF (1986) Equilibrium dialysis studies of the binding of radioprotector compounds to DNA. *Radiat Res* **107:** 194-204

Smoluk GD, Fahey RC, Ward JF (1988) Interaction of glutathione and other low-molecular-weight thiols with DNA: evidence for counterion condensation and coion depletion near DNA. *Radiat Res* **114:** 3-10

Steenken S (1989) One-electron redox reactions between radicals and molecules. Dominance of inner-sphere mechanisms. In: Minisci F (ed). *Free Radicals in Synthesis and Biology*. NATO ASI series, C260, Kluwer Academic

Publishers, Dordrecht, pp 213-231

Steenken S, Jagannadham V (1985) Reaction of 6-yl radicals of uracil, thymine, and cytosine and their nucleosides with nitrobenzenes via addition to give nitroxide radicals. OH^--catalyzed nitroxide heterolysis. *J Am Chem Soc* **107**: 6818-6826

Stratford MRL, Hodgkiss RJ (1986) Radioprotection by ascorbate *in vitro:* temperature dependent uptake. *Brit J Radiol* **59**: 83-84

Stratford MRL, Dennis MF, Watts ME, Watfa RR (1988) Radiosensitizer-DNA interactions. *Brit J Radiol* **61:** 860-861

Stratford MRL, Dennis MF, Watts ME, Watfa RR, Woodcock M (1989) Radiosensitizer-DNA interactions in relation to intracellular uptake. *Int J Radiat Oncol Biol Phys* **16:** 1007-1010

Symons MCR (1987) Application of electron spin resonance spectroscopy to the study of the effects of ionising radiation on DNA and DNA complexes. *J Chem Soc Faraday Trans I* **83**: 1-11

Varghese AJ (1975) Sensitization of thymine and uracil to ionizing radiation by *p*-nitroacetophenone. *Int J Radiat Biol* **28:** 477-484

Von Sonntag C (1987) The Chemical Basis of Radiation Biology. Taylor & Francis, London.

Ward JF (1981) Some biochemical consequences of the spatial distribution of ionizing radiation-produced free radicals. *Radiat Res* **86**: 185-195

Ward JF (1988) DNA damage produced by ionizing radiation in mammalian cells: identities, mechanisms of formation, and reparability. *Prog Nucleic Acid Res Mol Biol* **35**: 95-125

Wardman P (1984) Radiation chemistry in the clinic: hypoxic cell radiosensitizers for radiotherapy. *Radiat Phys Chem* **24:** 293-305

Wardman P (1987) The mechanism of radiosensitization by electron-affinic compounds. *Radiat Phys Chem* **30:** 423-432

Wardman P (1989a) Reduction potentials of one-electron couples involving free radicals in aqueous solution. *J Phys Chem Ref Data* **18:** 1637-1755

Wardman P (1989b) Molecular mechanisms of oxygen and 'electron-affinic' radiosensitizers. *Int J Radiat Oncol Biol Phys* **16**: 286-287

Wardman P, Clarke ED (1985) Electron transfer and radical-addition in the radiosensitization and chemotherapy of hypoxic cells. In: Breccia A and Fowler JF (eds). *New chemo and radiosensitizing drugs,* Lo Scarabeo, Bologna, pp 21-38

Wardman P, Dennis MF, White J (1989) A probe for intracellular concentration of drugs: delayed fluorescence from acridine orange. *Int J Radiat Oncol Biol Phys* **16:** 935-938

Transformation Reactions of Two Isomeric OH-Adducts of 2'-Deoxyguanosine

L. P. Candeias and S. Steenken
Max-Planck-Institut für Strahlenchemie, D-4330 Mülheim, Germany.

Introduction

It has been proposed (O'Neill 1983) that the OH radical reacts with 2'-deoxyguanosine in aqueous solution to give adducts. This suggestion is in line with the general property of $OH^\bullet$ to react by inner-sphere mechanisms (Steenken 1987). In the case of purines, molecular orbital calculations (Pullman 1963) indicate the C4-C5 bond and the C8 position as the most probable sites for the $OH^\bullet$ attack.

In agreement with this hypothesis and using the redox titration technique (Steenken 1989), it was shown (O'Neill 1983) that from the reaction of $OH^\bullet$ with 2'-deoxyguanosine at least two different radicals are formed and that the yields of oxidizing and reducing radicals are approximately equal. The same author (O'Neill 1985) has shown that the oxidizing radical is the same as that produced by the reaction of deoxyguanosine with one-electron oxidants. From its acid base properties and the use of 1-methylguanosine as model compound, it was later demonstrated (Candeias 1989) that the radical in question is the deprotonated radical cation (dGuo(-H)$^\bullet$), proposed to be formed by the addition of $OH^\bullet$ to C4 followed by water elimination, in a process identical to that found for the reaction of $OH^\bullet$ with N^6,N^6-dimethyladenosine (Vieira 1987a) and other 6-substituted purines (Vieira 1987b).

The addition of $OH^\bullet$ to the C8 position of 2'-deoxyguanosine can be inferred from product analysis work (Cadet 1985) in which 8-hydroxy-2'-deoxyguanosine, the product formed by <u>oxidation</u> of the C8 OH-adduct, and formamido-amino-pyrimidine compounds (FAPy) formed by opening of the imidazole ring of the C8-OH-adduct followed by <u>reduction</u>, were detected.

However, none of the two postulated reactions has so far been unambiguously demonstrated by time-resolved methods, the general belief being that they are either very slow or very fast. In this work we report the direct observation, by pulse radiolysis with optical detection, of the two transformation reactions of the OH-adducts of 2'-deoxyguanosine.

Results and Discussion

The reaction of $OH^\bullet$ with 2'-deoxyguanosine in N_2O saturated neutral aqueous solution yields transients with an uncharacteristic absorption spectrum, recorded immediately after completion of the reaction, with a maximum at 300 nm and extending until ≈ 750 nm. In longer times this spectrum transforms by two kinetically distinct first order processes: a fast ($k = 2 \times 10^5\ s^{-1}$) increase of the optical density in the range ≈ 290 - 500 nm and a slow ($k = 5 \times 10^3\ s^{-1}$) decay of the optical density at ≈ 550 - 680 nm. The rate of the latter can be

NATO ASI Series, Vol. H 54
The Early Effects of Radiation on DNA
Edited by E. M. Fielden and P. O'Neill

strongly increased by the addition of phosphate,[1] the rate constant for catalysis by phosphate being $6 \times 10^7\ M^{-1}s^{-1}$ at pH 7. This reaction is also H^+ and OH^- catalysed. The rate of the fast build-up of OD at 310 nm is not affected by the phosphate, but its rate increases with $[OH^-]$ for pH > 9.

The two reactions differ also with respect to the effect of oxygen. The fast reaction cannot be observed in the presence of 0.1 mM O_2, whereas the slow reaction is not affected even by 1 mM O_2. The transformation reactions are therefore concluded to involve radicals of different redox properties. The fast reaction is attributed to the ring opening reaction of the imidazole ring of the reducing OH-adduct to the C8 position (dGuo-C8-OH•) and the slow reaction to the water elimination from the oxidizing C4 adduct (dGuo-C4-OH•, see scheme).

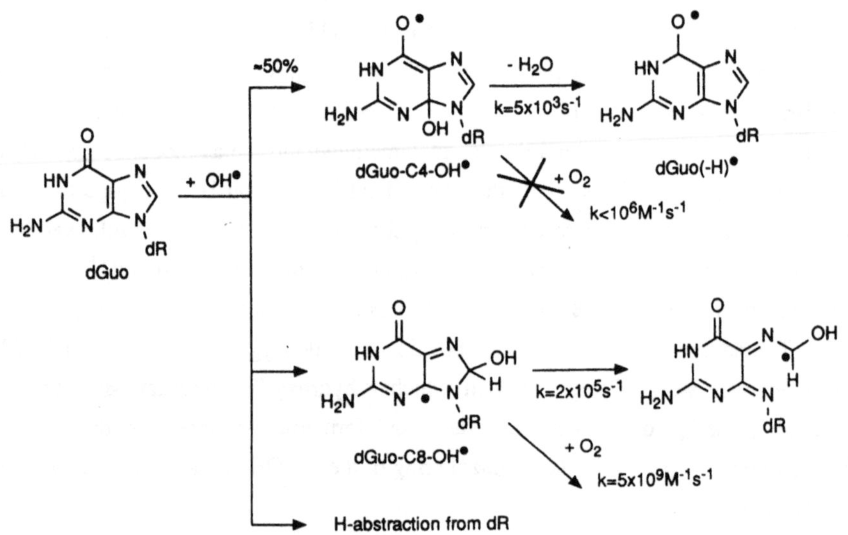

From the effect of the oxygen concentration on the plateau optical density at 300 nm it was determined that the half-scavenging point is at $[O_2] = 4 \times 10^{-5}$ M; from this value, the rate constant for the reaction of dGuo-C8-OH• with O_2 can be estimated as $5 \times 10^9\ M^{-1}s^{-1}$.

References

Cadet, J.; Berger, M. (1985) *Int. J. Radiat. Biol.* 47, 127
Candeias, L. P.; Steenken, S. (1989) *J. Am. Chem. Soc.* 111, 1094
O'Neill, P. (1983) *Int. J. Radiat. Biol.* 96, 128
O'Neill, P.; Chapman, P. W. (1985) *Int. J. Radiat. Biol.* 47, 71
Pullman, A.; Pullman, B. (1963) *Quantum Biochemistry* Interscience Publishers, New York
Steenken, S. (1987) *Faraday Trans. I* 83,113
Steenken, S. (1989) *Chem. Rev.* 89, 503
Vieira, A. J. S. C.; Steenken, S. (1987a) *J. Am. Chem. Soc.* 109, 7441
Vieira, A. J. S. C.; Steenken, S. (1987b) *J. Phys. Chem.* 91, 4138
Wardman, P.; Clarke, E. D. (1985) in New Chemo and Radiosensitizing Drugs Breccia, A.; Fowler, J. F. (eds) Ed. Scientifiche Lo Scarabeo

[1] The phosphate catalysis of the slow reaction has been previously reported (Wardman 1985).

THE RATES OF THE REACTION OF THIOLS WITH RADICALS DERIVED FROM OH ATTACK ON POLYMERIC NUCLEIC ACIDS AND ON DNA.

E. Bothe and D. Schulte–Frohlinde
Max–Planck–Institut für Strahlenchemie,
Stiftstraße 34–36, D 4330 Mülheim a.d. Ruhr

Aqueous, N_2O–saturated solutions of single–stranded calf–thymus DNA (ssDNA) containing different concentrations of dithiothreitol (DTT) were irradiated with ^{60}Co–γ–rays. At each concentration the G value of single–strand breakage (G_{ssb}) has been determined using the method of low–angle–laser–light–scattering (LALLS). The reciprocal of the G_{ssb} values thus obtained, after allowance for OH radical scavenging by DTT, should increase linearly with the DTT concentration for the case that DTT prevents ssb by reacting with one single type of DNA strand break precursor radical.

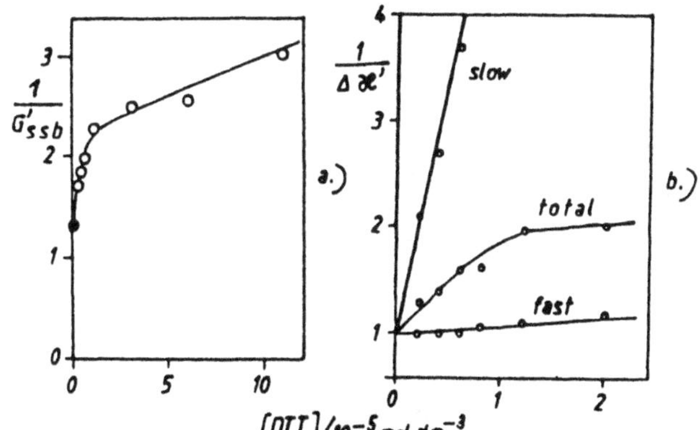

Figure 1. Inhibition of ssb formation in ssDNA with increasing DTT concentration observed a) by light scattering after ^{60}Co–γ–irradiation, b) by conductivity (Δκ) measurements in pulse radiolysis. The primed values are the measured values after allowance for OH radical scavenging by DTT.

The results presented in figure 1a demonstrate however that chemical repair by DTT with ssb formation in ssDNA is a two–component process with respect to DTT concentration. In experiments with DNA solutions containing, instead of DTT, ascorbate or trans 4,5–dihydroxy–1,2–dithiane, respectively, we found no protection by the additive except OH radical scavenging. This and other evidence shows that neither the DTT anion (DTT^-) nor the radical anion ($DTT^{-\cdot}$) is involved in the inhibition of strand breaks. Conductivity (Δκ) experiments in pulse radiolysis show that ssb formation in ssDNA in the absence of O_2 proceeds in two modes with different rates and with respective contributions of about 50%. The slow mode (overall half life about 1 s) reflects the decay of DNA base radicals and is quenched by small amounts of DTT ([DTT] $\leq 10^{-5}$ mol dm^{-3}, figure 1b). From the evaluation of the data obtained

NATO ASI Series, Vol. H 54
The Early Effects of Radiation on DNA
Edited by E. M. Fielden and P. O'Neill

by light scattering and by conductivity bimolecular repair rate constants k_r are obtained for the inhibition of ssb formation in ssDNA which are presented in table 1.

Table 1 Repair rate constants k_r for the inhibition of ssb in ssDNA by DTT.

Component	k_{ssb} / s^{-1}	k_r / $10^6\ dm^3\ mol^{-1}\ s^{-1}$	method
slow	0.7	0.55 ± 0.2	LALLS, G'_{ssb} decrease
		0.40 ± 0.15	cond., $\Delta\kappa'$ decrease
fast	100	0.55 ± 0.2	LALLS, G'_{ssb} decrease
		0.90 ± 0.2	cond., $\Delta\kappa'$ decrease
		0.6 ± 0.15	cond., k_{obs} increase

The influence of thiols on the decay rate of some polymeric nucleic acid radicals has been followed in pulse radiolysis by optical absorption spectroscopy at 450 nm. In the case of poly(A) (pH 7 and 9) with cysteine or 2-mercaptoethanol ($\leq 2 \times 10^{-4}$ mol dm^{-3}) the influence of the thiols on the decay rate of 80% of the transient absorption was small, corresponding to $k_r \leq 1 \times 10^5\ dm^3\ mol^{-1}\ s^{-1}$. It appears therefore that the N-centered radicals of adenine are not involved in ssb formation in ssDNA since the observed repair rate constants are larger (table 1). Likewise an involvement of the oxyl radical of guanine is ruled out since no influence of ascorbate was found (see above) which would have been expected to protect on the basis of the reactivity of guanosine radicals against RS^- (O'Neill, 1983). Using the same method, radicals derived from poly(C) and poly(U) exhibit higher rate constants with mercaptoethanol (k_r = 1–2 x $10^6\ dm^3\ mol^{-1}\ s^{-1}$ and 1.3 x $10^6\ dm^3\ mol^{-1}\ s^{-1}$ respectively). It appears therefore that possibly a significant extent of the slow mode of strand break formation arises from cytosine radicals. This is supported by the fact that poly(dC) shows a relatively high yield of ssb formation, one half of which is generated in slow components (Adinarayana et al, 1988). The fast component is also repaired relatively fast (k_r = 6–9 x $10^5\ dm^3\ mol^{-1}\ s^{-1}$, see table 1). This points to a reducing radical as precursor, which, for similar reasons as discussed for the slow mode, may be the cytosine OH adduct or the C(4') sugar radical.

Adinarayana M, Bothe E, Schulte-Frohlinde D (1988) Hydroxyl radical-induced strand break formation in single-stranded DNA in aqueous solution as measured by light scattering and by conductivity. Int J Radiat Biol 54:723–737

O'Neill P (1983) Pulse radiolytic study of the interaction of thiols and ascorbate with OH adducts of dGMP and dG. Implication of DNA repair process. Rad Res 96:198–210.

Radical Chemistry in Crystals, Matrices and in Aqueous Solution. Redox Properties and Proton Transfer Processes.

Steen Steenken
Max-Planck-Institut für Strahlenchemie
D-4330 Mülheim, Germany

The contributions in this session can be grouped into two categories, both by the methods employed and by their contents:

1. In the papers by a) Bernhard and b) Sagstuen et al., *electron spin resonance* (ESR) is used to identify the nature of the radicals formed initially on interaction of ionizing radiation with DNA bases, nucleosides, nucleotides, and polynucleotides at *low temperature in single crystals and matrices*. The emphasis is on the *structure* of the primarily formed species.

2. In the contributions of a) Fielden et al., b) Wardman, c) Candeias et al., and d) Bothe et al., the method of *pulse radiolysis* is employed to characterize radicals formed on irradiation of *aqueous solutions at room temperature,* and the emphasis is on *reactivity* of radicals which are mainly formed by addition of $OH^{\bullet}$ to the base moiety of nucleosides, nucleotides, and polynucleotides.

However, this distinction between structure-oriented and reaction-oriented radical chemistry is, of course, an oversimplification. E.g., it turns out that even at ca. 4 K protonation and *de*protonation reactions can be very rapid such that the situation resulting from an ionization cannot be understood without taking account of these proton transfer processes.[1] *De*protonation reactions are the results of the *increase in acidity* of molecules that occurs on *removal* of an electron. Such an increase can be very dramatic.[2] The driving force for *protonation* reactions is the *increase in basicity* on electron *addition* to a molecule. This effect can also be considerable.[2] It is obvious that for a proton to transfer from one site (a "transferable" proton is of course typically located at a hetero atom) to another site, the relative orientation of proton donor and proton acceptor should be "favorable". This condition is often fulfilled in crystals, it is *always* fulfilled in aqueous matrices (frozen or liquid), since water can serve as a proton donor *and* acceptor, and for many conceivable proton transfer reactions it is also fulfilled in DNA. With DNA, proton transfer can occur *between* the bases of a base pair ("intra-pair proton transfer")[2] and between a base and a water molecule in the DNA hydration shell ("extra-pair proton transfer)[2] . With the intra-pair case, the direction of proton transfer is orthogonal to the DNA axis (in which *electrons* may be transferred).

It is evident that for a molecule to undergo a proton transfer it is necessary that there be a partner in an appropiate distance and orientation. It is for this reason that proton transfer reactions

NATO ASI Series, Vol. H 54
The Early Effects of Radiation on DNA
Edited by E. M. Fielden and P. O'Neill

are very sensitive to "environment", and the cases presented and discussed by Sagstuen et al. and by Bernhard are beautiful examples.

The proton transfers observed in the crystalline state and in frozen glasses take place also in aqueous solution, where the reactive species are e^-_{aq} and $OH^•$. A method that has been very successful in establishing the nature of the primary radicals and their transformation reactions is time-resolved *redox titration*. With this technique, it has been established[3] that OH-adducts to the C4-C5 bond of the purines undergo a dehydration reaction by which are formed the same (oxidizing) radicals as those produced by ionization of the molecules. In these cases, from the point of view of the nature of the product radicals, a distinction between the direct and the indirect effect becomes impossible. These radicals, which are neutral entities at pH ≈ 7 and identical with those found in the crystalline state, can be repaired by electron donation from reductants such as ascorbate, N,N,N',N'-tetramethyl-para-phenylenediamine, or thiolates.[3] Oxidizing radicals are, however, not the only products of the interaction of $OH^•$ with purines,[2,3] and in the case of the pyrimidines the reducing radicals are even the predominant products. These reducing radicals react rapidly with oxygen and with organic oxidants such as quinones and nitroaromatics (=radiosensitizers), leading ultimately to the one-electron reduction of the oxidants and the one-electron oxidation of the radicals. (In many cases the *mechanism* of these redox processes is an inner-sphere one, i.e. involves addition/elimination.) The species resulting from the oxidation of the radicals have an oxidation state higher by *two* units as compared to their parent compounds. The reaction rates (and mechanisms) of reducing radicals with sensitizers depend critically on the reducing strength of the radicals and on the oxidizing power of the oxidants, i.e. on the driving power of the redox reaction. However, under biological conditions, other chemical properties of drugs have to be additionally taken into account such as solubility, lipophilicity and acid-base properties, in order to be of predictive value for therapeutic applications.[4]

To summarize, the *redox properties* of radicals are of very great importance in determining their *repairability* by protecting agents as well as the opposite, i.e. the possibility of *fixation of damage*. The redox (=electron transfer) properties of radicals are strongly influenced by their ability to *de*protonate or to *be* protonated, i.e. by proton transfer processes many of which depend critically on environment.

[1] W.A. Bernhard, *Adv. Radiat. Biol.* **1981**, *9*, 199.

[2] S. Steenken, *Chem. Rev.* **1989**, *89*, 503; *Free Rad. Res. Comm.* **1991**, in press.

[3] P. O'Neill, *Radiat. Res.* **1983**, *96*, 198.

[4] P. Wardman, In: G.E. Adams, A. Breccia, E.M.Fielden and P. Wardman (eds.) *Selective Activation of Drugs by Redox Processes*, Plenum, New York, **1991**, p. 11-24.

MODELS TO LINK DNA DAMAGE TO RBEs FOR FINAL CELLULAR EFFECTS

Dudley T Goodhead
Medical Research Council Radiobiology Unit
Chilton
Didcot OX11 0RD
United Kingdom

A major aim of research on the biological effects of radiation is to understand and be able to model the chain of processes from the initial radiation insult to the final observable effects in mammalian cells. It is widely believed that for many cellular effects of interest damage to DNA is a critical element in this chain. We may wonder, then, how this acts as an intermediate step in determining the relative biological effectiveness (RBE) of different ionising radiations.

1. SEQUENCE AND TIMESCALE OF PROCESSES AFTER IONISING RADIATION

A very large number of processes are involved between the initial passage of a radiation track in ~ 10^{-15} s to the observation of cellular phenotypic effects after some days (~ 10^{6} s) or longer. Some of these processes have been extensively studied and are quite well understood while others remain largely untouched. The degree of effort and understanding is not always in proportion to the likely importance of the process.

In a general way, the sequence of processes can be classified as described below, although there is considerable overlap between the classes, both in timescale and nature. Therefore the classifications are somewhat arbitrary and characteristic times are illustrative only.

The <u>initial physical damage</u> (~ 10^{-15} s) occurs with the passage of a charged particle from the radiation, causing predominantly ionisations and excitations in or near the cellular target material. Monte Carlo track structure computational methods allow this stage to be described in considerable detail for a water medium (see, for example,

NATO ASI Series, Vol. H 54
The Early Effects of Radiation on DNA
Edited by E. M. Fielden and P. O'Neill

Paretzke 1987). But only in the (near) future might it become possible to extend this to other materials of interest such as DNA and protein. It is reasonable to expect that many of the general features of track structure will be similar in these different media, but perhaps not those features that depend on the highest spatial and energy resolution. The relevance and nature of possible later physical processes, such as energy migration, are poorly understood.

Then follows the formation of the many <u>initial chemical species and their chemical reactions</u> with one another and other nearby molecules (~ 10^{-10} s). These processes have been widely studied, and are well understood, for water and for dilute aqueous solutions, especially for the more widely dispersed components of low-LET radiation tracks (see, for example, Magee & Chatterjee 1987). But less is known about reactions in the concentrated and structured molecular environment within cells. Even less effort seems to have been concentrated on the effects of direct interactions (ionisation and excitation) of the radiation with the atoms of the DNA target molecule itself or its closely associated protein or bound water. This is despite the large numbers of such interactions. Given the highly structured and clustered nature of initial radiation damage, a most notable lack may be in the field of non-linear chemistry within, and in the immediate environment of, the DNA (see §5 below).

Next comes the establishment of the <u>initial biochemical damage</u> to critical molecules and structures such as DNA (~ 10^{-4} s). A good deal is known about the simpler classes of these, such as DNA single-strand breaks (ssb) and double-strand breaks (dsb), for which direct experimental assays exist. Much less is known about more complicated forms of damage, including such things as combinations of complex breaks and cross-linking within DNA or with protein. It seems likely that many complex forms of damage have yet to be described.

<u>Biochemical modification</u> (~10^{0} s) proceeds via cellular processes such as repair, misrepair or fixation of the damage. Substantial information is available on the repair kinetics of the well-studied classes of DNA damage (notably ssb and dsb),

but there is very little information on what are the limitations to repair in normal cells, what determines the unrepaired or irreparable fractions of damage and what other types of damage may be less reparable. Yet it should be the unrepaired (and misrepaired) residues that are responsible for final cellular effects. Later consequences of unsuccessful biochemical modification can be observed in the form of chromosome aberrations, and more recently the technique of prematurely condensed chromosomes has allowed observations to be extended back in time to a few minutes after irradiation (Cornforth & Bedford 1983).

Finally there follows <u>cellular modification and expression</u> of damage, as for example loss of clonogenic ability or expression of mutation or transformation. There is very considerable data on, and understanding of, these effects in many different mammalian systems.

From this short summary it may be concluded that there is considerable knowledge of some of the relevant processes in the long chain from initial insult to final effect, but that many of the key elements are lacking.

2. MODELS TO LINK DNA DAMAGE TO RBE

Given the severe limitations in our current knowledge, how can we attempt to link the full chain of processes from initial physical radiation insult, to final cellular effects? A general approach may be first to seek unifying and simplifying features within the existing extensive body of experimental data. These can then be used to formulate models to clarify mechanistic options, reveal significant gaps in the data, make predictions and guide experimental design. These models may describe individual components of the chain or they may attempt to link major parts of the chain. Simultaneously the experimental database can be extended to develop and constrain the theoretical options, eliminate some models, guide the development of others and guide further experiments. In this circular way, substantial progress should be possible towards a comprehensive understanding.

In order to make the link between DNA damage and RBE for cellular effects, it is essential also to make the link between initial physical damage and DNA damage. This is because the radiation track structure defines the initial conditions of the entire system, that is the "quality" of the radiation. This initial physical state determines the RBE, within the boundary conditions of all the relevant intermediate processes.

3. FEATURES OF RADIATION TRACKS IN RELATION TO DNA DAMAGE

Ionising radiations differ fundamentally from other DNA-damaging agents, such as ultraviolet light or chemicals, because the initial atomic and molecular damage is inevitably in the spatial form of highly structured "tracks" of ionisations and excitations along the paths of the moving primary and secondary charged particles. Features of the track structures of different radiations in water have been described in considerable detail, as summarized for example in Paretzke (1980,1987,1990) and Goodhead (1987a,1989a,1990).

Some notable features in the present context include:

(a) Within the tracks the damage is by no means continuous at the atomic, molecular or DNA levels, because of the probabilistic nature of the individual ionisation and excitation processes. For all ionising radiations the spatial distribution is grossly inhomogeneous at the level of the DNA duplex, its higher structures and, except at high doses, at the level of the entire cell itself. Even for low-LET radiations, the distribution never approximates to a homogeneous distribution of single ionisations or of "spurs" (Paretzke 1987; Goodhead 1987a); individual localized clusters of ionisations and excitations of widely varying size occur with high frequency. High-LET radiations, too, can produce this full variation, as well as much larger clusters (Goodhead & Nikjoo 1989).

(b) For a given absorbed dose of radiation (i.e. overall average amount of energy deposited per unit mass), the total number of initial ionisations and excitations is approximately independent of LET. Because of the much larger number along a high-LET track, there are correspondingly fewer high-LET tracks for this given dose. Consequently, fewer target molecules will be hit by the tracks.

(c) If we consider targets of dimensions similar to a short, say 2 nm, segment of DNA (Goodhead & Nikjoo 1989; Nikjoo et al. 1989), it is seen that there is a full gradation of cluster sizes varying from single ionisations (or excitations) to tens of closely adjacent ionisations and excitations for low-LET radiations and many tens for high-LET radiations.

(d) For all radiations, very large numbers of the smallest clusters, including singlets, occur in the DNA volume. The yields after a given dose of radiation are quite similar to the experimentally measured yields of ssb (compare Tables 1 and 2). High resolution modelling of the energy depositions within the internal structure of the DNA, has suggested that the calculated absolute yields and the radiation-quality dependence of ssb match quite closely to experimental observations (Charlton et al. 1989).

(e) Despite the large numbers of these small events, the probability of two, from separate tracks, overlapping within the segment of DNA, or even within a considerable volume around it, is negligibly small except at excessively large doses ($\gtrsim 10^4$ Gy). Therefore, for all practical purposes, only clusters which occur <u>within</u> individual tracks need be considered. Similar conclusions apply to larger volumes of potential interest such as nucleosomes and chromatin fibre, with or without associated surrounding material.

(f) A moderate number of substantially larger clusters occur in the DNA for all radiations. For example, for clusters depositing >100 eV of energy in the form of say $\gtrsim 9$ ionisations and excitations in the DNA segment, the yield is ~20-100 cell^{-1} Gy^{-1} for low- and high-LET radiations. The yields are similar to those measured for dsb (cf Tables 1 and 2). Charlton et al. (1989) have shown that production of dsb, too, can be modelled with some success. These clusters should also be capable of causing DNA damage of complexity considerably greater than simple dsb. Such clusters occur predominantly near terminal ends of tracks of secondary electrons.

(g) Considerably larger clusters can be produced at meaningful frequencies by high-LET radiations. For example, clusters depositing $\gtrsim$250 eV ($\gtrsim$20 ionisations and excitations) in the DNA segment will occur at a rate of ~8 cell^{-1} Gy^{-1} for irradiation with slow α-particles and some will contain more than 30 ionisations and excitations. It is reasonable to expect that the biochemical damage which they cause may often be considerably more severe and complex than simple dsb. Of particular interest in respect of the unmodifiable, and presumably irreparable, cellular damage caused by high-LET radiations may be clusters of $\gtrsim$30 ionisations and excitations in target volumes similar to nucleosomes (Goodhead 1989a). It has been suggested previously that such a property may be the radiobiological characteristic of high-LET radiations, based on observed correlations with RBE for α-particles of different energies (Goodhead *et al*. 1985).

(h) In nucleosome-size, and larger, target volumes there can occur with high-LET radiations even much larger clusters than the above. These extend, for example, to clusters of $\gtrsim$ 120 ionisations and excitations ($\gtrsim$1300 eV) in the nucleosome target at a rate of ~1 cell^{-1} Gy^{-1}. These must be able to cause considerable complexities of damage to the DNA and surrounding proteins, but *perhaps* no greater yield of observable cellular effects if the damage described in (g) above is already irreparable.

(i) In general accordance with the above, it has been proposed (Goodhead 1989a) that the sparse ionisations and very small clusters described in (d) above, may be largely responsible for ssb and of little biological relevance in normal mammalian cells. Furthermore, the moderately-sized clusters (f) in DNA may be largely responsible for dsb and a minority of them, at least, be the cause of typical cellular damage from low-LET radiations. In this scheme, the unmodifiable and irreparable damage of (g), of complexity considerably greater than simple dsb, may be the main cause of cellular effects of high-LET radiation.

These concepts are very similar to those previously applied by Goodhead et al. (1980) in biophysical analysis of RBE-LET dependence of cell inactivation and mutation in mammalian cells irradiated with photons and a variety of heavy ions. They derive in turn from the analyses of Howard-Flanders (1958) on heavy-ion inactivation of microbes. These earlier analyses did not have the luxury or precision of Monte-Carlo track structure simulations and had to work with much cruder descriptions. It is perhaps interesting that the numbers describing radiobiologically relevant clustered damage have not been much altered by the greater precision of the current analyses.

Study of radiation track structure in relation to biological effectiveness strongly suggests that serious attention must be given to the variety and complexity of highly localized damage to DNA and related structures and that simple categorisation and measurement of dsb as a single group may not be adequate.

4. SOME UNIFYING AND SIMPLIFYING FEATURES OF RADIOBIOLOGICAL DATA

It is appropriate to seek unifying and simplifying features within the existing extensive body of experimental radiobiological data, so that these can lead us to realistic formulation of models.

- Damage to DNA is apparently implicated in most of the cellular effects of interest including mutation, chromosome aberrations, killing of many cell types and probably cell transformation.
- The vast majority of the damage produced by radiation in a cell is apparently unimportant to its final fate. This appears to be the case even for damage to the DNA itself (Table 2).
- Concentrating the energy deposition, as electron track-end clusters with ultrasoft X-rays or as high-LET particles, increases the biological effectiveness per unit dose despite the fact that fewer molecules will be hit and that single-radical yields should be reduced. This effect seems to apply to dimensions right down to 10 nm or less.

- The early damage predominantly responsible for the biological effects of high-LET radiations is apparently qualitatively, as well as quantitatively, different from that of low-LET damage. This is revealed by differences in the dose-dependence, time-dependence, reparability and modifiability of low-LET as compared to high-LET responses. In particular, cellular responses to concentrated high-LET radiations show no, or greatly reduced, abilities to be modified by factors which are known to modify DNA damage (oxygen, sensitizers, protectors), its cellular repair (dose-rate, dose-fractionation, delayed plating, inhibitors of enzyme repair, genetic repair defects) and final cellular effect after low-LET radiations.
- Diffusion of radiation-induced reactive free radicals in cells takes place only over very short distances, comparable to the DNA diameter, because of the highly scavenging intra-cellular environment. Therefore it is only the ionisations and excitations produced directly in the DNA, or in closely adjacent molecules, which are likely to damage the DNA. Such damage should be strongly dominated by effects of single radiation tracks because overlap of tracks within such distances is negligible at all doses of biological relevance for low- and high-LET radiations (Goodhead & Nikjoo 1989).

5. DNA DAMAGE IN MODELS OF RADIATION ACTION

It can be seen from Table 2 that the vast majority of the initial radiation damage to a cell is unimportant to its ultimate fate. Most of the damage is irrelevant, restored or repaired. But as the ionisation density (LET) is increased more complex damage is induced and this is less repairable. This appears to be the general basis of RBE. In order to understand and model this we need to know the nature of those minority components of DNA damage which are least likely to be repaired. Of prime relevance may be forms of damage more complex than simple dsb. But there is very little information on the chemistry or biochemistry of such complex damage. What follows from the highly localised clusters of initial physical damage which occur within radiation tracks, when say $\gtrsim 10$ ionisations and excitations are produced within a few adjacent base-pairs in DNA and perhaps with many tens of additional ionisations and excitations in the contiguous proteins and water?

In the absence of such information, models have tended to proceed by assuming rather simple damage whose molecular nature is usually unspecified. Present information from track structure simulations and experimental yields of DNA damage allow some constraints to be put on the interpretation of such models. This will be illustrated with a few models selected from those summarised schematically by Goodhead (1987b).

Consider, first, the theory of dual radiation action in its later "generalised" distance-model form (Kellerer & Rossi, 1978). In essence it assumes that irreparable lethal lesions arise from the interaction of two reparable "sublesions" and that when two sublesions occur close together they are much more likely to interact. The nature of the sublesions is unspecified, except that they are assumed to be produced from single interaction (ionisation or excitation) points with a probability proportional to the ionisation or excitation energy. Close proximity of sublesions is therefore much more likely to occur along the densely ionising tracks of high-LET particles which therefore have large RBE.

If we attempt to associate these sublesions and lesions with DNA damage, then the sublesions could probably be of complexity no greater than ssb because they need to be produced exclusively by single ionisations or excitations alone and they are allowed to have no inherent dependence on radiation quality or track structure. Then it might be most tempting to identify the full lethal lesions with dsb. One detailed attempt has been made to quantify the distances over which these sublesions interact (Brenner & Zaider 1984), by fitting the model to experimental data. This suggested a very high interaction probability for sublesions at distances apart of 0.1-0.3 nm (which is largely within a single atomic radius and therefore difficult to reconcile with any type of molecular damage), falling off to a few nanometres (consistent with dsb) and extending at much lower probability to many micrometers (which may be difficult to reconcile with current ideas on dsb formation, or may instead be considered as a totally different process of interaction between ssb's on different chromosomes). The model also allows for no qualitative differences between lesions so they would need to be simple, clean dsb for both high- and low-LET radiations with no differences in repair or modifiability.

A related model by Harder (1987) assumes that the ability of radiations to produce sublesions is characterised by the energy-restricted LET_{100eV} and that these sublesions can interact in pairs to produce final lesions. A distinction is made between fast, short-range (~nanometres) intratrack interactions and slow, distant-range intertrack interactions. Harder has suggested that the latter may be misrepair between two dsb.

The lethal-potentially-lethal (LPL) model of Curtis (1986) assumes that radiation produces two classes of initial damage. Potentially-lethal lesions may be repaired, or misrepaired by combining. If they are not repaired or if they are misrepaired, they are lethal. Lesions of the second class are lethal (irreparable) at the time of formation and it is these lesions which are required to be formed with greater efficiency by high-LET radiations so that the RBE is large. The model itself does not specify radiation, or track, parameters for the damage. Nor has it been applied quantitatively to RBE data. So the identity of the molecular damage remains open to choice with few constraints. Curtis has suggested that the potentially-lethal lesions may be due to individual localized clusters of ionisations and that the lethal lesions may be due to two of these occurring sufficiently closely along a track that they are effectively irreparable. It would appear that the potentially-lethal lesions could be simple dsb, or of somewhat greater complexity. But the requirement that low-LET radiations have a reasonable probability of producing two in close proximity along a track, suggests that the energy deposition within each cluster cannot be much more than about 100 eV (cf Table 1). If these lethal lesions from low- and high-LET radiations have similar identities to one another they should be similarly modifiable, or not, by experimental conditions. That is, the one-track component of the dose-response should behave similarly for all radiations.

The only quantitative RBE model which explicitly assumes specific DNA damage is the "molecular theory" of Chadwick & Leenhouts (1981). A single DNA dsb is assumed to be the critical lesion which can lead to a variety of cellular effects including a mutation, a chromosome aberration or cell death. A dsb is assumed to be formed by two adjacent ssb following attack by two radiation-induced OH radicals. The greater RBE of high-LET radiations is due to their greater chance of ionising two water molecules closely adjacent to the DNA by the passage of a single primary charged particle. A crucial feature of the model is that it requires low-LET radiations to be able to produce dsb by two-track as well as one-track action, because the model is built on the assumption that the curvature of observed dose-responses for cellular effects is due to similar curvature in the initial yield of dsb. For the two-track mechanism to be competitive despite the very low probability of tracks overlapping in small volumes, it is necessary for the model to allow an OH radical from the subsequent second track to diffuse a considerably greater distance to the DNA than is allowed for a single track. A critical test of the validity of this model relative to most other models is experimental determination of the true shape of the dose response for initial yield of dsb after low-LET irradiation.

The amorphous track structure model of Katz (Katz et al. 1971; Roth et al. 1976) considers only the average radial distribution of energy deposition along a track without any of the internal structure. This phenomenological model assumes that a cell is inactivated if sufficient energy is deposited in each of m radiation targets. The large RBEs of high-LET radiations arise from their ability to achieve this with a single track (acting by "ion-kill"). The formalism requires that these targets be large, typically a few micrometres in diameter, so they cannot be identified with realistic DNA structures. Additionally, in terms of the model a single electron track (including its secondaries) is totally incapable of inactivating a cell and is extremely unlikely, alone, to damage even a single target. This would imply that the required molecular damage is greater than would be produced by even the largest clusters of ionisations achievable by electrons.

This model is not intended for such mechanistic interpretations. However, because it provides one of the best parameterized fits to heavy ion radiobiological data, it is tempting to seek underlying interpretations. Accordingly, it has been suggested, from track-structure analyses, that a similar parameterization of the ion-kill mode could be achieved if the model were re-interpreted such that the cellular effect followed from the deposition of a local concentration of energy (~400 eV) in a single target of dimensions ~5-10 nm (Goodhead 1989b). This could be interpreted as producing characteristic high-LET DNA damage of complexity greater than a dsb. A different formalism would be required for the qualitatively different low-LET damage.

All the above models, with the possible exception of the suggested reinterpretation of Katz' model, allow for no qualitative differences between the predominant one-track damage produced by high- and low-LET radiations. Yet the radiobiological features of mammalian cells seem to require this (see Section 4). It has been proposed that this can be modelled in terms of a local threshold-energy concept (Goodhead _et al_. 1980, 1985; Goodhead 1989a). In these terms the probability of biological effectiveness of local damage is assumed to increase as a step function of the magnitude of the local damage, as an approximation to a more realistic monotonically increasing function. A minimum of two steps is required, one representing the rise from irrelevant minor damage to the predominant (and largely modifiable and/or reparable) damage from low-LET radiations and the other representing the further rise to the predominant (and largely unmodifiable and irreparable) damage from high-LET radiations. As working hypotheses these two classes of damage have been assigned the physical identities of deposition of $\gtrsim$ 100 eV within ~ 2 nm and $\gtrsim$ 400 eV within 5-10 nm (Goodhead 1989a). In terms of DNA damage these may correspond to dsb (or perhaps slightly greater complexity) and strand breaks plus substantial associated damage, respectively.

6. CONCLUSIONS

Due to the extreme complexity of the full sequence of processes which lead from initial physical insult, to DNA damage and finally to cellular effect, attempts to model and understand the mechanisms must make major simplifying assumptions. Experimental constraints are needed to limit the wide choice of options which are available and to lead the models through intermediate processes. A few of the critical questions in relation to DNA damage are:

- What are the features of that small fraction of DNA damage which remains unrepaired (or misrepaired) after many hours of cellular repair? Is it a random residue from some simple initial class of damage, such as dsb, or is it composed of some particular subclasses? It is presumably this fraction which is biologically important.
- Are simple dsb of much biological importance, or is it the reparability of more complex dsb, etc., which limits cellular repair and determines the effect?
- What is the nature of the damage from high-LET radiations which renders it less modifiable and less reparable?
- What relative contribution, in the cellular environment, do single-radical-produced dsb make to cellular effects? These should be most efficiently produced by sparse, as opposed to clustered, ionisations.
- What is the chemistry, and biochemical damage, which follows from different degrees of clustered initial physical damage in DNA and associated water and/or protein?
- Can the usual methods of "linear" chemistry adequately describe these processes in terms of each radical proceeding independently so that the final DNA damage is simply a linear sum of these independent products even if closely adjacent? A low-LET radiation track can cause up to about 20 ionisations and excitations within a distance of about 6 base pairs in DNA.

A slow α-particle track can cause even more in that target and can cause up to about 120 ionisations and excitations in a volume similar to a nucleosome (Goodhead 1990). Can linear chemistry describe the consequences or are there likely to be non-linear processes leading to radical-radical, or other, products which would not otherwise occur? Might irreparable high-LET damage be partly a consequence of such processes?
- Is there now a consensus that the relevant processes by which ionising radiations damage DNA in mammalian cells are highly local? In particular, is there only very short ranged ($\lesssim$ a few nanometres) diffusion of reactive radiation-induced radicals? In this case it should be reasonable to conclude that the properties of radiation tracks responsible for DNA damage are confined to the atomic interactions with the DNA itself and the very closely adjacent molecules of water and protein.

REFERENCES

Brenner DJ, Zaider M (1984) Modification of the theory of dual radiation action for attenuated fields. II. Application to the analysis of soft X-ray results. Radiat Res 99:492-501

Chadwick KH, Leenhouts HP (1981) The molecular theory of radiation biology, Springer Verlag, Berlin

Charlton DE, Goodhead DT, Wilson WE, Paretzke HG (1985) The deposition of energy in small cylindrical targets by high LET radiations. Radiat Prot Dosim 13:123-125

Charlton DE, Nikjoo H, Humm JL (1989) Calculation of initial yields of single- and double-strand breaks in cell nuclei from electrons, protons and alpha particles. Int J Radiat Biol 56:1-19

Cornforth MN, Bedford JS (1983) X-ray-induced breakage and rejoining of human interphase chromosomes. Science 222:1141-1143

Curtis SB (1986) Lethal and potentially lethal lesions induced by radiation - a unified repair model. Radiat Res 106:252-270

Goodhead DT (1987a) Relationship of microdosimetric techniques to applications in biological systems. In: Kase KR, Bjarngaard BE, Attix FH (eds) Dosimetry of ionizing radiations, Vol 2. Academic Press, New York, pp 1-89

Goodhead DT (1987b) Biophysical models of radiation action. In: Fielden EM, Fowler JF, Hendry JH, Scott D (eds) Radiation Research, Proc 8th Int Congr Radiat Res Edinburgh, July 1987, Vol 2. Taylor & Francis, London, pp 306-311

Goodhead DT (1989a) Initial physical damage produced by ionizing radiations. Int J Radiat Biol 56:623-634

Goodhead DT (1989b) Relationship of radiation track structure to biological effect: a re-interpretation of the parameters of the Katz model. Nucl Tracks Radiat Meas, Int J Radiat Appl Instrum D 16:177-184

Goodhead DT (1990) Radiation effects in living cells. Canadian J Phys (in press)

Goodhead DT, Nikjoo HN (1989) Track structure analysis of ultrasoft X-rays compared to high- and low-LET radiations. Int J Radiat Biol 55:513-529

Goodhead DT, Charlton DE, Wilson WE, Paretzke HG (1985) Current biophysical approaches to the understanding of biological effects of radiation in terms of local energy deposition. In: Schraube H, Burger G (eds) 5th Symp Neutron Dosim 17-21 Sept 1984. Commission of the European Communities, EUR9762, pp 57-68

Goodhead DT, Munson RJ, Thacker J, Cox R (1980) Mutation and inactivation of cultured mammalian cells exposed to beams of accelerated heavy ions. IV. Biophysical interpretation. Int J Radiat Biol 37:135-167

Harder D (1987) Pairwise lesion interaction - extension and confirmation of Lea´s model. In: Fielden EM, Fowler JF, Hendry JH, Scott D (eds) Radiation Research, Proc 8th Int Congr Radiat Res, Edinburgh, July 1987, Vol 2. Taylor & Francis, London, pp 318-324

Howard-Flanders P (1958) Physical and chemical mechanisms in the injury of cells by ionizing radiations. Adv Biol Med Phys 6:553-603

Katz R, Ackerson B, Homayoonfar M, Sharma SC (1971) Inactivation of cells by heavy ion bombardment. Radiat Res 47:402-425

Kellerer AM, Rossi HH (1978) A generalized formulation of dual radiation action. Radiat Res 75:471-488

Magee JL, Chatterjee A (1987) Track reactions of radiation chemistry. In: Freeman GR (ed) Kinetics of non-homogeneous processes. Wiley, New York, pp 171-214

Nikjoo H, Goodhead DT, Charlton DE, Paretzke HG (1989) Energy deposition in small cylindrical targets by ultrasoft X-rays. Phys Med Biol 34:691-705

Paretzke HG (1980) Advances in energy deposition theory. In: Thomas RH, Perez-Mendez V (eds) Advances in radiation protection and dosimetry in medicine. Plenum, New York, pp 51-73

Paretzke HG (1987) Radiation track structure theory. In: Freemen GR (ed) Kinetics of nonhomogeneous processes. Wiley, New York, pp 89-170

Paretzke HG (1990) Radiation physics: stochastic deposition of energy. Canadian J Phys (in press)

Roth RA, Sharma SC, Katz R (1976) Systematic evaluation of cellular radiosensitivity parameters. Phys Med Biol 21:491-503

TABLE 1. YIELDS† OF ENERGY CLUSTERS IN THE DNA STRUCTURES OF A SINGLE MAMMALIAN CELL AFTER A DOSE OF 1 Gy

	Number of clusters per cell per Gy						
	In a 2 nm segment of DNA				In a nucleosome		
Energy*	≳10 eV	>100 eV	>160 eV	>250 eV	≳10 eV	>400 eV	>800 eV
Radiation Low LET	1237	23	2	0	1624	4	0
Electron track-ends	972	46	7	0	609	11	0
Slow α-particles	667	103	38	8	255	70	20

* Dividing the values of energy deposition by ~11 provides an estimate of the corresponding average number of ionisations and excitations in the cluster

† Data extracted from Goodhead (1990), based on the track structure data and calculations of Charlton et al. (1985), Nikjoo et al. (1989) and Goodhead & Nikjoo (1989).

TABLE 2. YIELDS† OF DAMAGE IN A SINGLE MAMMALIAN CELL AFTER 1 Gy

Radiation:	Low-LET (x-, γ-rays)	Electron track-ends	High-LET (slow α-particles)
Tracks in nucleus	1,000	~ 1,100	~ 2
Ionisations:			
In nucleus	~100,000	~100,000	~100,000
In DNA	~ 1,500	~ 1,500	~ 1,500
DNA ssb	700-1000	n/a	300-600
DNA dsb: Initial	18-60	~40-130	70
After 8 h	6	n/a	30
PCC breaks:			
Initial	6	n/a	12
After 8 hr	~ 0.5	n/a	4
Lethal lesions	0.2-0.8	0.5-1.8	1.3-3.9
Cells inactivated	10%-50%	40%-80%	70%-95%

† Data extracted from Goodhead (1990), where references to experimental data are listed.

REPAIR OF DNA DAMAGE AND ITS EFFECT ON RBE - AN EXPERIMENTAL APPROACH

D. Frankenberg
Institut für Biophysikalische Strahlenforschung,
Gesellschaft für Strahlen-und Umweltforschung mbH München
Paul-Ehrlich-Str.20
D-6000 Frankfurt 70
FRG

Damage to DNA by ionizing radiation is essentially determined by the microscopic structure of the energy deposition of radiation at the nanometer scale. In eukaryotic cells a variety of DNA lesions can be detected: DNA protein crosslinks, base alterations and base detachments (base damage), sugar alterations, bulky lesions (i.e. clusters of base damage), DNA single- (ssb) and double- (dsb) strand breaks. For some of these lesions data are available about the yield per dose for a variety of radiation fields. For practical purposes values of the relative biological efficiencies (RBE) as a function of the linear energy transfer (LET) or lineal energy y can be calculated. These RBE-values are related to DNA damage at the end of the chemical phase of radiation action, i.e. when the processes of chemical restitution are completed and the DNA is in a *stable, however chemically altered* state.

Cell death, cell transformation and mutation are the ultimate cellular responses to the primary energy depositions by radiation. RBE-values of a variety of radiation fields for these biological effects are available. Again, for practical purposes RBE-values can be calculated in dependence of LET or y. These RBE-values are related to the *stable, however altered cellular* state when cells have completed their enzymatic repair processes evoked by the radiation-induced chemical changes in the DNA double helix. In the case of repair of all radiation-induced DNA damage, no cellular radiation effects would be detectable. Since this is not the case it has to be postulated that some of the DNA lesions are not repaired or misrepaired. It is suggested that for the cellular radiation effects, the RBE-values for unrepaired or misrepaired DNA lesions are responsible rather than those for the initial DNA lesions. The reasons why some of the DNA lesions are not repaired or misrepaired are threefold:

1. The structure of the energy deposition produced by a definite radiation field (*physical parameters).*

NATO ASI Series, Vol. H 54
The Early Effects of Radiation on DNA
Edited by E. M. Fielden and P. O'Neill

2. The ultrastructure of the whole genome of a cell, i.e. the subnuclear structure consisting of chromosomes, chromosomal fibers (25 nm), nucleosomes and the DNA double-helix (*target parameters).*
3. Enzymatic repair of radiation-induced damage to the genome (*biological parameters).*

The paper is mainly concerned with enzymatic repair of DNA damage, and discusses its effects on the RBE of cellular responses. Especially, it will be shown that the difference between the RBE-values for initial dsb (believed to be an important DNA lesion) and for cellular effects (cell death or cell transformation) can be explained when the enzymatic repair of dsb is taken into account.

RBE-values for initial DNA lesions

All initial DNA lesions so far as known are induced linearly with dose. However, for dsb both linear (Weibezahn and Coquerelle 1981, Van der Schans et al. 1982, Okayasu and Iliakis 1989) and non-linear (Radford 1985, Prise et al. 1987, Okayasu and Iliakis 1989) induction curves are reported when the neutral elution technique is applied. This is in contrast to the neutral sedimentation method which yields always a linear relationship between the number of dsb and dose even at doses of several thousands of Gy (Corry and Cole 1968, Lehman et al. 1970, Lehman and Stevens 1977, Frankenberg-Schwager et al. 1979, Blöcher 1982). In table 1 are listed the different DNA lesions with their RBE-values for densely ionizing radiations. It is reasonable to assume that for the induction of the initial DNA lesions only the ionization density at the *nanometer* scale is relevant. Sparsely ionizing radiations deposit approximately 70% of their energy through so called isolated spurs with only one or two ionizations. This is in contrast to densely ionizing radiations. Here, approximately 100% of the energy is deposited in volumes of some nanometer cord length containing at least 10 ionizations. Since a single ionization is sufficient to generate a base damage or a ssb, RBE-values < 1 are consistent with the energy deposition patterns of sparsely and densely ionizing radiations. The induction of dsb needs several (at least 2) ionizations within and/or in nanometer distance to the DNA double helix. Therefore, reported RBE-values > 1 for dsb induction are again in agreement with the track structures of sparsely and densely ionizing radiations. For the induction of DNA protein crosslinks and sugar alterations no RBE-values are available yet. It can be expected that for induced DNA protein crosslinks and sugar alterations the RBE-values will be < 1. The RBE-value of neutrons for induced bulky lesions is reported to be < 1 (Furuno et al. 1979). There is still an urgent need to determine RBE-values for this DNA lesion in dependence of ionization density, since its enzymatic repair is supposed to generate dsb (Geigl 1987).

Enzymatic repair of DNA lesions

For most of these DNA lesions, i.e. DNA protein crosslinks, base damage, ssb, dsb and bulky lesions, the kinetics of enzymatic repair has been determined. A detailed review of these repair kinetics is given elsewhere (Frankenberg-Schwager 1989, 1990a). Figure 1 summarizes the ranges of time constants ($t_{1/2}$ - values) for the removal of these DNA lesions. The shortest $t_{1/2}$ - values are observed for the removal of ssb and base damage , whereas DNA protein crosslinks, dsb and bulky lesions have $t_{1/2}$-values up to 14 h. The enzymatic repair of the initial DNA lesions shifts the irradiated cells from the chemically stable, however, altered

Table 1
RBE-values for initial DNA lesions in eukaryotic cells

DNA lesion	RBE of densely ionizing radiations
DNA protein crosslinks	?
Sugar alterations	?
Base damage	< 1
Single-strand breaks	< 1
Bulky lesions	< 1
Double-strand breaks	1.5 - 2.6

states either to the fully restored state or to altered cellular states, such as cell death, cell transformation and mutation. Enzymatic repair processes are dependent on the amount of repair enzymes, the time available for repair and energy supply. Mutants with a total (yeast cells) or partial (mammalian cells) deficiency in a specific repair pathway together with controlled variation of repair conditions during and/or after irradiation (low dose rate exposures, repair of potentially lethal damage (PLDR)) provide a powerful tool to help identify the DNA lesions responsible for cellular radiation effects.

RBE - values for residual DNA lesions

RBE - values are available only for residual ssb and dsb. The fraction of

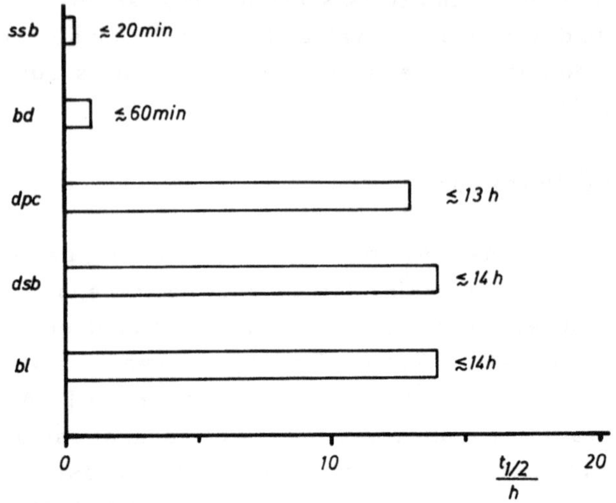

Figure 1: $T_{1/2}$-values in min or hours for the enzymatic repair of DNA lesions induced in eukaryotes by ionizing radiation. ssb: single-strand break, bd: base damage, dpc: DNA protein cross link, dsb: double-strand break, bl: bulky lesion

unrejoined ssb in Chinese hamster cells increases with the LET of heavy ions from 1.4 % (at 1 keV/μm) to 18% (at 1953 keV/μm) (Ritter et al 1977, Roots et al 1979). Compared to low LET radiation an increased yield of unrejoined ssb in Chinese hamster V79 cells (Ahnström and Edvardsson 1974) and in mouse cells (Sakai et al. 1987) is observed after irradiation with fast neutrons. For CHO cells irradiated with α-particles an RBE-value of about 2 is found for unrejoined ssb (Cole et al 1980). Measurements of ssb comprise always ssb originating from dsb. Since ssb are repaired much faster than dsb (see figure 1), ssb originating from dsb become gradually dominant with increasing repair and determine both the repair kinetics and RBE-values of residual breaks. It remains to be elucidated to which extent dsb contribute to the amount of unrejoined ssb (Dikomey and Franzke 1986, Radford 1986, Ritter et al 1977, Sakai et al 1987).

The fraction of unrejoined dsb is higher after high LET irradiation compared to sparsely ionizing radiations. Cole et al (1980) found an increase of the RBE of α-particles from 1.5 (initial dsb) to 6.0 (dsb left unrejoined in CHO cells kept in growth medium for 360 min). For diploid yeast, an increase of the RBE of 3.5 MeV α-particles from 2.6 (initial dsb) to 3.5 (non or misrepaired dsb after repair under non- growth conditions for 72 h) was reported (Frankenberg et al. 1981a, Frankenberg-Schwager et al. 1982).

A detailed analysis yielded an RBE-value of 5.7 and 3.0 for the linear and quadratic component, respectively, of the linear-quadratic relationship between unrejoined dsb and dose (Frankenberg-Schwager et al. 1990). Blöcher (1988) determined an RBE-value of 1.6 for initial dsb and 2.7 - 3.8 (depending on the dose from 100 Gy to 10 Gy of α-particles) for dsb left unrejoined in EAT cells kept in nongrowth medium for 24 h . In summary, an increase in RBE of α-particles due to dsb rejoining is observed not only at supralethal doses as used for mammalian cells but also at biologically relevant doses yielding high surviving fractions (> 10 %) as applied for yeast cells.

RBE-values for chromosome aberrations

At low doses the RBE-values of high LET radiations for chromosome aberrations are significantly higher than those observed for initial dsb (Vulpis 1973, Edwards et al. 1980, Purrott et al. 1980, Takatsuji and Sasaki 1984). With increasing dose the RBE-values approach those for initial dsb. There is experimental evidence that dsb may be the DNA lesions leading to chromosome aberrations (Bryant 1984, Natarajan and Obe 1984). Thus, the increased efficiency of high LET radiations for the formation of chromosome aberrations may be explained by their *increased* efficiency to induce not only total dsb but especially those dsb which are not repaired or are misrepaired. Dicentrics and chromosome fragments might be formed, the latter ones resulting from remaining dsb.

The effect of repair of DNA double-strand breaks on RBE for cell killing

It was first shown in yeast that repair proficient cells show higher RBE-values than cells which are not capable of post-irradiation recovery (i.e. liquid holding recovery (LHR) or PLD repair) (Petin 1979). The nature of the primary damage whose repair leads to an increased RBE-value has remained unknown. Three years later, Cox (1982) reported a very similar observation in the mammalian cell system, namely that normal human fibroblasts achieve higher RBE-values than those from ataxia-telangiectasia (AT) patients which carry a deficiency in the repair of PLD (Weichselbaum et al 1978, Cox 1981). Although AT cells show the same *rate* of dsb rejoining as normal cells (Paterson and Smith 1979), the elevated chromosomal damage points to a defect in the repair of strand breaks (Taylor 1978, Cornforth and Bedford 1985).

For yeast cells Ho (1975), Resnick and Martin (1976), and Frankenberg et al (1981b) have shown that about one dsb per cell corresponds to a lethal event. Moreover, it has been demonstrated that in comparison to sparsely ionizing radiation (30 MeV electron) densely ionizing α- particles (3.5 MeV) produce not only more dsb but also more DNA fragments after 72 h LH treatment (Frankenberg et al 1981a, Frankenberg-Schwager et al. 1982). These fragments may be composed of unrepaired dsb and of fragments formed in the course of misrepair of dsb. The effect of dsb repair on the RBE of densely ionizing α-particles was investigated in more detail with the help of the temperature conditional radiosensitive diploid yeast mutant rad54-3 (Frankenberg-Schwager 1984). This mutant is proficient in the repair of dsb at the permissive temperature of 23°C. When plated immediately (IP) after irradiation at this temperature a shouldered survival curve is observed. Furthermore, at the permissive temperature this mutant exhibits LHR of colony forming ability which in yeast cells is accompagnied by dsb repair (Frankenberg-Schwager 1980a). In contrast, at the restrictive temperature of 36°C this mutant is dsb repair-deficient (Budd and Mortimer 1982) and when plated immediately after irradiation at this temperature it yields an exponential survival curve where the D_o-value corresponds to about one dsb per cell. At this temperature LHR of colony forming ability is negligible (Frankenberg-Schwager et al. 1984). Therefore, in this mutant dsb repair function can be switched on and off by appropriate selection of temperature. At both temperatures this mutant grows to macrocolonies. Figure 2 shows the survival curves of rad54-3 cells for 30 MeV electrons and 3.5 MeV α-particles without (IP, 36°C) and with repair of dsb (IP, 23°C) as well as for various time periods of LHR (24 h, 72 h) at 23°C before plated on nutrient agar and incubated at this temperature for the macrocolony assay. Table 2 summarizes the RBE-values for increasing LHR periods available for dsb repair at selected surviving fractions. The RBE-value of 1.5 for initial dsb determined indirectly by the rad54-3 survival assay is smaller than the value of 2.6 reported when dsb were measured directly. The reason for this discrepancy is that per traversal of an α-particle a mean dose of about 10 Gy is deposited in the yeast nucleus (1.4 µm diameter) corresponding to 2.8 dsb per cell. Since this number is more than sufficient to kill a yeast cell, the RBE of α-particles for the induction of dsb as measured indirectly by the rad54-3 assay is smaller than that obtained by direct measurement of dsb. Table 2 indicates clearly that the RBE of 3.5 MeV α-particles increases gradually with increasing levels of dsb repair from 1.5 when dsb repair is prevented up to 4.5, when dsb repair is allowed for 72 h under LH conditions before cells were plated at 23°C (delayed plating, DP).

In figure 3 IP- and DP- survival curves of wild type (wt) yeast cells (strain 211) after irradiation with 30 MeV electrons (e^-) or 3.5 MeV α-particles (α) are compared with those of rad54-3 cells delayed plated after 72 h

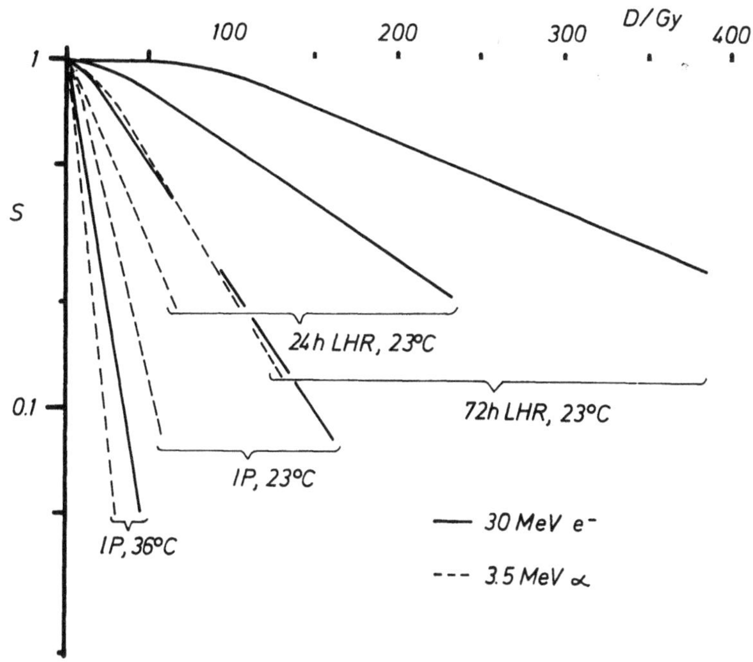

Figure 2: Survival curves of rad54-3 cells plated immediately after irradiation with 30 MeV electrons or 3.5MeV α-particles without repair of dsb (IP, 36°C) or with repair of dsb (IP,23°C) as well as with repair of dsb under LH conditions (24 h LHR, 23°C; 72 h LHR, 23°C) before plating and incubated at 23°C.

LHR (see figure 2). It can be seen that the intrinsic repair capacity of diploid rad54-3 cells at the permissive temperature for dsb repair is lower than that of wt cells. The survival curve of wt cells with 72 h LH treatment represents lethal, i.e. irreparable or misrepaired dsb, whereas the wt survival curve after IP comprises irreparable or misrepaired dsb as well as dsb which in principle are reparable, but under IP-conditions have not had the chance to become repaired (Frankenberg-Schwager et al 1987). The RBE-values of 3.5 MeV α-particles for survival of wt cells with IP and DP are given in the last two columns of table 2. Especially at high survival levels of wt cells the RBE of 3.5 MeV α-particles increases beyond that obtainable with rad54-3 cells, due to the higher repair capacity of wt cells. A dose dependence of RBE is clearly indicated.

Table 2: RBE-values of 3.5 MeV α-particles as a function of repair of dsb
S: surviving fraction

S	IP 36°C	IP 23°C	DP (24h LHR)	DP (48h LHR)	DP (72h LHR)	IP	DP (72h LHR)
0.9	1.5	2.8	3.7	3.9	4.5	7	10
0.5	1.5	2.8	3.7	3.8	4.4	4.5	4.5
0.1	1.5	2.8	-	-	-	3.3	3.4

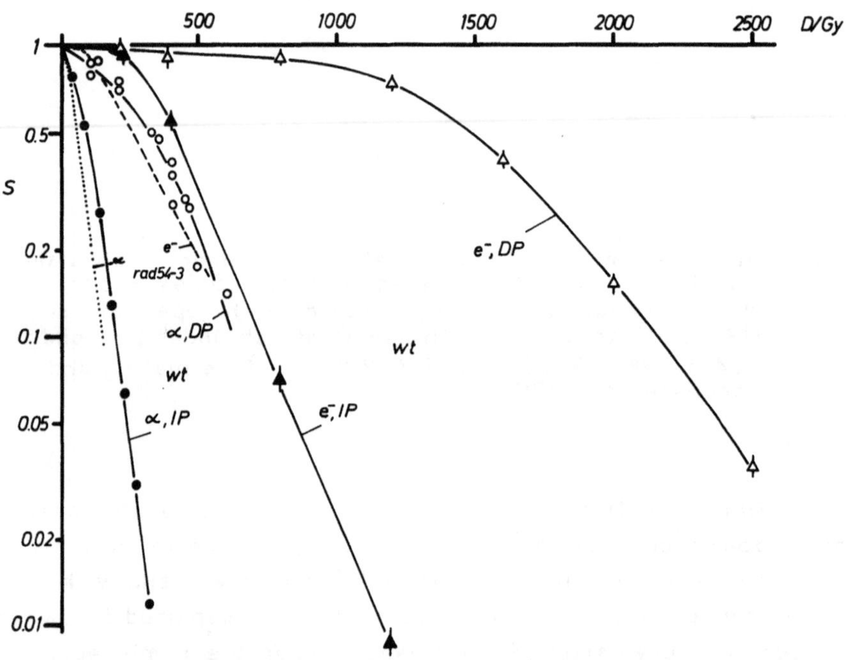

Figure 3: Survival curves of wild type (wt) yeast cells with immediate plating (IP) and delayed plating after 72h LHR (DP) after irradiation with 30 MeV electrons (e^-) or 3.5 MeV α-particles (α). For comparison are given the survival curves of the rad 54-3 cells delayed plated after 72h LHR (---- e^-, ······ α) at 23°C and incubated at 23°C.

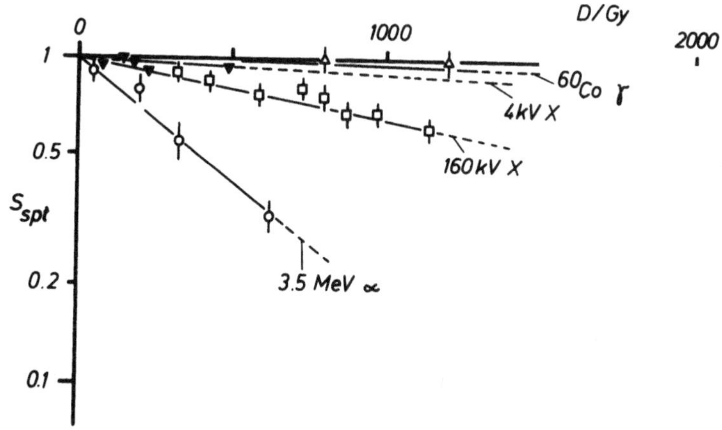

Figure 4: Exponential survival curves of wt yeast cells after irradiation with ^{60}Co γ-rays, 160 kV x-rays, 4 kV x-rays and 3.5 MeV α-particles at sufficently low dose rates. D: absorbed dose in Gy; S_{spt}: survival after irradiation at low dose rate; spt: single particle traversal through the cell nucleus.

Table 3: RBE-values of different radiations for S_{spt}, S_{acc}, dsb_{ind} and $S_{rad54-3(36°C)}$. For C_K x-rays no low dose rate exposures are possible.

	S_{spt}	S_{acc}	dsb_{ind}	$S_{rad54-3(36°C)}$
30 MeV electrons or ^{60}Co γ- rays	1	1	1	1
150 (160) kV X	8	1.3	1.3	1.3
4 kV X	2	2.1	1.8	1.8
C_K x-rays	–	–	3.8	2.6
3.5 MeV α	25	3.5	2.6	1.5

In contrast to high dose rate irradiations, survival curves of wt yeast cells are exponential at sufficiently low dose rate and optimum conditions for the repair of dsb during irradiation (Frankenberg 1984). Figure 4 presents cell survival S_{spt} (spt: single particle traversal through the cell nucleus) of wt cells after irradiation with ^{60}Co γ-rays, 160 kV X-rays, 4 kV X-rays and 3.5 MeV α-particles at sufficiently low dose rates. The RBE-values for S_{spt} relative to ^{60}Co γ-rays are given in table 3. Similar results were obtained at the dsb level. At high dose rate irradiation the RBE-value for the induction of dsb is dose independent. However, a dose dependence of the RBE-value for residual dsb is observed when repair of dsb is completed under LH-conditions and when linear-quadratic fits to experimental data where applied (Frankenberg-Schwager et al. 1990b). A similar observation was made in EAT cells (Blöcher 1988). In contrast, after irradiation at low dose rate a virtually linear relationship between residual dsb and dose has been observed (fig. 5) (Frankenberg-Schwager et al. 1981), suggesting that at low dose rate the RBE-value of 3.5 MeV α-particles for residual dsb might also be dose independent as in the case of survival. However, experimental data about residual dsb after irradiation with α-particles at low dose rates are not available yet.

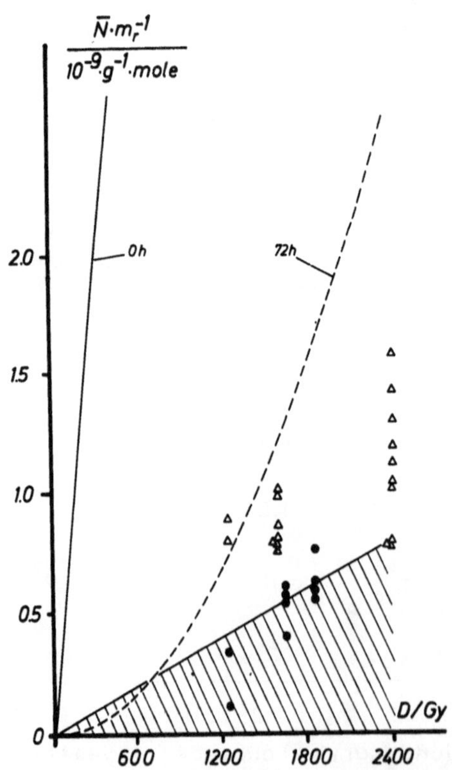

Figure 5:
The average number $\bar{N}$ of dsb per relative molar mass of DNA (m_r) induced by ^{60}Co γ-rays at the low dose rate of 0.55 Gy/min. Dsb measurements were performed either directly after irradiation (▲) or after LH periods up to 56h (●). In the latter case, the residual dsb approach or fall into the hatched area representing the putative linear component calculated for the relationship between the nonreparable or misrepaired dsb and dose. For comparison the results obtained by 30 MeV electrons at high dose rate without (0h) and with LH treatment for 72h are indicated.

Comparison of the survival curves of wt cells after irradiation at low dose rate (fig. 4) and at high dose rate followed by DP after 72 h LHR (fig. 3) shows that irradiation at sufficiently low dose rate induces irreparable and/or misrepaired dsb by spt, yielding an exponential survival curve whereas at high dose rate also binary misrepair of dsb occurs due to the accumulation (acc) of dsb resulting in the bending down of the survival curve. The surviving fraction S_{DP} obtained at high dose rate can therefore be interpreted as the product of two surviving fractions, the surviving fraction S_{spt} obtained at sufficiently low dose rate and the surviving fraction S_{acc} which is due to the accumulation of dsb during high dose rate exposures:

$$S_{DP} = S_{spt} \cdot S_{acc} = \exp(-k_{spt} \cdot D)\, S_{acc}$$

where k_{spt} is the inactivation coefficient for spt. Since S_{DP} and S_{spt} are known, S_{acc} as a function of dose can be calculated for 30 MeV electrons and 3.5 MeV α-particles (Frankenberg-Schwager et al 1984). Like for S_{spt} the RBE of 3.5 α-particles for S_{acc} is dose independent. Table 3 summarizes the RBE-values of 3.5 α-particles for cell death after irradiation at very low dose rate (S_{spt}) as well as after high dose rate exposure due to accumulation of dsb (S_{acc}), for radiation-induced dsb and for cell death of the dsb repair-deficient mutant rad54-3 (36 °C). In addition are given the corresponding RBE-values of 150 (160) kV x-rays and 4 kV x-rays. The agreement between the RBE-values for dsb_{ind} and $S_{rad54-3(36°C)}$ obtained with sparsely ionizing radiations indicates that about one dsb per cell is lethal. Furthermore, the agreement between the RBE-values for S_{acc} and dsb_{ind} supports the hypothesis that S_{acc} is caused by binary misrepair due to accumulation of dsb. The discrepancy between the RBE-values of 3.5 MeV α-particles for S_{acc} and $S_{rad54-3(36°C)}$ is due to the fact that more than one dsb is produced per traversal of an α-particle through the cell nucleus (see above).

Physical and target parameters suggested to lead to irreparable or misrepaired double-strand breaks

The RBE-values in table 3 may be analyzed by the following simplified concept of radiation quality in connection with the nucleosomal structure of the 25 nm chromosomal fibers. Three physical parameters describing the radiation quality are considered:

1. The ionization density at the level of the DNA double helix, i.e. the number of ionizations within a volume of only some nm cord length (local ionization density)

2. The number of local ionization densities sufficiently high to induce dsb which are generated by a single electron or particle traversing the cell nucleus (associated local ionization densities)
3. The fractions of the total energy imparted to matter under the conditions outlined under 1. and 2.

Radiation physics of "electron radiations" and the RBE-values of C_K and Al_K characteristic x-rays for dsb induction (Frankenberg et al. 1986) indicate that the electron track ends (electrons of energies from about 100 eV up to several hundred eV) generate local ionization densities sufficiently high to induce dsb. The increase of the RBE-values of x-rays with decreasing photon energies for dsb_{ind} and S_{acc} is due to the increase of the *fraction* of the total energy which is imparted to matter by electron track ends. 3.5 MeV α-particles generate *continuously* local ionization densities sufficiently high to to produce dsb. Since for 3.5 MeV α-particles and C_K x-rays 100% of the energy is imparted to matter by local ionization densities sufficiently high to induce dsb, and the local ionization density of 3.5 MeV α-particles is by a factor of about 3 higher compared to C_K x-rays, the higher RBE-value of C_K x-rays for dsb_{ind} indicates that there exists a saturation effect for the induction of dsb with increasing values of the local ionization density (see table 3).

It is useful to discuss the RBE-values for S_{spt} separately for electron radiations and particles, since electrons produce *discontinuously* associated local ionization densities, whereas particles do so either *discontinuously* or *continuously* depending on their energy. According to the secondary electron spectrum produced by the primary radiation field of electron radiations, two or more electron track ends may be produced by the same electron within a distance of 25 nm (i.e. the diameter of the chromosomal fiber). This may lead to irreparable dsb by loss of the DNA piece between two dsb. Alternatively, two dsb are induced in neighboured 25 nm fibers which by their interaction may lead to misrepaired dsb. Therefore, the RBE-values for S_{spt} are determined mainly by the associated ionization densities generated by a single electron as well as by the fraction of the total energy imparted to matter in such a way. The high RBE-value of 150 kV x-rays for S_{spt} is caused by their relatively high fraction of electrons with energies of some keV to about 20 keV. It is unlikely that for S_{spt} a saturation effect is achieved because of the relatively low cross sections of electrons to generate electron track ends (Paretzke 1976). The smaller RBE-alues of 4 kV x-rays for S_{spt} compared to 150 kV x-rays is simply due to the smaller fraction of electrons which generate more than one electron track end. Depending on their energy, particles generate *discontinuously* (for example 200 MeV protons) as well as *continuously* (for example 3.5 MeV α-particles) more than one local ionization density sufficiently high to induce dsb along their tracks through the cell nucleus. All mixtures between these two extremes are possible. Like for electrons, more than one dsb may be produced within a 25 nm fiber or, alternatively, in closely neighboured 25 nm fibers leading to irreparable or misrepaired dsb. Moreover, densely ionization particles may damage nucleosomal bodies in close vicinity to a single dsb so that repair of this dsb may be inhibited. However, the equal frequencies of dicentrics and acentrics obtaind with densely ionizing α-particles (Purrott et al 1980) suggest that this mechanism leading to irreparable dsb and in turn to cell death is of minor importance. High RBE-values for S_{spt} which are observed with densely ionizing particles may be explained by the small fraction of the total energy imparted to matter in form of associated ionization densities by ^{60}Co γ-rays (as reference radiation) as well as the higher local ionization densities achievable by densely ionizing particles. Saturation is observed when more than one irreparable or misrepaired dsb is produced by a single particle along its track through the cell nucleus. In contrast to wt cells, much lower RBE-values of densely ionizing particles are obtained for cell death of dsb repair-deficient mutants such as rad54-3 (36° C). Here, as well as for dsb induction the local ionization density determines the RBE. For ^{60}Co γ-rays the fraction of the total energy which is imparted to matter by local ionization densities sufficiently high to induce dsb amounts to around 25%. Therefore, considering the saturation effect achievable for the induction of dsb for densely ionizing particles, no higher RBE-values than about 4 have to be expected for dsb-deficient mutants.

The implications of yeast data for mammalian cells

The analyses of RBE-values for yeast cells can be applied to the data obtained with mammalian cells and their radiosensitive mutants, for example AT- cells or xrs-cells. Fig. 6 presents the survival curves of normal human fibroblasts and of AT strains after low dose rate exposure and after irradiation at high dose rate with or without postirradiation holding (PLDR). In table 4 are given RBE-values of α-particles for survival of normal cells and radiosensitive mutants. The cellular data may be summarized as follows:

1. With sparsely ionizing radiations exponential survival curves may be observed in contrast to normal cells (Weichselbaum et al. 1978, Cox et al. 1981, Thacker and Stretch 1985).

2. The radiosensitive cells lack recovery during postirradiation holding, i.e. they are not able to repair PLD (Weiselbaum et al. 1979, Cox et al. 1981).
3. No recovery during low dose rate exposure is observed in contrast to normal cells (Cox 1982, Thacker and Stretch 1985).
4. The dependence of RBE on ionization density is much less apparent in the radiosensitive cells (Cox 1982, Thacker and Stretch 1985).
The molecular and cytogenetic data which are available are the following:
5. After repair the residual levels of dsb are higher in the sensitive xrs mutants of Chinese hamster ovary cells compared to normal cells (Kemp et al. 1984).
6. AT cells show a lack of fidelity in the rejoining of dsb compared to normal cells (Cox et al. 1984).
7. Much higher levels of chromatid gaps or breaks are observed compared to those seen in normal cells (Taylor 1978, Kemp et al. 1984, Cornforth and Bedford 1985).

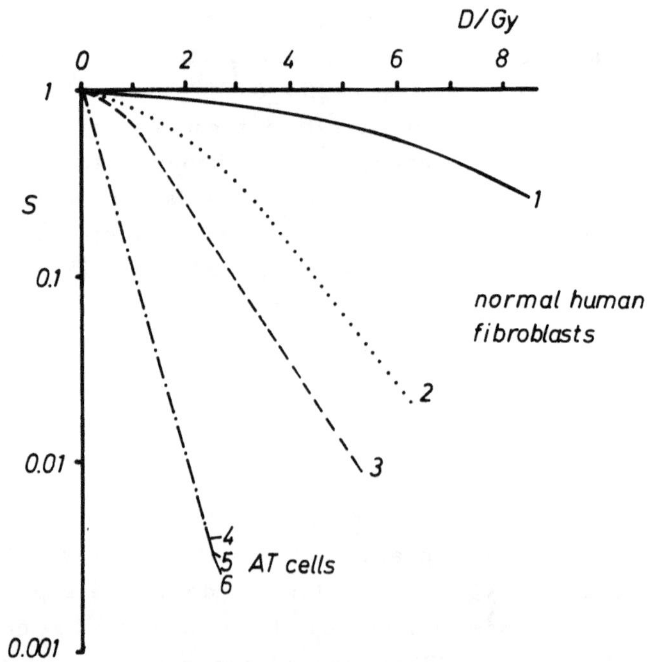

Figure 6: ^{60}Co γ-ray survival curves of normal human fibroblasts (HF 19) after low dose rate exposure (curve 1), with 6h postirradiation holding (curve 2) and without postirradiation holding (curve 3). For AT cells survival curves after low dose rate exposure (4), with (5) and without (6) postirradiation holding are identical within the experimental error (redrawn from Cox 1982).

Table 4: RBE-values of α-particles for normal and radiosensitive mammalian cells

Strain	LET (keV/μm)	RBE normal	RBE sensitive	Reference
Human fibroblasts	50	2.5	1.5	Cox, 1982
Human fibroblasts	100	4	2	
CHO cells	100	3.8	1 - 2	Thacker and Stretch, 1984

In contrast to the yeast system, radiosensitive mammalian cells are only partially defective in the repair of dsb, i.e. only a fraction of the initial dsb which are repaired in normal cells is not repaired or misrepaired (*no binary misrepair)* in these cells. Although a few dsb might be tolarated, the exponential survival curves indicate that a mutant cell is killed by a single unrepaired or misrepaired dsb. The cellular data summarized in 1.-4. can be explained consistently in the following way. The exponential survival curves obtained with sparsely ionizing radiations agree with a linear relationship between dsb induction and dose. The lack of PLDR and of the dose rate effect can be explained suggesting that the conditions of the normal survival assays provide sufficient time to repair all reparable dsb. Therefore, additional time given to the cells by postirradiation holding or by low dose rate exposures have no effect on cell survival. The smaller RBE-values of densely ionzing particles for killing of radiosensensitive cells compared to normal cells may be explained as follows: If a cell is killed by a single unrepaired or misrepaired dsb, the local ionization density with its RBE-values not higher than about 4 is responsible rather than the associated ionization densities by which much higher RBE-values are observed for cell killing.

The possible role of unrepaired or misrepaired DNA double-strand breaks for cell transformation

Besides cell killing, cell transformation is one of the major cellular effects of ionizing radiation. Several authors have shown that the RBE-values of densely ionizing particles for cell transformation increase with increasing LET or y (Robertson et al. 1983, Barendsen and Gaiser 1985, Hall and Hei

1985, Hieber et al. 1987, Hei et al. 1988). The RBE-values are dose dependent in a similar way as observed for chromosome aberrations and for cell death. They are high at low doses (RBE-values of up to about 12) and decrease (RBE-values of around 2) with increasing dose. Transformed cells are surviving cells and have repaired all their reparable radiation-induced DNA lesions. Thus, it is supposed that unrepaired or misrepaired DNA lesions are responsible for cell transformation. Since the RBE-values of densely ionizing particles for initial base damage, ssb and bulky lesions are < 1 (see table 1) and since the oxygen enhancement ratio (OER) of < 1 for DNA protein crosslinks does not match the OER for cell transformation (Borsa et al. 1984, Frankenberg-Schwager et al. 1990c), it is suggested that unrepaired or misrepaired dsb may be involved in cell transformation. This is supported by the fact that for both the un- or misrepaired dsb and cell transformation similar dose dependencies of their RBE-values are observed.

References

Ahnström G, Edvardsson K-A (1974) Radiation-induced single-strand breaks in DNA determined by rate of alkaline strand separation and hydroxylapatite chromatography: an alternative to velocity sedimentation. Int.J. Radiat.Biol. 26: 493-497

Barendsen GW, Gaiser JF (1985) Cell transformation in vitro by neutrons of different energies. Implications for mechanisms. Radiat.Protect. Dos. 13: 145-148

Blöcher D (1982) DNA double-strand breaks in Ehrlich ascites tumour cells at low doses of x-rays. Determination of induced breaks by centrifugation at reduced speed. Int.J. Radiat.Biol. 42: 317-328

Blöcher D. (1988) DNA double-strand break repair determines the RBE of α-particles. Int.J. Radiat.Biol. 54: 761-771

Borsa J., Sargent MD, Einspenner M, Azzam EI, Raaphorst GP (1984) Effects of oxygen and misonidazole on cell transformation and cell killing in C3H10T1/2 cells by x-rays in vitro. Radiat.Res. 100: 96-103

Bryant PE (1984) Enzymatic restriction of mammalian cell DNA using PvuII and BamH1: Evidence for the double-strand break origin of chromosomal aberrations. Int.J. Radiat. Biol. 46: 57-65

Budd M, Mortimer R.K. (1982) Repair of double-strand breaks in a temperature conditional radiation-sensitive mutant of Saccharomyces cerevisiae. Mutat.Res. 103: 19-24

Cole A, Meyn RE, Chen R, Corry PM, Hittelman W (1980) Mechanisms for cell injury. In: Radiation Biology in Cancer Research, pp 33-58. Meyn RE, Withers HR, edts., Raven Press, New York

Cornforth MN, Bedford JS (1985) On the nature of a defect in cells from individuals with Ataxia telangiectasia. Science 227: 1589-1591

Corry PM, Cole A (1968) Radiation-induced double-strand scissions of the DNA of mammalian metaphase cells. Radiat.Res. 36: 528-543

Cox R, Masson WK, Weichselbaum RR, Nove J, Little JB (1981) The repair of potentially lethal damage in x-irradiated cultures of normal and Ataxia telangiectasia human fibroblasts. Int.J. Radiat. Biol. 39: 357-365

Cox R (1982) A cellular description of the repair defect in Ataxia telangiectasia. In: Ataxia telangiectasia, pp 141-153. Bridges BA, Harnden DG, edts., John Wiley and Sons, Chichester

Cox R, Masson WK, Debenham PG, Webb MBT (1984) The use of recombinant DNA plasmids for the determination of DNA repair and recombination in cultured mammalian cells. Br.J. Cancer 49 (Suppl.VI): 67-72

Dikomey E, Franzke J (1986) DNA repair kinetics after exposure to x-irradiation and to internal β-rays in CHO cells. Radiat.Environm.Biophys. 25: 189-194

Edwards AA, Purrott RJ, Prosser JS, Lloyd DC (1980) The induction of chromosome aberrations in human lymphocytes by alpha radiation. Int.J. Radiat. Biol. 38: 83-91

Frankenberg D, Frankenberg-Schwager M, Blöcher D, Adamzcyk C (1981a) Initial and irreparable double-strand breaks in the DNA of irradiated eukaryotic cells in dependence of dose rate and LET: Implications for survival studies. In: Proc.7th Symp.Microdos., pp. 1033-1042, Booz J, Ebert HG, Hartfiel HD, edts., Harwood Acad.Publishers Ltd, London

Frankenberg D, Frankenberg-Schwager M, Blöcher D, Harbich R. (1981b) Evidence for DNA double-strand breaks as the critical lesions in yeast cells irradiated with sparsely or densely ionizing radiation under oxic or anoxic conditions. Radiat.Res.88: 524-532

Frankenberg D (1984) Interpretation of the dose and LET dependence of RBE-values for lethal lesions in yeast cells. Radiat. Res. 97: 329-340

Frankenberg D, Goodhead DT, Frankenberg-Schwager M, Harbich R, Bance DA, Wilkinson RE (1986) Effectiveness of 1.5 keV aluminium K and 0.3 keV carbon K characteristic x-rays at inducing DNA double-strand breaks in yeast cells. Int.J. Radiat.Biol. 50: 727-741

Frankenberg-Schwager M, Frankenberg D, Blöcher D, Adamzcyk C (1979) The influence of oxygen on the survival and yield of DNA double-strand breaks in irradiated yeast cells. Int.J. Radiat.Biol. 36: 261-270

Frankenberg-Schwager M, Frankenberg D, Blöcher D, Adamczyk C (1980a) Repair of DNA double-strand breaks in irradiated yeast cells under nongrowth conditions. Radiat.Res. 82: 498-510

Frankenberg-Schwager M, Frankenberg D, Blöcher D, Adamzcyk C (1980b) The linear relationship between DNA double-strand breaks and radiation dose (30 MeV electrons) is converted into a quadratic function by cellular repair. Int.J.Radiat. Biol. 37: 207-212

Frankenberg-SchwagerM, Frankenberg D, Blöcher D, Adamczyk C (1981) Effect of dose- rate on the induction of DNA double-strand breaks in eukaryotic cells. Radiat. Res. 87: 710-717

Frankenberg-Schwager M, Frankenberg D, Blöcher D, Harbich R, Adamczyk C (1982) Irreparable DNA double-strand breaks induced in eukaryotic cells by sparsely or densely ionizing radiation and their importance for cell killing. Mutat.Res. 96: 132 (Abstract)

Frankenberg-Schwager M, Frankenberg D, Harbich R (1984) Repair of DNA double-strand breaks as a determinant of RBE of alpha particles. Br.J.Cancer 49 (Suppl. VI): 169-173

Frankenberg-Schwager M, Frankenberg D, Harbich R (1987) Potentially-lethal damage repair is due to the difference of DNA double-strand break repair under immediate and delayed plating conditions. Radiat. Res. 111: 192-200

Frankenberg-Schwager M (1989) Review of repair kinetics for DNA damage induced in eukaryotic cells in vitro by ionizing radiation. Radiother.Oncol. 14: 307-320

Frankenberg-Schwager M (1990a) Induction, repair and biological relevance of radiation- induced DNA damage in eukaryotic cells. Radiat. Environm. Biophys., in press

Frankenberg-Schwager M, Frankenberg D, Harbich R, Adamczyk C (1990b) A comparative study of rejoining of DNA double-strand breaks in yeast irradiated with 3.5 MeV alpha particles or with 30 MeV electrons. Int.J. Radiat.Biol. 57: 1151-1168

Frankenberg-Schwager M, Frankenberg D, Harbich R (1990c) Do DNA double-strand breaks lead to radiation-induced cell transformation ? J. Cell. Biochem. Suppl. 14A: 80 (Abstract)

Furuno I, Yada T, Matsudaira H, Maruyama T (1979) Induction and repair of DNA strand breaks in cultured mammalian cells following fast neutron irradiation. Int.J. Radiat. Biol. 36: 639-648

Geigl EM (1987) Untersuchung von S1-Nuclease-sensitiven Stellen im Genom der Hefe Saccharomyces cerevisiae nach in vivo-Bestrahlung mit Gamma-Strahlen. GSF-Bericht 18/87, GSF München

Hall EJ, Hei TK (1985) Oncogenic trnaformation in vitro by radiation of varying LET. Radiat. Protect.Dos. 13: 149-151

Hei TK, Komatsu K, Hall EJ, Zaider M (1988) Oncogenic transformation by charged particles of defined LET. Carcinogenesis 9: 747-750

Hieber L, Ponsel G, Roos H, Fenn S, Fromke E, Kellerer AM (1987) Absence of a dose- rate effect in the transformation of C3H10T1/2 cells by α-particles. Int.J. Radiat.Biol. 52: 859-869

Ho KS (1975) Induction of DNA double-strand breaks by x-rays in a radiosensitive strain of the yeast Saccaromyces cerevisiae. Mutat.Res. 30: 327-334

Kemp LM, Sedgwick SG, Jeggo PA (1984) X-ray sensitive mutants of Chinese hamster ovary cells defective in double-strand break rejoining. Mutat.Res. (DNA Repair Reports) 132: 189-196

Lehman AR, Ormerod MG (1970) Double-strand breaks in the DNA of mammalian cells after x-irradiation. Biochim.Biophys.Acta 217: 268-277

Lehman AR, Stevens S (1977) The production and repair of double-strand breaks in cells from normal humans and from patients with Ataxia telangiectasia. Biochim. Biophys.Acta 474: 49-60

Natarajan AT, Obe G (1984) Molecular mechanisms involved in the production of chromosomal aberrations. III. Restriction endonucleases. Chromosoma (Berlin) 90: 120-127

Okayasu R, Iliakis G (1989) Linear DNA elution dose response curves obtained in CHO cells with non-unwinding filter elution after appropriate selsction of the lysis conditions. Int.J. Radiat. Biol. 55: 569-581

Paretzke HG (1976) An appraisal of the relative importance for radiobiology of effects of slow electrons. In: Proc. 5th Symp.Microdos. pp. 41-58, Booz J, Ebert HG, Smith BGR, edts., CEC Luxemburg

Paterson MC, Smith PJ (1979) Ataxia telangiectasia. An inherited human disorder involving hypersensitivity to ionizing radiation and related DNA-damaging chemicals. Ann.Rev.Genet. 13: 291-318

Pètin VG (1979) Effect of gamma and alpha irradiation on survival of wild-type and sensitive mutants of yeast. Mutat. Res. 60: 43-49

Prise KM, Davies S, Michael BD (1987) The relationship between radiation-induced DNA double-strand breaks and cell killing in hamster V79 fibroblasts irradiated with 250 kVp x-rays, 2.3 MeV neutrons or Pu-238 alpha particles. Int. J. Radiat.Biol. 52: 893-902

Purrott RJ, Edwards AA, Lloyd DC, Stather JW (1980) The induction of chromosome aberrations in human lymphocytes in vitro irradiated with alpha particles from plutonium-238. Int. J. Radiat. Biol. 38: 277-248

Radford IR (1985) The level of induced DNA double-strand breakage correlates with cell killing after x-irradiation. Int.J. Radiat.Biol. 48: 45-54
Radford IR (1986) Effect of radiomodifying agents on the ratios of x-ray-induced lesions in cellular DNA: use in lethal lesions determination. Int.J. Radiat.Biol. 49: 621-637
Resnick MA, Martin P (1976) Repair of double-strand breaks in the nuclear DNA of Saccharomyces cerevisiae and its genetic control. Molec.Gen. Genet. 143: 119-129
Ritter MA, Cleaver JE, Tobias CA (1977) High LET radiations induce a large proportion of non-rejoining DNA breaks. Nature 266: 653-655
Robertson JB, Koehler A, George J, Little JB (1983) Oncogenic transformation of mouse BULB/3T3 cells by plutonium-238 alpha particles. Radiat.Res. 96: 261-274
Roots R, Yang TC, Craise L, Blakely EA, Tobias CA (1979) Impaired repair capacity of DNA breaks induced in mammalian cellular DNA by accelerated heavy ions. Radiat.Res. 78: 38-49
Sakai K, Suzuki S, Nakamura N, Okada S (1987) Induction and subsequent repair of DNA damage by fast neutrons in cultured mammalian cells. Radiat.Res. 110: 311-320
Takatsuji T, Sasaki MS (1984) Dose-effect relationship of chromosome aberrations induced by 23 MeV alpha particles in human lymphocytes. Int.J. Radiat. Biol. 45: 237-243
Taylor AMR (1978) Unrepaired DNA strand breaks in irradiated Ataxia telangiectasia lymphocytes suggested from cytogenetic observations. Mutat.Res. 50: 407-418
Thacker J, Stretch A (1985) Responses of four x-ray sensitive CHO cell-mutants to different radiations and to irradiation conditions promoting cell recovery. Mutat.Res. (DNA Repair Reports) 146: 99-108
Van der Schans GP, Centen HB, Lohman PHM (1982) DNA lesions induced by ionizing radiation, In: Progr.Mutat.Res., Vol.4, pp. 285-299, Natarajan AT, Obe G, Altmann H, edts., Elsevier, Amsterdam
Vulpis N (1973) Chromosome aberrations induced in human peripheral-blood lymphocytes using heavy particles from $^{10}B(n,\alpha)$ ^{7}Li reaction. Mutat.Res.18: 103-111
Weibezahn KF, Coquerelle T (1981) Radiation-induced DNA double-strand breaks are rejoined by ligation and recombination processes. Nucl. Acids Res. 9: 3139-3150
Weichselbaum RR, Nove J, Little JB (1978) Deficient recovery from potentially lethal radiation damage in Ataxia telangiectasia and Xeroderma pigmentosum. Nature 271: 261-262

DNA DOUBLE-STRAND BREAK INDUCTION IN YEAST BY HEAVY ION IRRADIATION

T.C. Akpa, K.J. Weber, J. Kiefer, M. Frankenberg-Schwager*, R. Harbich*, D. Frankenberg*

Strahlenzentrum der Justus-Liebig-Universität, Giessen and
* Abteilung für Biophysikalische Strahlenforschung der GSF, Frankfurt

DNA double-strand breaks are regarded as fundamental lesions in the genome of irradiated cells. In yeast lethal damage from sparsely ionizing radiation could conclusively be identified with this lesion type whereas for other systems the situation is less clear. Although there are data for some high LET radiations, mainly α-particles, it is still an open question whether DSB are also the determinants of cellular effects from heavy ion exposure. The first preliminary results on DSB induction in wild-type yeast (strain 211*B) exposed to accelerated charged particles are presented here. The irradiations were performed at the Gesellschaft für Schwerionenforschung (GSI, Darmstadt) using the UNILAC-accelerator. Exposure conditions and dosimetry were as described by Schöpfer et al. (1980). DSB were assessed by neutral sucrose sedimentation according to Frankenberg-Schwager et al. (1979). Measurements are so far available for neon, argon and lead ions of various energies up to about 20 MeV/u (see table 1).
DSB induction was found to depend linearly on particle fluence, the respective production rates are given in table 1. The parallelism between DSB induction and inactivation cross sections is striking, the similarity of numerical values is, of course, coincidental.

Table 1. Physical parameters of ions used, DSB induction and inactivation cross section

ion	energy MeV/u	LET keV/μm	DSB cross sect. $\mu m^2/10^9$ g/mol	inact. cross sect. μm^2
He	0.88	130	0.31	
Ne	4.9	661	0.54	0.57
Ne	14.4	315	0.39	0.37
A	8.8	1296	0.47	0.51
A	12.9	1020	0.66	0.69
A	14.0	963	0.62	0.52
Pb	12.7	11080	2.08	2.08

Figure 1 summarizes DSB-induction cross sections as a function of LET for a number of ions obtained by other groups in various biological systems. Despite large variations in cellular DNA-contents and considerable scatter of the data there is an indication of a common behaviour: Cross sections increase linearly in the low-LET region and approach

NATO ASI Series, Vol. H 54
The Early Effects of Radiation on DNA
Edited by E. M. Fielden and P. O'Neill

some kind of saturation (with the exception of lead) after passing a range where the rise is supralinear with LET. This points to similar induction mechanisms at the molecular level.

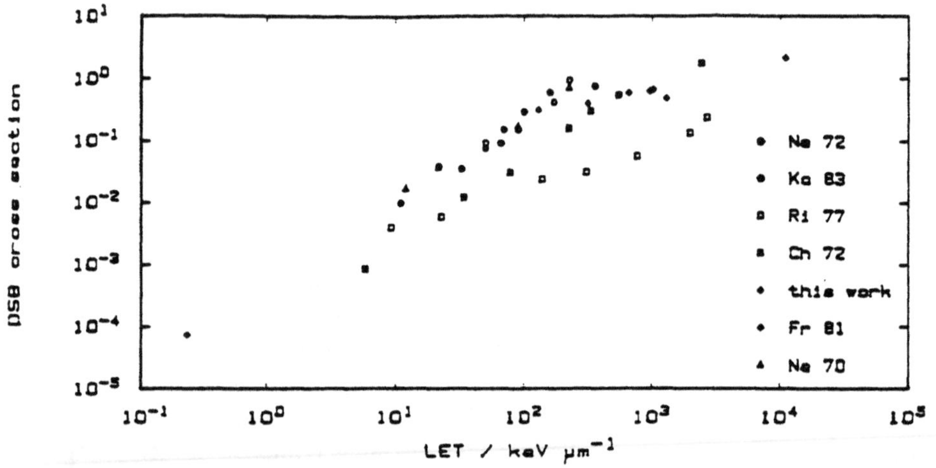

Fig. 1. DSB production rates in different systems. The data were normalized to DSB per 10^9 g/mol. Cross sections are in square micrometer.
Key: Ne 72: Neary et al. 1972 (phage); Ka 83: Kampf and Eichhorn 1983 (mammalian cells), Ri 77: Ritter et al. 1977 (mammalian cells); Ch 72: Christensen et al. 1972 (phage); Fr 81: Frankenberg et al. 1981 (yeast); Ne 70: Neary et al. 1970 (phage)

These studies are supported by the "Gesellschaft für Schwerionenforschung", Darmstadt,Germany. We thank the accelerator crew and Dr. G. Kraft for help with the irradiations and Beate Barth for technical assistance.

REFERENCES

Christensen RC, Tobias CA, Taylor WD (1972) Heavy-ion induced single- and double-strand breaks in ϕX-174 replicative form of DNA. Int J Radiat Biol 22:457-477

Ritter MA, Cleaver JE, and Tobias CA (1977) High-LET radiation induce a large proportion on non-rejoining DNA breaks. Nature 266:653-655

Neary GJ, Horgan VJ, Bance DA, Stretch A (1972) Further data on DNA strand breakage by various radiation qualities. Int J Radiat Biol 22:525-537

Neary GJ, Simpson-Gildemeister VFW, Peacocke AR (1970) The influence of radiation quality and oxygen on strand breakage in dry DNA. Int J Radiat Biol 18:25-40

Kampf G, Eichhorn K (1983) DNA strand breakage by different radiation qualities and relations to cell killing: Further results after the influence of alpha-particles and carbon ions. Stud Biophys 93:17-26

Frankenberg-Schwager M, Frankenberg D, Blöcher D, Adamczyk C (1979) The influence of oxygen on the survival and yield of DNA double-strand breaks in irradiated yeast cells. Int J Radiat Biol 36:261-270

Frankenberg D, Frankenberg-Schwager M, Blöcher D, Harbich R (1981) Evidence for double-strand breaks as the critical lesions in yeast cells irradiated with sparsely or densely ionizing radiation under oxic and anoxic conditions. Radiat Res 88:524-532

Schöpfer F, Kiefer J, Schneider E, Kraft G (1980) Inactivation of dried yeast cells by accelerated heavy ions of very high LET. Radiat Res 82:235-243

THE USE OF DNA PRECIPITATION ASSAY FOR EVALUATING DSB INDUCED BY HIGH AND LOW LET RADIATIONS: COMPARISON WITH SEDIMENTATION RESULTS

M. Belli#, D. T. Goodhead$, F. Ianzini#, T. J. Jenner$, G. Simone*, M. A. Tabocchini#.
*Istituto FRAE - CNR
Via de' Castagnoli, 1
40126 Bologna, ITALY

The assays used for measurements of DNA double strand breaks (dsb), such as neutral sucrose gradient or filter elution, need sophisticated and expensive equipments as well as a great deal of time. Recently, Olive (1988) has proposed a relatively simple method for detecting dsb-related damage: a DNA precipitation assay in non-denaturing conditions. The principle of this method is that the treatment of cells with sodium dodecyl sulphate and KCl causes cell proteins to precipitate, trapping long pieces of DNA. If breaks are introduced by radiation, DNA fragments remain in the supernatant. The yield of dsb induced at dose D can be related to the ratio: $Y(D) = \ln[F_p(D)/F_p(O)]$, where $F_p(D)$ and $F_p(O)$ are the percentage of precipitated DNA at dose D and in the control, respectively.

In this paper we have compared the yield of initial dsb induced in Chinese hamster cells by ionizing radiations, using low speed sedimentation on neutral sucrose gradient and DNA precipitation assays.

V79-4 cells, grown as monolayer on Hostaphan bases of glass walled dishes, were exposed to protons and alpha particles with LET of 20 and 23 keV/µm. Irradiations were carried out using the Variable Energy Cyclotron at the Harwell Laboratory. The dependence of the initial yield of dsb on the dose was determined and compared with that for X-rays and ^{238}Pu alpha particles. The dose range was 30-200 Gy.

Linear fits of the data were performed by the least square method using: $1/M = \alpha D + \beta$ (with the common intercept β = 1.84

ISS, Rome, Italy; $ MRC, Chilton, United Kingdom

NATO ASI Series, Vol. H 54
The Early Effects of Radiation on DNA
Edited by E. M. Fielden and P. O'Neill

TABLE 1

RADIATION	($\alpha \pm$s.e.)10^{-9} (sedimentation)	RBE$\pm$s.e. (sedimentation)	($\alpha \pm$s.e.)10^{-2} (precipitation)	RBE$\pm$s.e. (precipitation)
X-rays	0.43$\pm$0.05		0.19$\pm$0.01	
20 keV/μm protons	0.37$\pm$0.04	0.88$\pm$0.15	0.14$\pm$0.01	0.76$\pm$0.06
" " alpha	0.51$\pm$0.05	1.21$\pm$0.18	0.13$\pm$0.02	0.68$\pm$0.09
23 keV/μm protons	0.43$\pm$0.04	1.01$\pm$0.15	0.16$\pm$0.01	0.82$\pm$0.07
" " alpha	0.50$\pm$0.04	1.17$\pm$0.16	0.13$\pm$0.02	0.67$\pm$0.10
124 keV/μm 238Pu	0.39$\pm$0.02	0.91$\pm$0.12		

x 10^{-10} $\pm$ 0.23 $dalton^{-1}$) for sedimentation and Y(D) = αD for precipitation. The α parameter (expressed as $dalton^{-1}.Gy^{-1}$) and the RBE values (α/α_x) are reported in Table 1.

The sedimentation results show little differences among radiations of different quality. This finding is not consistent with the inactivation and mutation induction data (Thacker et al, 1979, 1982), and indicate that the amount of initial dsb is not directly related to cellular damage. A possible explanation could be that different types of radiation may produce dsb with different biological consequences.

The DNA precipitation results confirm that the initial yield of dsb for protons and alpha particles of the same LET are similar, while low LET X-rays appear to be more effective than these particles, as is shown by the RBE values lower than 1. This apparent discrepancy may reflect a difference in the type of lesions scored by the two methods.

On the whole, our data are consistent with the hypothesis that differences in cellular response can be due not only to the amount but also to the spectrum of lesion induced by different types of radiation.

REFERENCES

Olive P L (1988) DNA precipitation assay: A rapid and simple method for detecting DNA damage in mammalian cells. Env Mol Mut 11:487-495

Thacker J, Stretch A, Goodhead D T (1982) The mutagenicity of α particles from plutonium-238. Rad Res 92:343-352

Thacker J, Stretch A, Stephens M A (1979) Mutation and inactivation of cultured mammalian cells exposed to beams of accelerated heavy ions 11. Chinese hamster V79 cells. Int J Rad Biol 36:137-148

DISCUSSION OF THE SESSION ON RBE FROM THEORY TO EXPERIMENT

K.H. Chadwick
Radiation Protection Programme
Commission of the European Communities
Rue de la Loi 200
B - 1040 Brussels

Introduction

The different biological effectiveness per unit dose of different types of radiation has always been intriguing and a challenge to explain. The physics of radiation interaction with matter clearly demonstrates the different spatial patterns of energy deposition by the different radiations and is obviously very relevant. Calculations show that the specific spatial patterns of energy deposition diffuse and are no longer specific after a time interval of less than a microsecond so that the induction of the biologically effective damage is very fast. The typical dependence of biological effectiveness with radiation quality implies the importance of two or more energy deposition events within a discrete volume and might be interpreted to imply a structured target such as the DNA double helix or the chromatin. The shape of the dose effect relationships changes with increasing radiation quality from a curvi-linear relationship for sparsely ionising radiation to a linear relationship for more densely ionising radiation. This might be interpreted to imply different types of critical lesions being induced by the different radiations or a different effectiveness of repair of the lesions because densely ionising radiation might cause multiply damaged sites which are less likely to be correctly repaired than the lesions induced by sparsely ionising radiation. All these topics are admirably dealt with in the presentations but several relevant points were raised in the discussion, namely the relationship between DNA double strand breaks and lethality, the repairability of lesions induced by densely ionising radiation, the modification of the effect induced by densely ionising radiation, the comparison of effects in radiation sensitive and normal mammalian cells, the comparison of effects in yeast and mammalian cells and recombinational repair of DNA damage.

NATO ASI Series, Vol. H 54
The Early Effects of Radiation on DNA
Edited by E. M. Fielden and P. O'Neill

The relationship between DNA double strand breaks and lethality

Several current biophysical models associate DNA double strand breaks and cell lethality. Some assume one double strand break as the critical lesion, others assume pairs of double strand breaks as the lethal lesion. Measurements of DNA double strand break induction by different types of radiation have not yet resolved the question of how critical the double strand break lesion really is. At low LET, correlations have been made between cell survival and DNA damage, purported to be double strand breaks, measured by neutral elution techniques by several groups although the measurements and correlations are disputed by other groups. The dispute is fuelled by the fact that the dose relationship for the damage measured by neutral elution differs in shape from that measured using neutral sucrose gradient sedimentation and the fact that the RBE for the damage measured using neutral elution appears to be close to or less than one as radiation LET increases in contradiction to the biological effect. The RBE of the damage, presumably double strand breaks, measured using neutral sucrose gradient sedimentation in yeast cells does increase as LET increases. While there may be several reasons for the discrepancies such as different dose ranges covered by different techniques, use of different techniques in different cells, the influence of multiple double strand breaks per track traversal of a cell nucleus on the efficiency of the damage measurement techniques, and even the way in which the biological effect is analysed, there is no doubt that the resolution of the disputes and discrepancies between the different measurement techniques will lead to considerable progress in the understanding of the relationship between DNA double strand breaks and biological effect.

The repairability of lesions induced by densely ionising radiation

Some biophysical models ascribe the curvi-linear dose relationship for biological effects induced by sparsely ionising radiation to a restricted or saturated repair process and the linear dose relationship for densely ionising radiation to the lack of repair or a different type of repair. Although there is little direct evidence for a saturation of a biochemical repair pathway in mammalian cells and none in yeast cells, it is not possible as yet to discount totally its existence. The repair of double strand breaks, although only measured at very high doses, does appear to have the same kinetics after sparsely and densely ionising radiation, although the total amount of repair may not be the same. If prematurely condensed chromosomes can be considered indicative of double strand breaks, and this is still debatable, then similar repair kinetics but relatively more unrepaired damage would be indicated after densely ionising radiation. This obviously leads to the question of whether completely different lesions are induced by densely ionising radiation compared with sparsely ionising radiation or whether the nature of the lesion might be different. For instance, sparsely ionising radiation might induce a simple double strand break whilst densely ionising radiation, depositing more energy near the DNA, might induce a double strand break associated with other damage to the DNA or chromatin structure close to the break making it more difficult to repair. These could be the local multiply damaged sites. In this situation the RBE effect as a function of radiation quality would be considered to arise as a consequence of a radiation quality effect on the lesions induced and a radiation quality effect on the repair of the lesions induced. It would be quite complicated to try to unravel these different components of the effect of radiation quality unambiguously.

The modification of the effect induced by densely ionising radiation

The use of the terms 'direct' and 'indirect' radiation action has always caused confusion among the physicists who consider 'direct' action to imply the energy deposition event taking place in the DNA molecule itself and 'indirect' action to imply any DNA damage arising from energy deposition events outside the DNA molecule, and the chemists who consider 'direct' action to be all damage which cannot be scavenged or modified. An attempt was made at the meeting to resolve this difference in interpretation as it was clear that the two concepts about 'direct' action were probably not identical and energy depositions in the water within one or two molecular layers from the DNA will be very difficult to scavenge. It was suggested that the terms 'unscavengeable' and 'scavengeable' damage be adopted and though this might be an improvement, it might also be difficult to define in a totally unambiguous way. It does however represent the difference in radiation action which can be determined phenomenologically.

The effect of densely ionising radiation is often thought to be unmodifiable compared with the effect of sparsely ionising radiation and the oxygen enhancement ratio decreases to unity as LET increases. It was suggested that the modification of the sparsely ionising radiation effect would alter the ratios of the relative RBE for different radiation and that this might influence the conclusion that an energy deposition of 100 eV in the 2nm cube was critical for the biological effect. It seems reasonable to anticipate this change but it was felt that this would not alter by very much the level of energy deposition required for the biological effect. The modification of mutation frequency induced by fission neutrons by thiols was reported but this was not really considered to be too surprising and it seems to be the effects of the very densely ionising radiation which are not, or only slightly modifiable. In this case most of the damage will arise from energy deposition in or very close to the DNA molecule and the role of saturation caused by the induction of multiple lesions in a cell nucleus by the passage of a single particle track may also influence the measurement of any modifying effect. It is perhaps worth noting here that in a linear-quadratic analysis of the effect of oxygen on cell survival after sparsely ionising radiation, the oxygen gave a much greater modification of the quadratic term than of the linear term of the dose effect relationship.

The comparison of effects in different cells

Some questions were raised about whether it was entirely valid to make comparisons of radiation effects arising in different cells, such as the comparison between yeast and mammalian cells or between radiation sensitive, repair deficient cells and normal cells. Although comparisons are sometimes difficult the results in general are encouraging and work in yeast points to the importance of the DNA double strand break lesion and its repair and the work in repair deficient mammalian cells also shows that cells which are defective in the repair of double strand breaks are also radiation sensitive. Increasing evidence shows that there has been a remarkably strong preservation of genes for the repair of DNA damage with evolution and a high level of homology of repair genes or proteins can be found extending from the bacteria E.Coli through yeast and drosophila to mammals and man. This would surely suggest that some comparisons between the different cells can be cautiously made bearing in mind the differences between the cells which may influence the radiation effect, such as single or multi-chromosome nucleus, mitosis or budding, the presence of both exons and introns in the genes, the presence of repetitive DNA, the length of chromosomes, the ability of cells to undergo both homologous and non-homologous recombination. That being noted, it remains an interesting and intriguing observation that the effect of modifiers and of different radiation qualities is in general smaller in repair deficient radiation sensitive cells when compared with normal cells.

Another question raised in connection with the comparison of different cells concerned the recombinational repair of DNA double strand damage and whether this process occurred in mammalian cells as well as yeast and bacteria. Recombinational repair is a complicated enzymatically controlled process which involves an undamaged stretch of DNA closely homologous to the damaged region and which can, at least theoretically, lead to either complete, perfect repair, rejoining of the originally damaged DNA molecule with some loss of bases or incorrect rejoining involving the exchange between the damaged and homologous DNA molecules. This sort of repair can also be invoked to explain the formation of chromosomal aberrations. Evidence for this sort of repair has been found in bacteria and yeast cells and some DNA structures associated with the recombination process have been identified in mammalian cells. The strong preservation of repair processes throughout the evolutionary process would also argue in favour of recombinational repair also being present in mammalian cells.

Summary

Many aspects of the variation of relative biological effect as a function of radiation quality remain to be solved and if as is currently suggested the variation is made up of both a track structure/target structure effect and a repair effect, it will be difficult to resolve the problems unambiguously. Even so, considerable progress has been made over the past few years and more progress can be anticipated if a good interdisciplinary co-operation can be established to attack specific aspects of the questions which remain.

THE DEVELOPMENT OF CHEMICAL DAMAGE OF DNA IN AQUEOUS SOLUTION

D. Schulte–Frohlinde and E. Bothe
Max–Planck–Institut für Strahlenchemie
Stiftstrasse 34–36, 4330 Mülheim a.d. Ruhr, FRG

Abstract. A summary of results obtained in our laboratory is given concerning the formation of free radicals produced in DNA and components of DNA following ionization or reaction with OH radicals in aqueous solution and the rate of their conversion and decay. In the case of laser excitation biphotonic ionization of the nucleobases takes place and the radical cations primarily formed are converted to neutral base radicals by deprotonation or by reaction with water. The neutral base radicals have their highest free spin densities at the nitrogens and oxygens of the nucleobases (N– and O–radicals). In contrast reaction with OH radicals leads primarily to carbon radicals by addition to the double bonds of the nucleobases and to sugar radicals by H abstraction. The reactions, the rate of the conversion and the decay of these radicals in DNA in aqueous solution is discussed with special regard to the radicals leading to strand break (sb) formation. In cells the decay is dominated by the reaction of the DNA radicals with endogenous thiols.

Introduction. In order to help to understand the chemical reactions of DNA in an irradiated cell the chemical status of a DNA molecule in aqueous solution a few microseconds after an electron pulse will be examined. In the literature two pathways to damage of DNA in a cell, called either the direct or the indirect effect are discussed. The direct effect leads to damage of DNA by ejection of an electron from the DNA itself producing DNA radical cations and electrons, whereas by the indirect effect first OH and other radicals are generated in the neighborhood of DNA which diffuse to DNA and react with it.

Radicals in DNA and nucleosides formed by the direct effect within a few μs after an electron pulse. The ejection of an electron (ionization) induced by high–energy irradiation with low LET occurs only during irradiation. After the pulse no further ionization takes place. Since the electron is not considered to cause much damage to DNA only the fate of the radical cations is considered. Knowledge of the behavior of the radical cations of the bases in DNA stems from laser studies and from studies with $SO_4^{\bullet-}$ radical ions. The latter compound reacts with bases and in many cases leads to radical cations of these bases (Rustgi and Riesz, 1978, O'Neill and Davies, 1987, Hazra

NATO ASI Series, Vol. H 54
The Early Effects of Radiation on DNA
Edited by E. M. Fielden and P. O'Neill

and Steenken, 1981). Both methods for the generation of radical cations have been employed. The laser method in addition to ionization by a biphotonic process (eq 1) produces various photoproducts, many of them by monophotonic processes (for a review see Nikogosyan, 1990).

$$(1) \qquad DNA + 2\,h\nu \longrightarrow DNA^{+} + e^{-}$$

What are the rates and the chemical reactions of the base radical cations in model systems?

The radical cation of thymidine has been produced by reaction with $SO_4^{\cdot -}$. Reaction Scheme I is a complete presentation of the reactions of the thymidine radical cation with all rate constants at pH 7 as measured by Deeble at al. (1990) in our laboratory. The radical cation disappears with a half life of less than a few μs and is converted into N and C radicals by elimination of a proton either from the NH–3–group or from the methyl group. The methyl radical is produced only in low yield. The reactivity of the thymine–N–3–radical is not known. In aqueous solution the N–3 radical of thymidine disappears bimolecularly. It should be noted that many of the reactions are reversible, under certain conditions this leads to a behavior of the system which is difficult to predict. In DNA, due to the long lifetime of DNA radicals, the N–3 radical should be converted into the OH adduct radical faster than formation of strand breaks. This conclusion is based on the rate constants given in Reaction Scheme I under the assumption that the N–3 radical reacts only slowly with sugars.

Reaction Scheme I

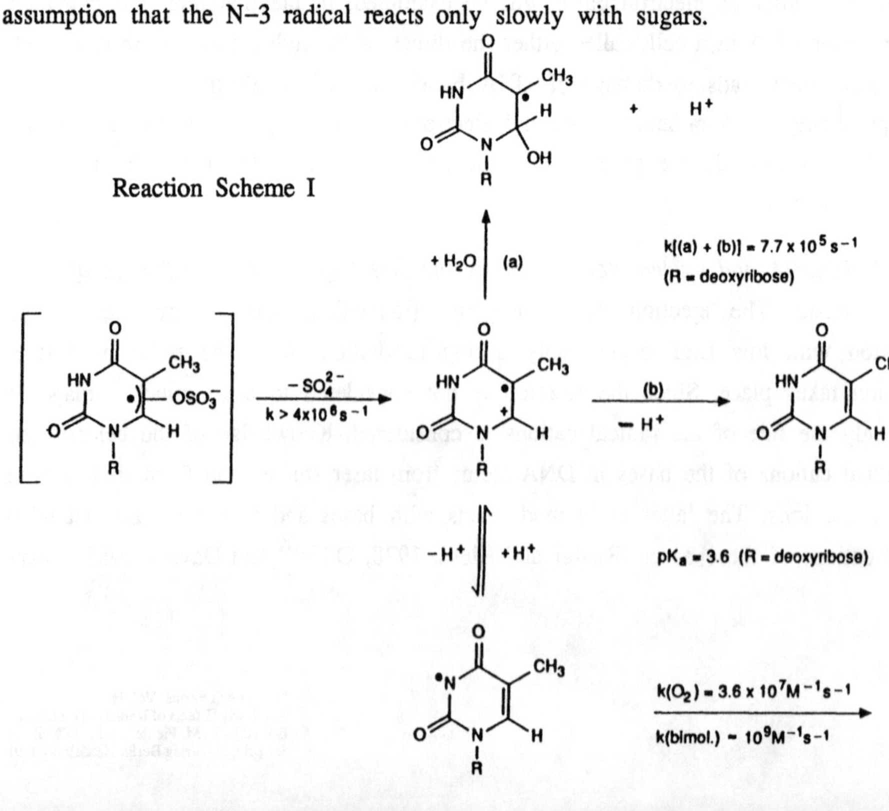

The cytidine radical cation possibly loses a proton from the amino group in a few μs and is converted into a nitrogen radical (eq 2).

(2)

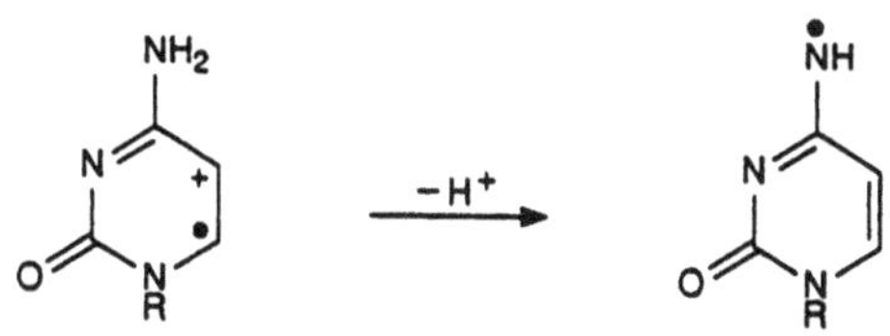

Esr studies of cytidine (R=ribose) in aqueous solution show that the N radical formed as in eq (2) is so reactive that only sugar radicals can be observed (Hildenbrand, Behrens, Schulte-Frohlinde, and Herak, 1989).

The guanosine radical cation (R=ribose) loses a proton at pH 7 and forms an oxyl radical (O radical) (eq 3) (Willson et al., 1974). This reaction is in equilibrium with the back reaction (eq 3). The pK_a value is 3.9. No reaction of the radical cation of guanine with water leading to addition at position 8 is observed since no 8-OH-guanosine is formed as product (Steenken, private communication).

(3)

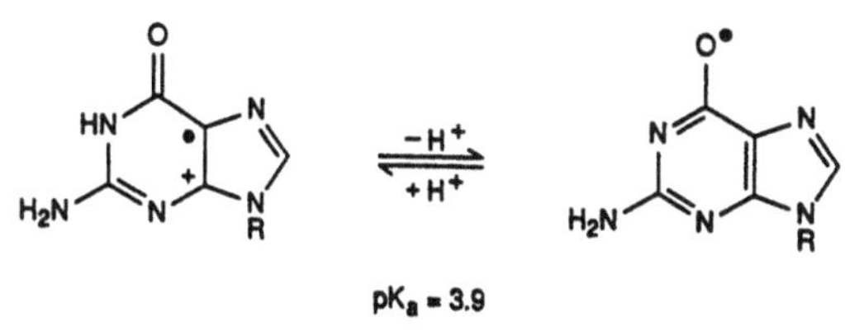

The guanosine oxyl radical is not very reactive. It has been tentatively identified in dsDNA in aqueous solution with a lifetime of about 5 s (Hildenbrand and Schulte-Frohlinde, 1990). In single-stranded DNA, however, the oxyl radical of guanine could not be observed by esr.

The adenosine radical cation decays by proton release and formation of an N-radical (eq 4) (Vieira and Steenken, 1987).

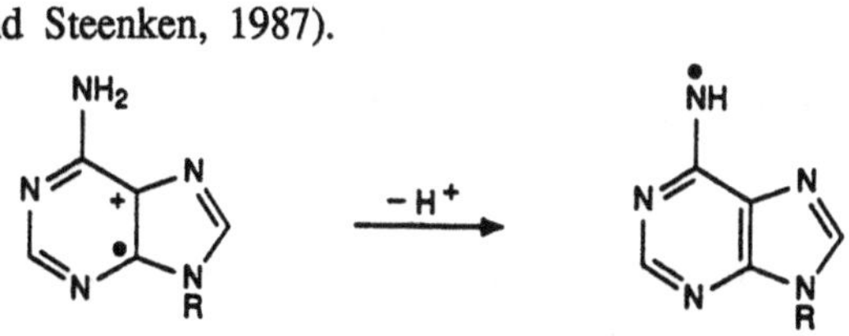

(4)

This H^+ elimination takes place in less than a few µs. No water addition occurs at position 8 of the radical cation as inferred from the absence of 8–OH adduct formation (Steenken, private communication).

If it is assumed that the above described reactions of the radical cations of the bases of nucleosides occur also in DNA it follows that neutral N and O base radicals are produced within a few µs in DNA and it is only with thymine that carbon radicals are also formed. Table 1 summarizes those radicals which are considered to be formed by ionization of the bases of DNA followed by deprotonation.

Table 1. Radicals produced in nucleotides following ionization of the bases and deprotonation. Data from O'Neill and Chapman, 1985, Willson, Wardman and Asmus, 1974, Vieira and Steenken, 1987, Deeble et al. 1990.

base	radical
Adenine	N–6–radical
Guanine	O–6–radical
Cytidine	N–4–radical*
Thymine	N–3–radical
	$-\dot{C}H_2$–radical

* probable but not certain

In DNA in addition to deprotonation the decay of a radical cation may occur by charge transfer. In frozen solution of DNA X–irradiation leads to guanine radical cations as the only positively charged radicals observed (Gräslund et al., 1975). This may be interpreted as indicating that charge transfer from the other base radical cations to guanine takes place. Charge transfer (or excitation transfer) from one base to another has been discussed as an explanation for the difference in the ssb yields of ssDNA induced by ·OH radicals or by radical cations in contrast to equal yields for poly(A) and poly(C) (see Table 2) (Wala et al., 1990).

The conversion of the radical cations into the neutral radicals in dsDNA has been discussed by Steenken (1989). The rate constants in DNA, however, are not well known. In the special case of polyuridylic acid (poly(U)), the uracil radical cation reacts with water and forms the 5–OH–6–yl radical (Schulte–Frohlinde et al., 1985).

An interesting question is whether or not the direct effect induced by high–energy absorption leads to the formation of sugar radical cations and phosphate radicals. An answer is difficult to obtain. In frozen aqueous solution of DNA X–irradiation leads solely to guanine radical cations and thymine radical anions (and cytidine radical anions as was shown recently by Bernhard (1989) and by Hüttermann (private communication, 1989)). It is intriguing that no sugar radicals are observed although sugar radicals are formed on warming showing that the sugar radicals can be observed by esr (Cullis et al., 1986). The sugar radical formation on warming is likely to be due to H abstraction from the sugar by base radicals and this sugar radical formation is therefore no indication of a direct ionization of the sugar. The formation of sugar and phosphate radical cations may be inhibited by the rigidity of the system due to low temperatures, or by the neighborhood of the bases. If the latter is the main cause then at room temperature sugar and phosphate radicals should not be formed directly as a consequence of ionization by the direct effect of high–energy absorption.

What is the fate of the N and O radicals of the bases of DNA in aqueous solution and what is the time dependence of their decay? In DNA two kinds of decays of DNA radicals have to be distinguished. The first is the disappearance of the radicals by a final, bimolecular reaction (either by dimerization, disproportionation or reaction with additives e.g. a metal ion). The second is the decay of a DNA radical by conversion into another DNA radical. It is clear that this conversion must be faster than the bimolecular reaction in order to compete. Insight into the conversion of DNA base radicals into DNA sugar radicals can be obtained from strand break measurements.

Table 2. Yield of single–strand break formation of various polynucleotides and DNA in aqueous solution at pH ~ 7. Yields are given in percent per reacting OH and per radical cation produced in the nucleic acid. Data from Wala et al. (1990).

	radical cation	·OH
poly(U)	–	40
poly(C)	43	40
poly(A)	6	8
ssDNA	5	15
dsDNA	9*	22*

* double–strand break with a mainly quadratic dependence on total dose

The results available from laser excitation of DNA are yields for single– and double–strand break formation per radical cation (Table 2). Table 2 shows that the yield for ssb formation is low for poly(A) but high for poly(C) and that the yields induced by OH and by radical cations are similar. This has led to the suggestion that the effect of the direct and the indirect effect on strand break formation is similar (Wala et al., 1990, Schulte–Frohlinde, 1986). The main message obtained from the laser results is that base radicals can lead to strand break formation. This follows from the fact that the ns laser light (λ 248 nm) is absorbed by the bases only and not by the sugar and water at the intensities and concentrations used in these experiments. That base radicals can lead to ssb formation has already been shown earlier for poly(U) as model system by other methods (Lemaire et al., 1984). Not very much is known concerning the rate of disappearance of neutral radicals induced in DNA by laser excitation. For poly(U) the rate constant for the conversion of the base radical into the sugar radical is identical to that induced by OH radicals (Schulte–Frohlinde et al., 1985).

Radicals in DNA formed by the indirect effect and present a few μs after an electron pulse. In cells the reaction of OH radicals with DNA occurs in a few nanoseconds since the diffusion length is very short due to the high concentration of organic material which act as OH scavengers (Roots and Okada, 1975). What kind of radicals are produced in DNA by OH radicals? The OH radicals add mainly to the C=C and C=N bonds of the nucleobases.

Table 3. Estimates of radicals produced in ssDNA by reaction with OH radicals based on results and rate constants measured with nucleotides in aqueous solution. Rate constants from various authors as cited in the review by von Sonntag, 1987, p. 117.

Adenine:	4–OH adduct	10 %
	8–OH "	10 %
Guanine:	4–OH "	17 %
	8–OH "	17 %
Thymine:	5–OH "	12 %
	6–OH "	6 %
	$-{}^{\bullet}CH_2$	2 %
Cytidine:	5–OH "	18 %
	6–OH "	2 %
Sugar:		6 %
		100 %

Experimental results of Ward and Kuo (1978) show that OH radicals abstract H atoms from the sugar of dsDNA to a considerable extent in aqueous solution. Scholes et al. (1969) have concluded that the distribution of the OH adduct base radicals in DNA is similar to that expected on the basis of reaction rate constants of OH with nucleosides and nucleotides. There is practically no difference between the absorption spectra of ss and dsDNA radicals following reaction with OH (Bothe, 1990). These results allow to make an estimate of the distribution of radicals obtained with OH in DNA. In Table 3 the percentages are given for the yield of these radicals in ssDNA if they would be formed according to the rates with which the mononucleotides react with OH. The real distribution will deviate from percentages given in Table 3, however, the figures give an indication of the distribution under the assumptions made.

Reactions of the neutral radicals present a few μs after the electron pulse in DNA. Not very much is known about the chemical reactions of the neutral radicals in DNA leading to their final disappearance in aqueous solution. Organic radicals can only disappear bimolecularly in the absence of metal ions. Since with dilute DNA solutions (10^{-5} - 10^{-4} mol dm^{-3} in nucleotide units) and doses of a few Gray with DNA of 10^6 Dalton several tens of radicals are formed per DNA molecule, the decay of the radicals may occur intramolecularly or intermolecularly. Washino and Schnabel (1982) have considered this situation for poly(A). In any case the DNA radicals have to encounter in order to disappear but the high negative charge of the DNA leads to repulsion between the radicals and to an unusually long lifetime of the order of seconds (Michaels and Hunt, 1977, Schulte–Frohlinde et al. 1986). Different from that is the conversion of DNA base radicals into other DNA radicals which occur by first order reactions. From studies of nucleosides the following is known concerning the decay of purine radicals. According to the work of different authors as cited by Steenken (1989), the 4–OH adduct radicals of the purine derivatives eliminate water and form N radicals (adenine) and O radicals (guanine). These are identical with those formed by laser excitation. This may be one reason why the yields of ssb formation induced by OH and by radical cations in poly(A) are similar (Table 2). In the case of poly(C), however, although the yields of ssb formation are similar (Table 2) the radicals which lead to H abstraction from the sugar may be different in the two cases. The corresponding 8–OH adduct radicals ring open.

Rate of decay of DNA radicals in aqueous solution. The decay of DNA base radicals in aqueous solution has been followed by absorption spectroscopy (Michaels and Hunt, 1977). The decay shows complex kinetics (Fig. 1). This indicates that different DNA base radicals decay with different rates. The decay includes the conversion of base radicals into sugar radicals. The latter are difficult to identify by absorption spectroscopy due to their low extinction coefficients.

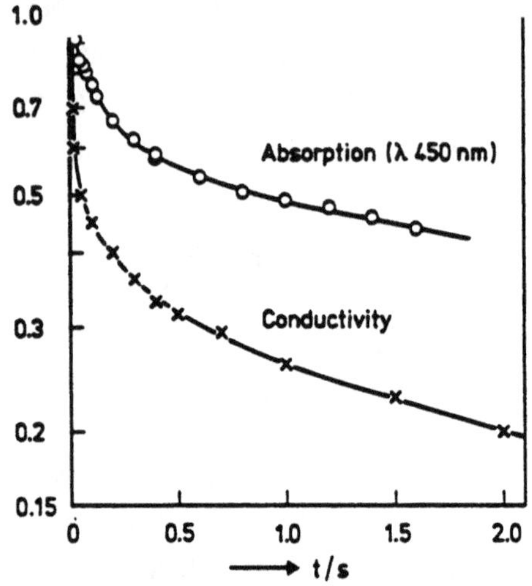

Figure 1. Decay of base radicals of ssDNA radicals following a 1 μs electron pulse in anoxic, N_2O–containing solution as measured by the decay of the optical absorption. Concentration of DNA 4 x 10^{-4} mol dm^{-3}. Data from Bothe et al. (1990). After 20 s the absorption due to the radicals has almost disappeared.

By means of esr it can be seen that different radicals have different lifetimes in DNA which depend on the gaseous conditions, on single– or double–strandedness of the DNA and on the presence of additives (Table 4).

Table 4. DNA radicals identified by esr in single– and double–stranded DNA in aqueous solution and their decay rate constants under various gaseous conditions. Data from Hildenbrand and Schulte–Frohlinde (1990), Schulte–Frohlinde et al. (1986), Schulte–Frohlinde and Hildenbrand (1989).

		$k(s^{-1})$	identified radicals
ssDNA	N_2	0.44, 1.8	6–OH–5–yl thymine and 5–methylenuracil
	O_2	4.7	peroxyl radicals
dsDNA	N_2	0.14 n.d.	$G(-H)^{\cdot}$ (guanine(–H) radical) 6–OH–5yl thymine
	O_2	2.0	peroxyl radicals

From a comparison of the rate constants in Tables 4 and 5 it follows that the radicals identified in Table 4 are not those which are responsible for the fast and larger portion of strand break formation. However, it can be concluded that the rate constants are the upper limit for those radicals which finally disappear bimolecularly. The conversion of some base radicals of DNA occurs much faster. Bothe et al. (1990) found that the rate constant for the water elimination reaction of the 4–OH adduct of adenine in poly(A) and dsDNA is very similar to that found by Vieira and Steenken for mononucleotides (1987).

Rate of strand break formation. A free radical–induced strand break formation can only occur when the sugar is damaged. In the absence of oxygen, the sugar 4' radical in DNA leads to sb formation. The intramolecular conversion of a base radical into a sugar radical and the demonstration that this is the rate determining step for ssb formation is clearly indicated for poly(U) in the absence of O_2 on the basis of esr and conductivity measurements. The k value of this conversion is 0.2 s^{-1} at 4 ^{o}C and identical with that measured for ssb formation (Hildenbrand and Schulte–Frohlinde, 1989, Schulte–Frohlinde, 1989). For dsDNA no clear evidence is available as yet for the extent of the involvement of base radicals in ssb formation. The results with ssDNA are presented by Bothe and Schulte–Frohlinde elsewhere in this meeting.

For DNA in the presence of oxygen it was shown that the rate of disappearance of the DNA peroxyl radical coincides with strand break formation (Liphard et al., 1990) and that the mechanism of sb formation involves a reaction of base peroxyl radicals with the sugar of the DNA.

Table 5. First order rate constants for single–strand break formation (k_{ssb}) of nucleic acids measured at pH 7.5 ± 0.5 in the absence and presence of oxygen, respectively, following an electron pulse at room temperature in the presence of N_2O. The k values are obtained from the first half life. Sources are given in the footnotes.

	k_{ssb} / s^{-1} (absence of oxygen)	k_{ssb} / s^{-1} (presence of oxygen)
poly(U)	0.6[a]	0.21[h]
poly(dU)	200, 2.3[b]	
poly(C)	10[c]	0.3[i]
poly(A)	10, 1000[d]	
poly(dA)	60[e]	
ssDNA	100[f]	

Footnotes to Table 5: [a] determined by conductivity measurements under salt–free conditions, Bothe and Schulte–Frohlinde, 1982. [b] two components of about equal contribution as estimated from the results of conductivity measurements, Adinarayana, Bothe and Schulte–Frohlinde, 1988. [c] determined by conductivity under salt–free conditions, Müller, 1983. [d] determined by conductivity under salt–free conditions. The two rate constants represent the rates of two components of strand breakage, with an approx. 30% contribution of the fast mode, Bothe and Schulte–Frohlinde, unpublished data. [e] determined by conductivity under salt–free conditions, Adinarayana, Bothe and Schulte–Frohlinde, 1988. [f] for the fast mode, as determined by conductivity under salt–free conditions, Bothe, Qureshi and Schulte–Frohlinde, 1983. [h] calculated from the overall half–lifes $t_{1/2}$, Bothe, Behrens, Böhm, Sethuram and Schulte–Frohlinde, 1986. [i] determined by conductivity under salt–free conditions, Bothe and Schulte–Frohlinde, unpublished data.

From the k values for ssb formation compiled in Table 5 it follows that the nature of the sugar has a strong influence on the rate constants, since ssDNA, poly(dA) and poly(dU) have much higher k values than poly(U) and poly(A). Poly(dU) and poly(dA) show k values which are close to the values measured for ssDNA. This may be explained by assuming that the rate determining step is H abstraction from the sugar by base radicals in which case the reactivity of the base radicals is a function of the stereochemical mobility of the bases in the macro molecules. However, the yields of ssb formation (Adinarayana et al., 1988) do not correspond to the rate constants measured so far (poly(U) (G_{ssb} = 2.4), with poly(dU) (G_{ssb} = 0.4) and of poly(A) (G_{ssb} = 0.4) with poly(dA) (G_{ssb} = 2.3)).

Rate constants of the reaction of DTT with DNA radicals in aqueous solution. The rate constants of reactions of DTT (dithiothreitol) with various ss and ds DNA radicals have been determined by different methods (Table 6).

Table 6. Rate constants for the reaction of DNA radicals with DTT at room temperature and neutral pH measured by different methods. Sources are given in the table.

	gas	$k(dm^3mol^{-1}s^{-1})$	method
ssDNA	N_2O	0.6×10^6	ssb measured by light scattering[a] and by conductivity[a]
	O_2	$(3.5 \pm 50\%) \times 10^4$	decay of the esr signal of the DNA peroxyl radical[b]
dsDNA	N_2O	2.0×10^6 (±50%)	ssb measured by the supercoil→open circle conversion of M–13DNA[c]
		1.4×10^6	disappearance of the supercoiled form of pBR322 DNA[d]
	O_2	$(5 \pm 50\%) \times 10^4$	decay of the esr signal of the DNA peroxyl radical[b]

Footnotes to Table 6:
[a] Prakash, Bothe and Schulte–Frohlinde (1990). [b] Schulte–Frohlinde, Behrens and Önal (1986). [c] Ludwig, Rettberg, and Schulte–Frohlinde (1990). [d] with 2–mercaptoethanol which is expected to react with similar rate constants as DTT since both thiols are neutral (Fahey et al., 1989). However it may be less reactive due to the presence of only one RSH group in 2–mercaptoethanol in comparison to the two RSH groups in DTT. For the influence of the charge of the thiol on repair rate see Fahey et al. (1990).

The reaction of DTT with DNA radicals in the absence of oxygen prevents sb formation. The sb formation can be quenched nearly completely (± 10%) by DTT. With which DNA radicals does DTT react in order to prevent sb formation? Table 1 offers 5 and table 3 offers 10 radicals which could contribute to sb formation. Since esr studies have excluded the route to sb formation via the 5–OH–6–yl radical of thymine and via the oxyl radical of guanine in dsDNA only the ring–opened radicals of the purines, the N–radical of adenine (however see the contribution of Bothe to this meeting), the cytidine radicals and the sugar 4'–radicals should contribute to sb formation. From studies of the reaction of the model systems poly(C) and poly(A) and others it follows that sb formation in poly(C) can be prevented by DTT with a rate constant of $k = 8 \times 10^6$ dm^3 mol^{-1} s^{-1} (Müller, 1983) and in poly(A) with a rate constant of $k \sim 8 \times 10^5$ dm^3 mol^{-1} s^{-1} (Bothe and Önal, unpublished). A more detailed study shows that DTT prevents by two different modes ssb formation in ssDNA (Prakash, Bothe and Schulte–Frohlinde, 1990). Liphard et al. (1990) have shown that the rate constant of the reaction of glutathione with ssDNA radicals preventing ssb formation is 8×10^4 dm^3 mol^{-1} s^{-1}. This value may be compared with $\sim 1.6 \times 10^5$ dm^3 mol^{-1} s^{-1} found for protection against inactivation of transforming DNA in aqueous solution (Held et al., 1984). With transforming DNA an unrepaired ssb is the main lethal event (K.H. Held, personal communication).

Rate constants of the reaction of DTT with DNA radicals as measured by survival. The fate of DNA radicals in E. coli is determined by the rate of reaction of the DNA radicals with endogenous thiols. In Table 7 some rate constants are listed which have been obtained from survival measurements in the presence of DTT. A comparison is made between transforming DNA, M–13 phage DNA, and E. coli. No comparison with mammalian cells will be made here. The data presented in Table 7 have large error margins.

Table 7. Rate constants for the repair reaction of DTT with radicals of biologically active DNA in aqueous solution and in E. coli cells based on survival measurements.

DNA	$k_r(dm^3\ mol^{-1}s^{-1})$	authors
transforming DNA	2×10^6	Held et al. (1984)
phage DNA (M–13)	2.8×10^6 (+50%)	Ludwig et al. (1990)
E. coli AB1157	4×10^6 (±50%)	Schulte–Frohlinde et al. (1990)[a]
AB2480	3×10^6 (±50%)	Schulte–Frohlinde et al. (1990)[a]

[a] k_r values calculated from the half value concentration of DTT measured by survival at a constant oxygen concentration (air).

Although in the range of our error margins, the k_r values for protection of survival seems to be higher than the k_r values for protection against ssb in agreement with conclusions drawn elsewhere (Schulte–Frohlinde and Bothe, 1990). It is not expected that the k_r values for the repair of the precursors of ssb and of inactivation are identical since Lafleur et al. (1980) have already shown that the pH dependence of inhibition of ssb formation differs from that for inactivation for biologically active DNA (ϕX174) γ–irradiated in aqueous solutions. The cause is proposed to be slow rearrangements of labile chemical structures, probably associated with the bases. These rearrangements depend on pH and temperature. Furthermore, Michael et al. (1990) has shown that the rate constant for repair by thiols of clustered radicals in DNA have higher rate constants. Another interesting result is that the rate for ssb formation increases with the degree of superhelicity of the dsDNA (Ye et al., 1990). This effect should also have an influence on the rate of chemical repair by thiols.

In the first part of the present paper the various radicals produced in DNA by the direct and the indirect effects are discussed. Do all the different radicals in relaxed DNA react with DTT with a similar rate constant? Michael et al. (1981) have found with the oxygen explosion technique two thiol dependent repair rate constants in bacteria. The two rate constants may represent two groups of DNA radicals which could be the N and O base radicals on the one hand and the carbon base radicals on the other, or they could be base radicals and sugar radicals. Another distinction may be based on reaction of the pyrimidine versus the purine radicals, or the two rate constants reflect different damages (Schulte–Frohlinde and Bothe, 1990). A further possibility, however, is that the higher k_r values measured in E. coli are due to the repair of clustered radicals (Michael et al., 1990). Studies of the repair rates of DNA radicals by thiols in aqueous solution will be presented by Bothe in this meeting.

Biological relevance of the results. The biological relevance of a damage in DNA depends on its enzymatic reparability. It is known that different damages in DNA are repaired by different enzymatic repair systems with different efficiences. This might be the reason why E. coli cell lines with different enzymatic repair capacities show different rate constants for chemical repair by DTT. The rate constant may apparently be shifted by a small factor due to a shift in the importance of the damages for survival. This explanation for the difference found in k_r for the two E. coli strains leads to the thought that the mutant might be inactivated by a different class of damages than the wild type. This idea was already considered by Bresler et al. (1975). However, Sargentini and Smith (1986) have shown that the survival of E. coli strains differing in their capacity for enzymatic dsb repair, depends solely on the number of unrepaired dsb and not on other damages. Therefore, we prefer to believe that the small difference between the k_r values of the strains AB1157 and AB2480 is in the range of the error margin.

A further question may be posed considering the fact that the lethal event for the biologically active DNA (baDNA) irradiated in aqueous solution are a few single-strand damaged sites. For E. coli cells on the other hand one unrepaired dsb has been proposed to be the lethal event. This difference is not easily understood since in both cases the same set of enzymatic rapair systems should repair the damaged DNA as long as survival is measured in the same E. coli cells and therefore the lethal damage should be similar. We believe that this phenomenon can be explained on the basis of the large difference in the molecular weight of the chromosomal DNA of E. coli and of the plasmid DNA as well as the higher lethality of clustered damaged sites as produced by spur effects. The importance of spur effects has been pointed out by Ward (1981). It seems as if these differences do not lead to differences larger than a factor of 2 – 4 between chemical repair rate constant of baDNA radicals in aqueous solution and chemical repair of DNA radicals in E. coli with DTT as external thiol.

We should like to thank Professor Dr. C. von Sonntag for his permission to present his unpublished data (Reaction Scheme I).

References

Adinarayana M, Bothe E, and Schulte-Frohlinde D (1988) Hydroxyl radical-induced strand break formation in single-stranded polynucleotides and single-stranded DNA in aqueous solution as measured by light scattering and by conductivity. Int J Radiat Biol 54:723-737

Bernhard W A (1989) Sites of electron trapping in DNA as determined by ESR of one-electron-reduced oligonucleotides. J Phys Chem 93:2187-2189

Bothe E, Behrens G, Böhm E, Sethuram B, and Schulte-Frohlinde D (1986) Hydroxyl radical-induced strand break formation of poly(U) in the presence of oxygen. Comparison of the rates as determined by conductivity, ESR and rapid-mix experiments with a thiol. Int J Radiat Biol 49:57-66

Bothe E, Hankiewicz E, and Schulte-Frohlinde D (1990) Pulsradiolysis of poly(U). In Preparation

Bothe E, Qureshi G A, and Schulte-Frohlinde D (1983) Rate of OH radical induced strand break formation in single stranded DNA under anoxic conditions. An investigation in aqueous solutions using conductivity methods. Z Naturforsch 38c:1030-1042

Bothe E, Schulte-Frohlinde D (1982) Release of K^+ and H^+ from poly U in aqueous solution upon γ and electron irradiation. Rate of strand break formation in poly U. Z Naturforsch 37c:1191-1204

Bresler S E, Sverdlov A G, Noskin L A, and Stepanova I M (1975) Influence of genotype of bacterial cells on radioprotective action of indolilealkylamines. stud biophys 53:163-166

Cullis P M, Symons M C R, Sweeney M C, Jones G D D, and McClymont J D (1986) The effect of ionizing radiation on deoxyribonucleic acid. Part 4. The role of hydrogen peroxide. J Chem Soc Perkin Trans II 1671-1676

Deeble D J, Schuchmann M N, Steenken S, and von Sonntag C (1990) Direct evidence for the formation of thymine radical cations from the reaction of $SO_4^{\cdot -}$ with thymine derivatives: a pulse radiolysis study with optical conductance detection. J Phys Chem submitted

Fahey R C, Newton G L, Aguilera J A, and Ward J F (1990) Thiol uptake and radioprotection for V-79 cells as a function of the net charge of the thiol. Abstracts of papers, Ep-6. Thirty-eight Annual Meeting of the Radiation Research Society, New Orleans, Louisiana, April 7-12

Fahey R C, Prise K, Davies S, Michael B, Stratford M, and Watfa R (1989) The effect of counterion condensation and coion depletion on the rates of repair of pBR322 DNA radicals by thiols. In: Gray Laboratory Annual Report (Michael B and Hance M eds), London: Cancer Research Campaign pp 127-129

Gräslund A, Ehrenberg A, Rupprecht A, Ström G, and Crespi H (1975) Ionic base radicals in γ-irradiated oriented non-deuterate and fully deuterated DNA. Int J Radiat Biol 28:313-323

Hazra D K, Steenken S (1983) Pattern of OH radical addition to cytosine and 1-, 3-, 5- and 6-substituted cytosine. Electron transfer and dehydration reactions of the OH adducts. J Am Chem Soc 105:4380-4386

Held K D, Harrop H A, and Michael B D (1984) Effect of oxygen and sulphydryl-containing compounds on irradiated tranforming DNA. III Reaction rates. Int J Radiat Biol 45:627-636

Hildenbrand K, Behrens G, Herak J N, and Schulte-Frohlinde D (1989) Comparison of the reaction $^{\cdot}OH$ and $SO_4^{\cdot -}$ radicals with pyrimidine nucleosides. An electron spin resonance study in aqueous solution J Chem Soc Perkin Trans 2 283-289

Hildenbrand K, Schulte-Frohlinde D (1989) ESR studies on the mechanism of hydroxyl radical-induced strand breakage of polyuridylic acid. Int J Radiat Biol 55:725-738

Hildenbrand K, Schulte-Frohlinde D (1990) ESR spectra of radicals of single-stranded and double-stranded DNA in aqueous solution. Implication for $^{\cdot}OH$-induced strand breakage. Free Rad Res Commun, accepted

Lafleur M V M, Woldhuis J, and Loman H (1980) Effects of sulphydryl compounds on radiation damage in biologically active DNA. Int J Radiat Biol 37:493–498

Lemaire D G E, Bothe E, and Schulte–Frohlinde D (1984) Yields of radiation–induced main chain scission of poly U in aqueous solution: strand break formation via base radicals. Int J Radiat Biol 45:351–358

Liphard M, Bothe E, and Schulte–Frohlinde D (1990) The influence of glutathione and single–strand breakage in DNA irradiated in aqueous solution in the absence and presence of oxygen. Int J Radiat Biol, in press

Ludwig C, Rettberg P, and Schulte–Frohlinde D (1990) The influence of thiol–repaired damaged sites on the survival of M_{13}–DNA. In preparation

Michael B D, Harrop H A, and Held K D (1981) Timescale and mechanisms of the oxygen effect in irradiated bacteria. In: Oxygen and Oxy–Radicals in Chemistry and Biology. (Rodgers M A J and Powers E L eds.) New York: Academic Press

Michael B D, Prise K, Fahey R, and Davies S (1990) Chemical repair of DNA single and double strand break precursor radicals: implactions for models of chemical modification. Abstract of papers, ARR Biennal Meeting, Clarendon Laboratory and St. Johns College, Oxford, April 4–6

Michaels H B, Hunt J W (1977) Radiolysis of the double–stranded polynucleotides poly(A+U) and DNA in the presence of oxygen. Radiat Res 72:32–47

Müller M (1983) Untersuchungen zum ·OH–induzierten Strangbruch von Poly(C) in wässriger Lösung. Dissertation, Universität Bochum

O'Neill P, Davies S E (1986) A pulse radiolytic study of the interaction of nitroxyls with free–radical adducts of purines: consequences for radiosensitization. Int J Radiat Biol 49:937–950

O'Neill P, Davies S E (1987) Pulse radiolytic study of the interaction of $SO_4^{\cdot-}$ with deoxynucleotides. Possible implication for direct energy deposition. Int J Radiat Biol 52:577–587

Nikogosyan D N (1990) Two–quantum UV photochemistry of nucleic acids: comparison with conventional low–intensity UV photochemistry and radiation chemistry. Int J Radiat Biol 57:233–299

Prakash Rao P J, Bothe E, and Schulte–Frohlinde D (1990) Reaction of dithiothreitol with radical precursors of strand breakage in single–stranded DNA in anoxic aqueous solution: A study by pulse radiolysis and by light scattering. In preparation

Prise K, Davies S, Michael B, and Conlan J (1989) Fast chemical repair of DNA strand break precursors: the role of protein sulphydryls. In: Gray Laboratory Annual Report. London: Cancer Research Campaign

Roots R, Okada S (1975) Estimation of life times and diffusion distances of radicals involved in X–ray–induced DNA strand breaks or killing of mammalian cells. Radiat Res 64:306–320

Rustgi S N, Riesz P (1978) An esr and spin–trapping study of the reaction of the $SO_4^{\cdot-}$ radical with protein and nucleic acid constituents. Int J Radiat Biol 34:301–316

Sargentini N J, Smith K C (1986) Quantitation of the involvement of the recA, recB, recC, recF, recJ, recN, lexA, radA, radB, uvrD, and umuC genes in the repair of X–ray–induced DNA double–strand breaks in Escherichia coli. Radiat Res 107:58–72

Scholes G, Willson R L, and Ebert M (1969) Pulse radiolysis of aqueous solutions of deoxyribonucleotides and of DNA: Reaction with hydroxy–radicals. Chem Commun 17–18

Schulte–Frohlinde D (1986) Comparison of mechanisms for DNA strand break formation by the direct and indirect effect of radiation. In: Mechanisms of DNA Damage and Repair. (Simic M G, Grossman L, and Upton A D eds.) Basic Life Science Vol. 38 New York: Plenum Press pp 19–27

Schulte–Frohlinde D (1989) Studies of radiation effects on DNA in aqueous solution. ICRU News 4–15

Schulte–Frohlinde D, Behrens G, and Oenal A (1986) Lifetime of peroxyl radicals of poly(U), poly(A) and single– and double–stranded DNA and the rate of their reaction with thiols. Int J Radiat Biol 50:103–110

Schulte–Frohlinde D, Bothe E (1990) Determination of the constants of the Alper formula for single– strand breaks from kinetic measurements on DNA in aqueous solution and comparison with data from cells. Int J Radiat Biol accepted

Schulte–Frohlinde D, Hildenbrand K (1989) Electron spin resonance studies of the reaction of ·OH and $SO_4^{\cdot -}$ radicals with DNA. In: Free radicals in synthesis and biology. (Minicsi F ed.) NATO ASI series C, Vol. 260, Bardolino, Italy 1988. Dordrecht: Kluwer pp 335–359

Schulte–Frohlinde D, Opitz J, Görner H, and Bothe E (1985) Model studies for the direct effect of high–energy irradiation on DNA. Mechanism of strand breakformation induced by photoionization of poly U in aqueous solution. Int J Radiat Biol 48:397–408

Schulte–Frohlinde D, Rettberg P, Hartke A, and Jeremic L (1990) Relationship between thiol induced repair and repair by enzymes. Contribution to: 3rd International Conference on Anticarcinogenesis and Radiation Protection, held at Dubrovnik, Yugoslavia, October 15–21, 1989

Steenken S (1989) Purine bases, nucleosides and nucleotides: Aqueous solution redox chemistry and transformation reactions of their radical cations, e– and OH adducts. Chem Rev 89:503–520

Vieira A J S C, Steenken S (1987) Pattern of OH radical reaction with N^6,N^6–dimethyladenosine. Production of three isomeric OH adducts and their dehydration and ring opening reactions. J Am Chem Soc 109:7441–7448

von Sonntag, The Chemical Basis of Radiation Biology (1987) London: Taylor and Francis

Wala M, Bothe E, Görner H, and Schulte–Frohlinde D (1990) Quantum yields for generation of hydrated electrons and single–strand breaks in poly(A), poly(C) and single stranded DNA in aqueous solution upon 20–ns laser excitation at 248 nm. Photochem Photobiol, to be published

Ward J F (1981) Some biochemical consequences of the spatial distribution of ionizing radiation produced free radicals. Radiat Res 86:185–195

Ward J F, Kuo I (1978) Radiation damage to DNA in aqueous solution: a comparison of the response of the single–standed form with that of the double stranded form. Radiat Res 75:278–285

Washino K, Schnabel W (1982) OH Radical–induced main chain scission of poly(ribo–adenylic acid). Macromol Chem Rapid Commund 3:427–430

Willson R L, Wardman P, and Asmus K–D (1974) Interaction of dGMP radical with cysteamine and promethazine as possibel model of DNA repair. Nature 252:323–324

Ye M, Nelson J M, and Miller J H (1990) Superhelicity increases the sensitivity of plasmid DNA to strand break induction by X–rays. Abstract of papers, Ei–22. Thirty–eight Annual Meeting of the Radiaton Research Society, New Orleans, Louisiana, April 712

RADICAL MULTIPLICITY IN RADIATION-INDUCED DNA STRAND BREAKS: IMPLICATIONS FOR THEIR CHEMICAL MODIFICATION

Barry D. Michael, Kevin M. Prise, Robert C. Fahey[1], and Melvyn Folkard

Cancer Research Campaign Gray Laboratory
P O Box 100 Mount Vernon Hospital
Northwood Middlesex HA6 2JR UK.

SUMMARY

Chemical repair or restitution of DNA break precursors reduces the number of breaks detected from hypoxically irradiated cells by approximately 3-fold. Under hypoxia it occurs on a millisecond timescale and is accessible to kinetic measurement. Although a large part of the chemical repair seen in eukaryotic systems is attributable to non-protein thiols, there is a residual component which appears not to be. Evidence from fast mixing studies with mammalian cells and with plasmid DNA in the presence of a thiol shows a difference between the chemical repair kinetics of DSB and SSB precursors. The radical precursors of DSB appear to be repaired about twice as fast as those of SSB. This finding supports the concept that DSB's derive from *paired* radical sites and indicates that current models of chemical modification, which are based on competing repair and fixation at a *single* radical site may be inappropriate. Also, a consideration of this phenomenon offers a new mechanism for the dependence of OER on LET based on the multiplicity of radical sites induced in each lesion.

---

[1]Department of Chemistry
University of California San Diego
La Jolla CA 92093 USA.

NATO ASI Series, Vol. H 54
The Early Effects of Radiation on DNA
Edited by E. M. Fielden and P. O'Neill

INTRODUCTION

Radiation damage to DNA is initiated by direct and indirect processes in times that are generally very much less than 10^{-6} s. The DNA radicals so produced undergo subsequent reactions some of which lead to the formation of strand breaks (Schulte-Frohlinde *et al.* 1985). These DNA strand break precursor radicals are susceptible to chemical modification. Either the radical sites may be reduced by endogenous thiols, such as glutathione, resulting in their chemical repair and the prevention of strand breakage, or they may be oxidized leading to damage fixation and the eventual formation of strand-breaks. It is the antagonistic competition between reductive repair by thiols and oxidative radiosensitisation by dissolved oxygen which determines the extent of DNA damage remaining at the end of this stage of radiation action. Typically, these competing processes take place on a timescale of the order of 10^{-3} s. The lesions remaining at the end of this stage are relatively stable chemically and they are then susceptible to enzymatic repair processes leading to repaired or misrepaired DNA molecules. This paper describes the timescale of some of the relevant chemical repair and fixation reactions, both in intact cells and in DNA preparations. In particular, it examines experimentally and theoretically the underlying kinetics of processes involving more than one radical site on the DNA and implications that follow for cellular response and its chemical modification.

Previous work from this Laboratory has shown that the timescale of modification of radiosensitivity by oxygen is approximately 10^{-3} s in a variety of cell types (Michael *et al.* 1979). We have also shown that analogous processes studied using bacterial transforming DNA occur on approximately the same timescale and with similar reaction kinetics (Held *et al.* 1984). A comparative study of the chemical repair kinetics of precursors of lethal lesions and DNA strand breaks in V79 Chinese hamster cells, showed that both endpoints have similar lifetimes (Michael *et al.* 1986); essentially similar results were obtained in studies with *Saccharomyces cerevisiae* yeast cells (Frankenberg *et al.* 1990).

Our studies with V79 cells and with yeast both indicate that the major non-protein thiol, glutathione, present in these cells is responsible only for a part of the observed chemical repair. The source of the residual repair has not yet been identified. Recent work with pBR 322 plasmid DNA has compared the chemical repair of its radicals by glutathione and other thiols of differing net charge (Fahey *et al.* in preparation). These studies explained our earlier observations that glutathione is a relatively poor chemical repair agent for DNA radicals (Held *et al.* 1984) due to its negative charge. The methods developed with pBR 322 system have been applied in the present work to a comparison of repair kinetics of single and multiple radical sites in irradiated DNA.

CHEMICAL REPAIR KINETICS IN V79 CELLS

Using the gas explosion technique (Michael *et al.* 1973; Michael *et al.* 1979), we have compared the timescales for chemical repair of the radical precursors of lethal lesions, single- and double-strand breaks, in V79-379A fibroblasts irradiated on membrane filters. The methods of scoring survival and DNA strand breaks by filter elution have been described previously (Prise *et al.* 1989). Figure 1 compares the oxygen effect fast

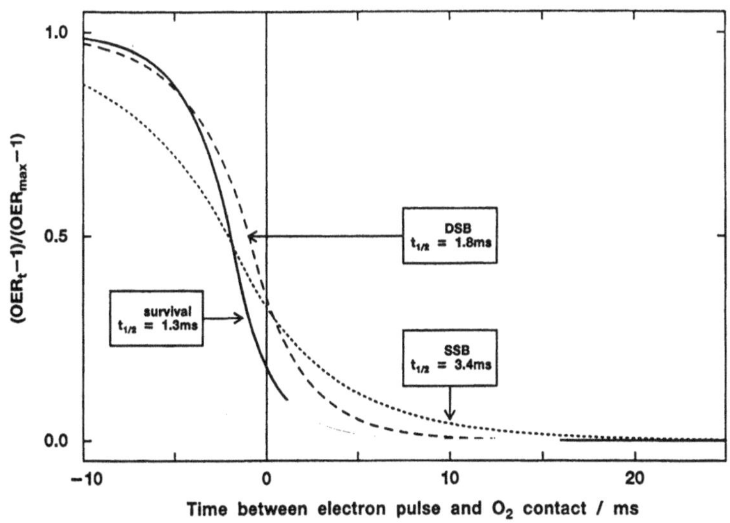

Figure 1. Fast kinetics of oxygen effects on cell kill, DSB and SSB in pulse-irradiated V79 cells, determined using the gas explosion technique. The vertical axis has been normalized.

kinetics for the three different endpoints. For clarity, the data points have been omitted and the best fit kinetic profiles are shown normalised on a scale of 0 – 1, these values representing the fully hypoxic and fully oxic responses respectively. In this diagram, single-pulse electron irradiation takes place virtually instantaneously at $t = 0$ and oxygen contact with the previously hypoxic cells occurs at the times indicated. Thus on the right-hand side of the diagram where $t > 0$ significant post-irradiation sensitization by oxygen can be seen for all three endpoints. As in previous studies, time exponential fits to these post-effect regions are used to determine the first-order rates of chemical repair of their radical precursors. Fitted repair rates (in s^{-1}) are: cell survival, 530; DSB, 380; SSB, 210.

Thus the radical precursors of all three endpoints show generally similar rates of chemical repair and the reasonably close agreement between the values for cell survival and DSB illustrate the importance of the latter in relation to lethal effect. However, the repair rate associated with SSB precursors appears somewhat slower and the significance of this will be discussed later.

RESIDUAL CHEMICAL REPAIR

Glutathione is the most abundant naturally occurring thiol. The effect on chemical repair kinetics of depleting GSH levels by treatment of V79 cells with buthionine sulfoximinine (BSO), are illustrated in Figure 2.

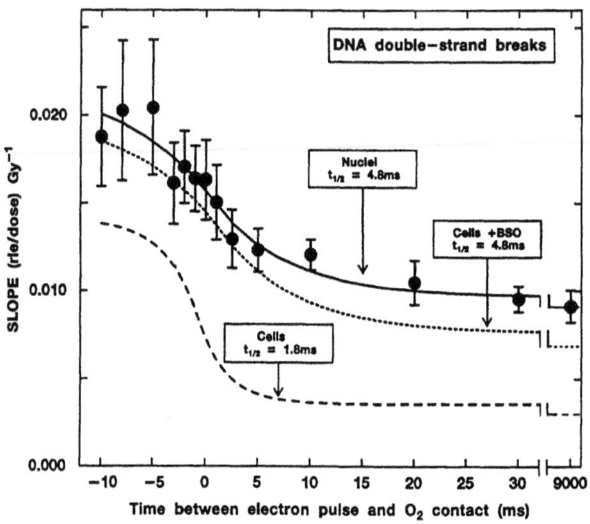

Figure 2. Oxygen effect kinetics for radical precursors of DSB in V79 cells ± BSO treatment and in isolated nuclei. Yield of DSB per unit dose is plotted *vs.* time of O_2 contact relative to radiation pulse.

Here oxygen effect kinetics using the gas explosion technique are shown for DSB induction. Treatment with BSO depleted the GSH content to < 1% of that in control cells. However, as shown in Figure 1, this depletion only resulted in a 2.7-fold extension of the chemical repair half-life from 1.8 to 4.8 ms. Similar observations with intact V79 cells have been described previously (Michael *et al.* 1986). Also shown in Figure 2 are data for V79 cell nuclei prepared as a monolayer. These have been washed and are severely depleted of non-protein thiols. The nuclei show DSB repair kinetics very similar to those of intact BSO-treated cells. Also the OER for the isolated nuclei is 2.5

compared with a value of 3.4 for DSB measured in control cells. Thus, even in isolated nuclei where the GSH content is < 1% of that present in control cells, there remains a substantial proportion of the control cells' chemical repair capacity. Similar small dependencies on GSH content were found for OER and fast kinetics measurements in GSH-proficient and -deficient strains of yeast (Frankenberg *et al.* 1990). This residual chemical repair is not evident in purified DNA systems (Held *et al.* 1984; Fahey *et al.* in preparation) and we propose that it is attributable to protein thiols present in chromatin, which contribute appreciably to the thiol content of the cell nucleus (Loh *et al.* 1990). However, further work is required to elucidate the nature of this residual repair capacity.

CHEMICAL REPAIR OF PLASMID DNA RADICALS

Although our previous studies of DNA radical repair using bacterial transforming DNA (Held *et al.* 1984) gave useful kinetic information, it was clear that plasmid DNA systems offered improved specificity in the DNA damage scored. Adaptation of the gas explosion technique to studies with pBR 322 plasmids, has allowed us to determine the reactivities of a range of thiol compounds with SSB precursor radicals (Fahey *et al.* in preparation). These studies showed that, even at physiological ionic strength, the phenomena of counter-ion condensation and co-ion depletion strongly influence the apparent reactivity of charged thiols with radicals formed on polyanionic DNA. We have extended these studies to a comparison of the chemical repair kinetics of single- and double-strand break precursors. The object of this study was to see whether a predicted difference in their repair kinetics could be observed. pBR 322 plasmids, mounted on membrane filters, were irradiated in the gas explosion apparatus (Fahey *et al.* in preparation). After irradiation, plasmids were washed from the filters and loaded onto agarose gels for analysis by electrophoresis. The gels were stained with ethidium bromide and were read directly using a scanning fluorimeter (Chromascan, Joyce Loebl Ltd, Newcastle, England). The proportions of supercoiled, relaxed and linear plasmid forms were measured corresponding to plasmids with no damage, SSB and DSB respectively. In this way, for each irradiated sample the extent of single- and double-strand break induction could be determined. These are shown as functions of oxygen contact time in Figure 3. In the upper panel are the kinetics of chemical repair of SSB precursors scored in terms of percentage supercoiled plasmid. Loss of the supercoiled form as shown corresponds closely to production of the relaxed form because the relative proportion of the linear form is small. The lower panel shows chemical repair kinetics for precursors of double-strand breaks expressed as production of

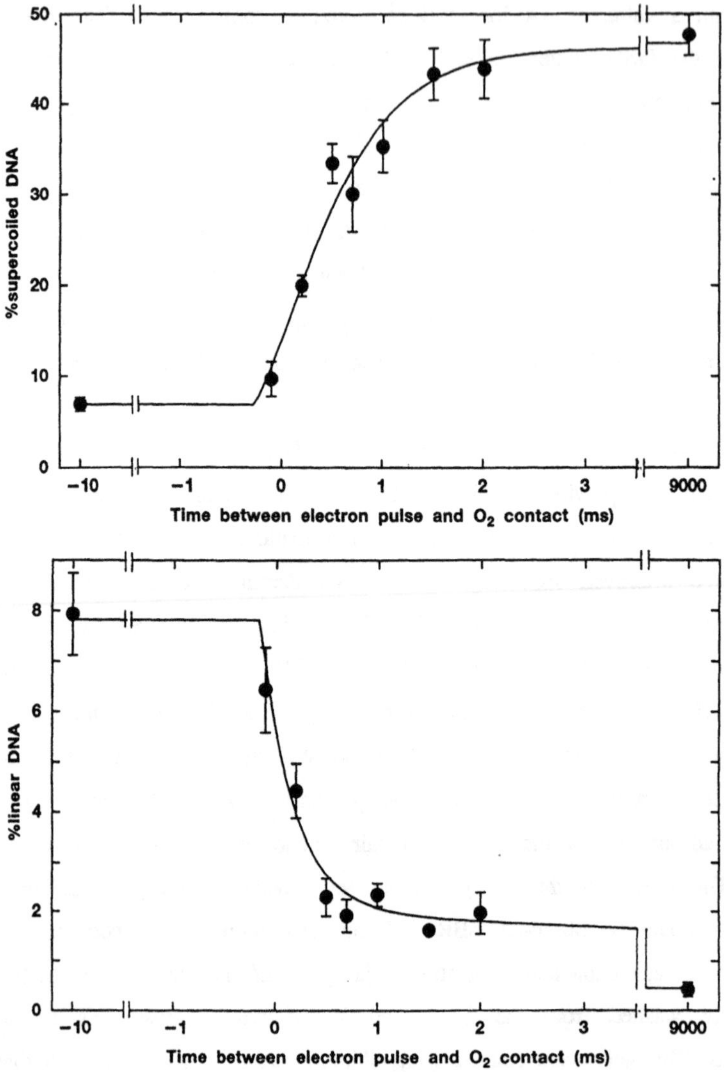

Figure 3. Fast chemical repair of radical precursors of SSB (upper panel) and DSB (lower panel) in pBR 322 plasmids measured using the gas explosion technique.

the linear form of the plasmid. The fitted kinetic profiles show chemical repair rates of 1800 and 3000 s^{-1} for the repair of single- and double-strand break precursor radicals respectively. The ~2-fold difference in these repair rates is similar to that found also for SSB and DSB precursor radicals in mammalian cells, as shown in Figure 1. The significance of this is discussed below.

PAIRED RADICAL SITES IN RELATION TO CHEMICAL MODIFICATION

The proposal that DSB's result from locally multiply damaged sites which occur when clusters of ionisation overlap the DNA (Ward, 1981) implies that the precursors of these lesions consist of pairs of independent radical sites formed one on each strand. The more frequent induction of SSB's occurs when fewer ionisations occur close to the DNA, resulting in the induction of a single radical site on one or other strand. There are important differences between the kinetic schemes applying to the chemical repair of these single and paired radicals and these are illustrated in Figure 4.

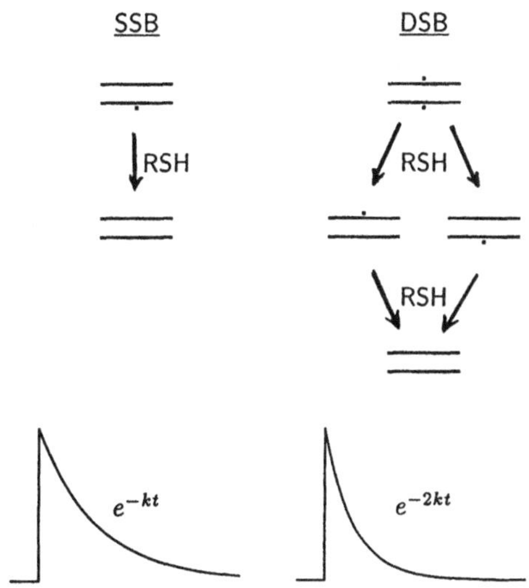

Figure 4. Repair kinetics for radical precursors of SSB and DSB.

The chemical repair of SSB precursor radicals will occur in a simple one-step pathway as shown. However, the repair of DSB precursor radicals will proceed in two steps with two parallel pathways. Only the first step is required to prevent the formation of a DSB and this would therefore be seen experimentally as repair, regardless of whether the second repair step occurs or whether a SSB is formed at that point. However, because the crucial first step of repair can take place in two ways as shown, the overall instantaneous probability of this process will be twice that for the chemical repair of a single-strand break precursor radical. This is illustrated by the kinetic profiles shown in Figure 4 where, for first-order repair of SSB precursor radicals with a rate constant of k, the corresponding repair kinetics of DSB precursors will also appear first-order but the rate constant will be 2k. The predicted 2-fold difference in chemical repair rate is clearly consistent with our observations described above for single- and double-strand break precursor repair in V79 cells and in pBR 322 plasmid DNA in the presence of 10 mM GSH. This finding is of importance in understanding the nature of chemical repair kinetics in DNA systems, but it also has implications for models of chemical modification of radiation response and these are discussed further below.

MODELS OF CHEMICAL MODIFICATION WITH PAIRED RADICAL SITES

Numerous studies point to the importance of DSB induction in radiation effects. Thus cell kill and chromosome aberrations are thought to derive from DSB's. The kinetic data shown above support the view that chemical repair of DSB's occurs at paired radical sites. Most models of chemical modification of radiation response are, however, based upon the single-radical model proposed by Alper and Howard-Flanders (1956). The Alper and Howard-Flanders equation (1) below

$$OER_{ssb} = \frac{m[O_2] + K}{[O_2] + K} \tag{1}$$

describes oxygen enhancement of response as a function of oxygen concentration in terms of a maximum OER value, m, and a half-effect oxygen concentration equivalent to K. The corresponding expression modified for paired radical sites is

$$OER_{dsb} = \left[\frac{m[O_2] + K}{[O_2] + K}\right]^2 \tag{2}$$

where m and K, as before, apply to each of the SSB precursor radicals which, as a pair, constitute the precursor of the DSB. A consequence of the relationships shown in equations (1) and (2) is that, at high oxygen concentration, the maximum OER's for SSB and DSB should be m and m^2 respectively. We have compared the OER values for SSB and DSB in the pBR 322 + 10 mM GSH system, mounted on membrane filters, and the value for SSB is 5.1 ± 0.4 and for DSB 11.3 ± 2.7. Although the observed DSB OER is not as high as the square of that for SSB, the difference is sufficient to lend some support to equation (2) being applicable to DSB lesions and therefore also to cell killing effects of ionising radiations. Also, it is possible that, even with low-LET radiation, some of the DSB involve more than two radical sites (e.g. those from electron track ends) and that this lowers the observed OER for DSB's. Such radical multiplicity effects are described in the next section. A similar modification to the Alper and Howard-Flanders formula has been proposed previously (Ward, 1987).

K-curves, relating OER to oxygen concentration, have been derived from equations (1) and (2) and are shown in Figure 5.

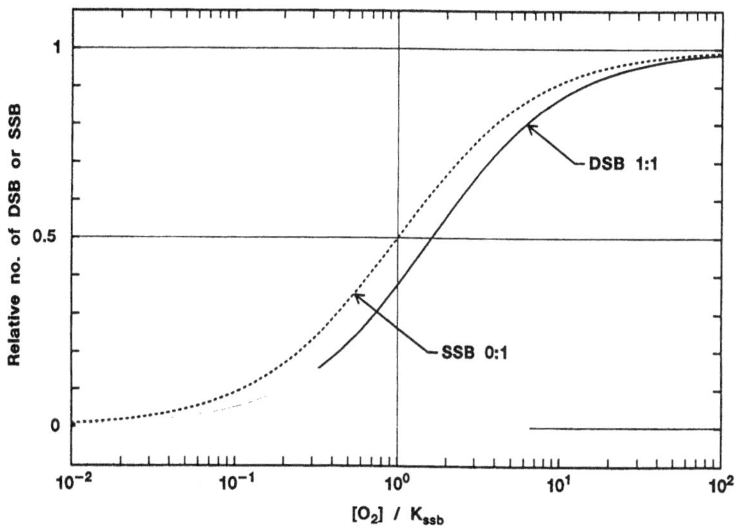

Figure 5. Theoretical oxygen K-curves for SSB and DSB.

As in Figure 1, to allow comparison, the OER's have been normalised on a scale of 0 – 1. Similarly, oxygen concentration has been shown relative to the K-value for a single-radical SSB precursor lesion. The nomenclature used for SSB precursor radicals is 0:1, representing one such radical on one strand only and the corresponding expression for DSB precursors consisting of one radical on each strand is 1:1. Comparison of the normalised K-curves in Figure 5 shows that their shapes differ and that the K-value for DSB's is, in theory, higher than that for SSB's. The differences illustrated in Figure 5 would be difficult to detect in practice because of experimental errors, although the underlying effects on chemical repair can be seen by fast kinetics methods as shown in Figures 1 and 3. Clearly, however, these factors should be considered in the analysis of K-curves and in relating parameters so obtained to data from fast kinetics experiments, e.g. relationships between K-values and rates for chemical repair and fixation reactions.

MULTIPLE RADICAL SITES INDUCES IN DNA BY HIGH LET RADIATION

The above analysis for single and paired radical sites led us to consider the more general case for DNA breaks resulting from more than one radical site on each strand as are expected to occur with high LET radiations. Monte Carlo track structure calculations show that, with increasing LET, the probability of induction of such multiple lesions increases (Charlton *et al.* 1989; Holley *et al.* 1990; Holley and Chatterjee, 1990). Figure 6 illustrates how the DNA radical multiplicity resulting from DSB induction by low LET radiation (e.g. Ward, 1981) might differ from that resulting from, say, a 3

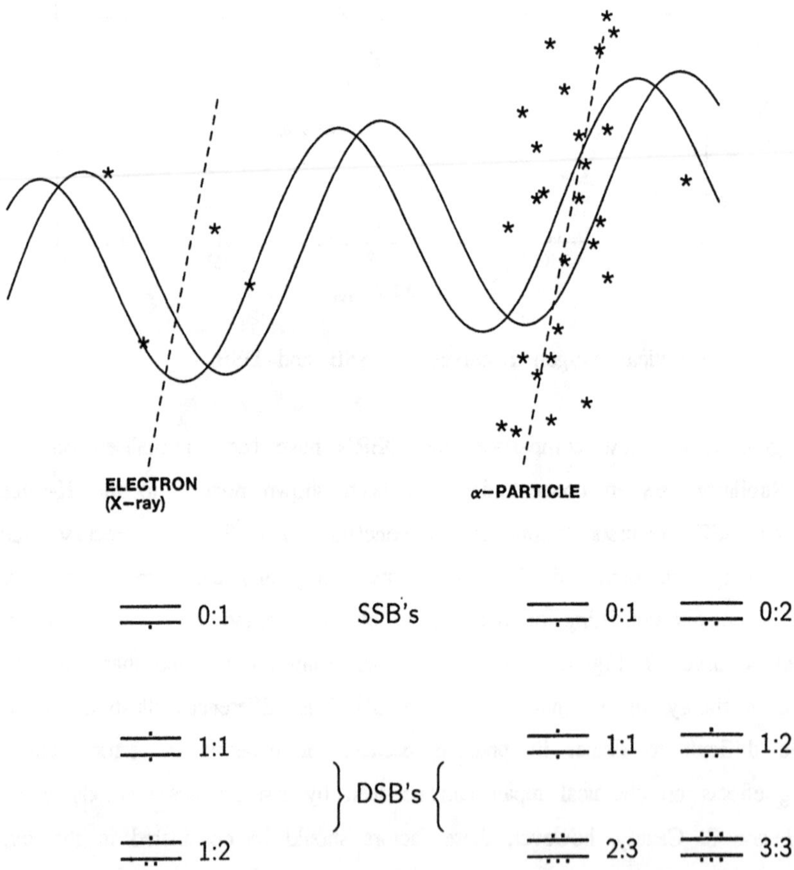

Figure 6. Strand break radical precursor multiplicity after electron (left) and α-particle (right) irradiation.

MeV α-particle track (Goodhead and Nikjoo, 1989). In this diagram, the numbers of strand-break precursor radicals induced on each strand are shown as 1:2, 2:3, etc., and in general the designation j:k will be used to represent DSB's consisting of respectively j and k strand-break sublesions on either strand. The general expression relating OER to oxygen concentration for a DSB of j:k radical multiplicity is shown in equation (3).

$$OER = \frac{\left[\left[\frac{(m-1)K}{m([O_2]+K)}\right]^j - 1\right]\left[\left[\frac{(m-1)K}{m([O_2]+K)}\right]^k - 1\right]}{\left[\left[\frac{m-1}{m}\right]^j - 1\right]\left[\left[\frac{m-1}{m}\right]^k - 1\right]} \tag{3}$$

Here, the values of m and K apply, as in equations (1) and (2), to the underlying single-radical SSB precursor. Equation (2) is a special case of the general equation (3) where j = 1 and k = 1. K-curves derived using equation (3) are shown in Figure 7.

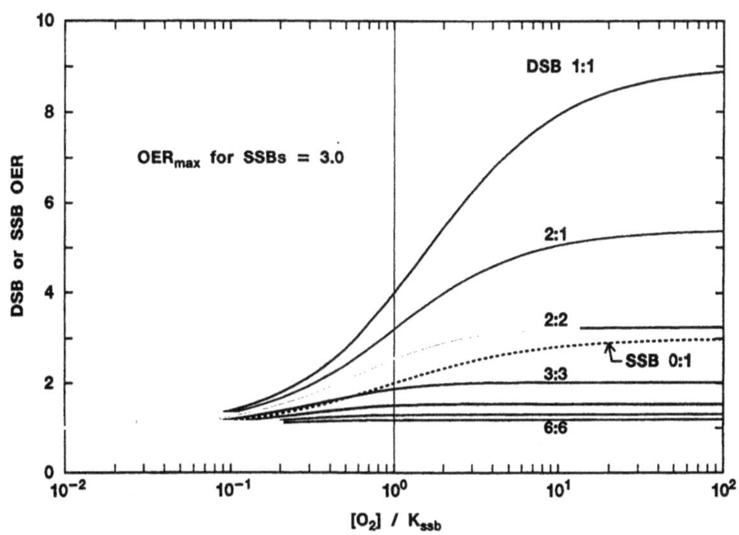

Figure 7. Theoretical oxygen K-curves for DSB of varying multiplicity, as indicated.

As in Figure 5, oxygen concentrations are shown on a relative scale which is normalised to the oxygen K-value for a single-radical SSB precursor. The curves have all been calculated assuming that the value of m for a 0:1 SSB is equal to 3.0. Thus, in this illustrative diagram, the maximum OER values for various DSB are: 1:1, 9.0; 2:1, 5.40; 2:2, 3.24; 3:3, 2.01; 6:6, 1.20. The dependence of maximum OER on multiplicity shown in these examples suggests a mechanism for the observed relationship between OER and LET. Thus, as values of j and k increase with LET, the maximum OER's derived using equation (3) decrease. This is because as j and k increase, so does the probability that, even under hypoxia, at least one radical site on each strand will become a strand break, leading to the formation of a DSB. This invokes the concept of a hypoxic fixation pathway for strand break precursor radicals which either competes with or by-passes their chemical repair. Such a pathway is required in order to explain the inability of thiols to protect completely under hypoxia. We propose the lesion multiplicity concept outlined above to explain the relationship between OER and LET, as an alternative to the 'interacting radicals' and 'oxygen-in-the-track' hypotheses (see Alper, 1979).

There is little information on the underlying multiplicity of strand break events described above. Some data are available from studies of DNA damage due to decays of incorporated ^{125}I (Martin and Haseltine, 1981; Charlton, 1986) and there is information from results of Monte Carlo track structure calculations, as cited above. The conventional methods of assaying DSB's give no indication of their underlying multiplicity, but clearly the complexity and consequent reparability of DSB's varies widely with the LET of the radiation used to induce them (Prise *et al.* these proceedings). In future studies, we hope to develop methods to measure the underlying multiplicity of DSB's induced by various radiations.

CONCLUSIONS

1. Chemical repair of the radical precursors of lethal events, SSB's and DSB's occur on similar timescales. In eukaryotic cells some of the observed repair is attributable to GSH, but there is a substantial residual component which may be due to thiol groups in the proteins present in chromatin.
2. In mammalian cells, chemical repair kinetics of lethal damage and DSB precursor radicals are similar and their rates are approximately double those for SSB's.
3. In pBR 322 plasmids plus GSH, the chemical repair rates associated with DSB and SSB formation also differ by a factor of ~2. This is consistent with DSB precursors that consist of paired radical sites, as also is the observation of a substantially higher OER for DSB than for SSB in this system.
4. Current models of chemical modification based on single-radical chemical kinetics, therefore, may not be appropriate for endpoints involving DSB's or cell kill.
5. With increasing LET, the multiplicity of radical sites involved in each induced DSB will increase. This leads to a proposed mechanism for the observed decrease of

OER with increased LET which is related to the concept of hypoxic fixation of strand-break radical precursor sites.

6. Improved assay methods for strand breakage need to be developed to score their changing multiplicity with LET.

ACKNOWLEDGEMENTS

The authors wish to acknowledge that this work was supported by the Cancer Research Campaign and to thank Mrs D.M. Brown for preparing the manuscript.

REFERENCES

ALPER, T. (1979), *Cellular Radiobiology* (Cambridge University Press, Cambridge), pp. 133-145.

ALPER, T., HOWARD-FLANDERS, P. (1956), Role of oxygen in modifying the radiosensitivity of *E. coli B*. *Nature*, **178**, 978-979.

CHARLTON, D.E. (1986), The range of high LET effects from ^{125}I decays. *Radiat. Res.*, **107**, 163-171.

CHARLTON, D.E., NIKJOO, H. AND HUMM, J.L. (1989), Calculation of initial yields of single- and double-strand breaks in cell nuclei from electrons, protons and alpha-particles. *Int. J. Radiat. Biol.*, **56**, 1-19.

FAHEY, R.C., PRISE, K.M., STRATFORD, M.R.L.S., WATFA-CHALLAND, R.R. AND MICHAEL, B.D. (In preparation), Rates for repair of pBR 322 radicals by thiols as measured by the gas explosion technique: Evidence that counter-ion condensation and co-ion depletion are significant at physiological ionic strength (for submission to *Int. J. Radiat. Biol.*).

FRANKENBERG, D., MICHAEL, B.D., FRANKENBERG-SCHWAGER, M. AND HARBICH, R. (1990), Fast kinetics of the oxygen effect of DNA double-strand breakage and cell killing in irradiated yeast. *Int. J. Radiat. Biol.*, **57**, 485-501.

GOODHEAD, D.T. AND NIKJOO, H. (1989), Track structure analysis of ultrasoft X-rays compared to high- and low-LET radiations. *Int. J. Radiat. Biol.*, **55**, 513-529.

HELD, K.D., HARROP, H.A. AND MICHAEL, B.D. (1984), Effects of oxygen and sulphydryl-containing compounds on irradiated transforming DNA. III. Reaction rates. *Int. J. Radiat. Biol.*, **45**, 627-636.

HOLLEY, W.R., CHATTERJEE, A. AND MAGEE, J.L. (1990), Production of DNA strand breaks by direct effects of heavy charged particles. *Radiat. Res.*, **121**, 161-168.

HOLLEY, W.R. AND CHATTERJEE, A. (1990), Clustering of damage around the site of a radiation-induced DNA strand-break. Abstract Co-1 from *38th Meeting of Radiation Research Society,* New Orleans, April 1990.

LOH, S.N., DETHLEFSEN, L.A., NEWTON, G.L., AGUILERA, J.A. AND FAHEY, R.C. (1990), Nuclear thiols: Technical limitations on the determination of endogenous nuclear glutathione and the potential importance of sulfhydryl proteins. *Radiat. Res.*, **121**, 98-106.

MARTIN, R.F. AND HASELTINE, W.A. (1981), Range of radiochemical damage to DNA with decay of Iodine-125. *Science*, **213**, 896-898.

MICHAEL, B.D., ADAMS, G.E., HEWITT, H.B., JONES, W.B.G. AND WATTS, M.E. (1973), A post-effect of oxygen in irradiated bacteria: a submillisecond fast mixing study. *Radiat. Res.*, **54**, 239-251.

MICHAEL, B.D., DAVIES, S. AND HELD, K.D. (1986), Ultrafast chemical repair of DNA single and double strand break precursors in irradiated V79 cells. In *Mechanisms of DNA Damage and Repair*, edited by M.G. Simic, L. Grossman and A.C. Upton (New York: Plenum Press), pp. 89-100.

MICHAEL, B.D., HARROP, H.A. AND MAUGHAN, R.L. (1979), Fast response methods in the radiation chemistry of lethal damage in intact cells. In *Radiation Research*, edited by S. Okada, M. Imamura, T. Terashima and H. Yamaguchi (Japanese Association for Radiation Research), pp. 288-296.

PRISE, K.M., FOLKARD, M., DAVIES, S. AND MICHAEL, B.D. (1989), Measurement of DNA damage and cell killing in Chinese Hamster V79 cells irradiated with aluminium characteristic ultrasoft X-rays. *Radiat. Res.*, **117**, 489-499.

PRISE, K.M., FOLKARD, M., DAVIES, S. AND MICHAEL, B.D. (these proceedings). The lethality or radiation-induced DNA double-strand breaks for radiations of differing LET.

SCHULTE-FROHLINDE, D., OPITZ, J., GÖRNER, H. AND BOTHE, E. (1985), Model studies for the direct effect of high-energy irradiation on DNA. Mechanism of strand break formation induced by photoionization of poly U in aqueous solution. *Int. J. Radiat. Biol.*, **48**, 397-408.

WARD, J.F. (1981), Some biochemical consequences of the spatial distribution of ionizing radiation-produced free radicals. *Radiat. Res.*, **86**, 185-195.

WARD, J.F. (1987), Abstract, *Fourth International Congress on Oxygen Radicals (4-ICOR)*, July 1987 at University of California San Diego, La Jolla.

EFFECT OF DOSE MODIFIERS ON RADIATION-INDUCED CELLULAR DNA DAMAGE

G.P. van der Schans
TNO Medical Biological Laboratory
Lange Kleiweg 139
2288 GJ Rijswijk
The Netherlands

Summary

In research on the modification by chemical agents of the effects of ionizing radiation, most attention has been directed towards application of radiosensitizing or radioprotective drugs in cancer radiotherapy.

Studies on radiosensitization have led to the development of a broad class of compounds which cause pronounced differential radiosensitization of cells under hypoxic conditions but show no activity in oxic cells. Radiobiological studies on the general occurrence of the oxygen effect led to the conclusion that the hypoxity of tumor cells makes them relatively radioresistant. This is a major factor in the failure of radiotherapeutic treatment of some human tumors. The large majority of radiosensitizers established as such are of the oxygen-mimetic type, the activity of which is governed largely by their electron-affinity. Other mechanisms involved in sensitization are changes in intracellular thiol levels, inhibition of cellular DNA repair processes, and/or inhibition of oxygen utilization.

The development of radioprotective drugs has shown much less progress. In the past, thousands of compounds were synthesized and screened for their protective capacity against damage arising from whole body irradiation, with only limited success. One compound, however, the thiophosphate WR2721, appeared sufficiently promising to warrant further investigation of its potential as a protective agent against destruction of normal tissue during radiotherapy of cancer.

In this paper, special attention is given to the influence of oxygen on the induction of radiation damage in DNA of mammalian cells after sensitization by intracellular glutathione depletion. Furthermore, data are presented about the modification of radiation response at the DNA level after *in vivo* irradiation.

NATO ASI Series, Vol. H 54
The Early Effects of Radiation on DNA
Edited by E. M. Fielden and P. O'Neill

1. INTRODUCTION

For several years, there has been increasing interest in the possibility to apply radiosensitizing or radioprotective drugs in the treatment of cancer by radiation. Therefore, much effort has been directed to the research on the modification of the effects of ionizing radiation by these chemical agents.

Studies on radiosensitization have led to the development of a broad class of compounds which cause pronounced radiosensitization of cells under hypoxic conditions but show no activity in oxic cells. Radiobiological studies on the general occurrence of the oxygen effect provided evidence to support the hypothesis that the relative radioresistance of hypoxic tumor cells is a major factor in the failure of radiotherapeutic treatment of some human tumors. The large majority of radiosensitizers established as such are of the oxygen-mimetic type, the activity of which is governed largely by their electron-affinity. For one reason or another, most of these could not be developed for clinical use, but one series of compounds, the nitroimidazoles, showed promise in this respect. One compound of this class, misonidazole, was subject to widespread clinical investigation, also because of its apparently favorable pharmacological properties. However, it became clear that the clinical potential of this drug is severely restricted by its neurotoxic properties. The search for a clinically superior drug is aimed either at improving sensitization efficiency or reducing toxicity. Other topics that receive growing attention in sensitization research include (*i*) changes in intracellular thiol levels, (*ii*) inhibition of cellular DNA repair processes and (*iii*) inhibition of oxygen utilization.

The development of radioprotective drugs has shown much less progress. In the past, thousands of compounds were synthesized and screened for their protective capacity against damage arising from whole body irradiation, with only limited success. One compound, however, the thiophosphate WR2721, appeared sufficiently promising to warrant further investigation of its potential as a protective agent against destruction of normal tissue during radiotherapy of cancer. There is some evidence that the high hydrophilicity of WR2721 influences its passage through cellular membranes and that tumor cells may present a stronger membrane barrier than normal tissue cells. However, notwithstanding the differential effect observed in experimental systems, radioprotection by WR2721 has been observed in some tumors. The use of this radioprotector is compromised, however, by the need to increase radiation dosage by such an extent that any tumor protective effect is abolished.

In this paper, the effects of radiosensitizing and radioprotecting drugs on radiation-induced cellular DNA damage will be discussed. Special attention will be given to the influence of oxygen on the induction of radiation damage in DNA of mammalian cells after sensitization by intracellular glutathione depletion. Furthermore, some preliminary results, obtained with recently developed methods, will be presented on the modification of radiation response at the DNA level after *in vivo* irradiation.

2. RADIOSENSITIZATION BY OXYGEN

One of the most common radiation-dose modifiers is oxygen. The effect of oxygen on the induction of different types of lesions has been extensively studied. Nevertheless, there are many conflicting data about the extent of the effect of oxygen. For example, the oxygen enhancement ratio (OER) - which is the ratio between the effects measured under anoxic *vs* oxic conditions - for damage to the sugar moiety, measured as single-strand breaks (SSB) or double-strand breaks (DSB), ranges from 2 to 10. For base damage, detected in the γ-endonuclease assay with *Micrococcus luteus* extracts, OER's of 0.5-1.4 are reported. Most of the controversies can be explained by the fact that different amounts of oxygen are still present under the hypoxic rather than anoxic conditions used in practice. Only under strictly anoxic conditions the highest values for the OER can be obtained. The OER for sugar damage is generally higher than that for base damage. To explain this observation, one has to keep in mind that the effect of oxygen can not be considered separate from the effects of other constituents in the cell, e.g. thiol-compounds. The effect of thiol-compounds, like intracellular glutathione (GSH) is usually explained by their ability to reconstitute chemically the target free radical in critical molecules by hydrogen donation, in competition with the reaction of oxygen with the same free radical (see e.g. Alper and Howard-Flanders 1956). The difference observed in OER of sugar damage and base damage, supports the notion that the oxygen-thiol competition model may be valid for some types of DNA damage, such as sugar-type radicals, but not for others, e.g. base radicals (O'Neill 1983). This can be explained by the fact that reconstitution of base damage owing to the presence of endogenous SH-compounds leads in some cases to other damage (modified base), whereas reaction of the SH-compound with the sugar leads to a complete reconstitution. Most of the damage of which the level is *enhanced* by oxygen can be prevented by DMSO, a scavenger of OH·-radicals: the amount of SSB is reduced by 80%, that of ESS by 75% (Skov, 1984).

3. RADIOSENSITIZATION BY NITROIMIDAZOLE COMPOUNDS

Most investigations on radiosensitizers dealt with their effects on cell survival. Radiosensitization by nitroimidazole compounds such as misonidazole (MISO; 1-(2-nitro-1-imidazonyl)-3-methoxy-2-propanol) can occur in two ways.

(*i*) Oxygen-mimetic radiosensitization due to the electron-affinity of the NO_2 group results in a change in the slope of the radiation-survival curve when hypoxic cells are irradiated *immediately* after the addition of the drug.

(*ii*) A reduction in the shoulder region of the radiation-survival curve occurs when *prolonged* exposures to the drug *in vitro* under hypoxic conditions precede the irradiation (Wong *et al.* 1978). This type of radiosensitization does not require the presence of the nitro compound during the irradiation and is observed for cells that are irradiated after exposure, under oxygenated as well as hypoxic conditions. Presumably, the radiosensitization is a result of alterations which occur as a consequence of the metabolism of MISO under hypoxic conditions (nitroreduction).

The enhancement ratio for MISO amounts to about 2.3 (Skov *et al.* 1982). However, in the presence of another radiosensitizer, TAN (described in section 4), the enhancement by MISO is reduced to 1.5.

Skov (1984) showed that the presence of MISO during irradiation under hypoxic conditions resulted not only in an enhanced induction of SSB (3x), but also of base damage sensitive to *M.luteus*-extracts (ESS; 1.9x). The enhancement of ESS-induction can almost completely be eliminated by the presence of an OH·-scavenger (DMSO), whereas the enhancement of SSB-induction was reduced to about 30%.

In the presence of TAN, the sensitizing effect of MISO on SSB-induction is totally eliminated, the enhancement of ESS-induction only partly.

Taylor *et al.* (1983) applied the alkaline elution assay (Kohn *et al.* 1976) to assess DNA damage after treatments with MISO, graded doses of X-rays and the combination of MISO-pretreatment with X-rays. Unlike X-rays, sublethal doses of MISO produce very little increase in the elution rate of DNA. When a marginally toxic MISO-pretreatment (2h, 5 mM, survival>67%) is combined with various X-ray doses, more strand breaks are measured immediately after the irradiation than can be accounted for by the sum of the MISO-induced and radiation-induced breaks. The rate of repair of strand breaks is not influenced by MISO-pretreatment.

MISO-pretreated cells show survival levels and DNA elution rates which suggest that they received an extra dose of 1.5 Gy. The results indicate that sensitization by pretreatment with MISO followed by irradiation, both under hypoxic conditions, could be accounted for by the interaction of MISO-induced damage and radiation at the DNA level. The sensitization is not a consequence of inhibition of the repair of X-ray-induced strand breaks by MISO-pretreatment.

4. RADIOSENSITIZATION BY NITROXYL COMPOUNDS

The most widely studied compound of this group is TAN (2,2,6,6-tetramethyl-1,4-piperidone-*N*-oxyl). An enhancement ratio for cell death of 1.4 has been observed under hypoxic conditions (Skov *et al.* 1982).

TAN has no effect on radiation-induced formation of SSB whereas the induction of ESS is enhanced by a factor of 1.5. The explanation is that TAN forms an adduct with the radicals on one of the bases (Willson and Emmerson, 1970), which gives rise to an ESS and to a decrease in survival. To explain the protective effect of TAN in the presence of MISO, it is suggested that TAN acts as a radical and "repairs" the MISO-enhanced damage or converts it into damage that is more easily repaired.

5. CHANGES IN INTRACELLULAR THIOL LEVELS

It has been known for many years that various simple chemical compounds can reduce intracellular non-protein-bound sulfhydryl (NPSH) levels. N-ethylmaleimide is one of the best known examples of this type of reagent. Some of the sensitizers first characterized in bacterial systems (Adams and Cooke, 1969) can reduce GSH concentration but these suppressors of GSH are also active as electrophilic sensitizers in free-radical mechanisms. Examples include various electrophors such as conjugated diketones, pyruvates and esters of maleic and fumaric acid. Of particular interest at the present time is diethylmaleate (DEM; Bump *et al.* 1982). This agent can enhance considerably the sensitizing action of misonidazole both *in vitro* and *in vivo*. The effect on cellular DNA is still not known.

Cellular GSH levels can also be depleted via biochemical processes. The compound buthionine sulphoximine (BSO) inhibits the activity of γ-glutamylcysteine synthetase, an enzyme which catalyses the GSH biosynthesis (Griffiths and Meister, 1979).

Both approaches to intracellular GSH depletion were used in our own study of the role of oxygen during irradiations under conditions of GSH depletion (Van der Schans *et al.* 1986). In two cell lines, CHO and HeLa, the effects of the oxygen concentration on the induction of SSB and DSB and on cell killing were investigated. Great care was taken to obtain a really low oxygen content of the medium (Chapman *et al.* (1970). In the cells we observed no, or only a marginal, radiosenstization due to GSH depletion both under aerobic and anoxic conditions; at intermediate concentrations of oxygen, however, radiosensitization values were much higher. These results can be explained by the simplified model described by Koch *et al.* (1984) taking into account the competition between endogenous sensitizers and protectors and an exogenous sensitizer, although there is not a strictly first-order dependence of radiosensitization on oxygen concentration.

5.1. *The induction of SSB and DSB as a function of the concentration of oxygen*

Detailed studies on the effect of oxygen on sensitization by DEM and BSO were performed with CHO and HeLa cells grown on Mylar. In figure 1, the induction of SSB in DNA in CHO cells, calculated from dose-effect curves, is plotted as a function of the concentration of oxygen in the surrounding medium. Without pretreatment with BSO or DEM the maximum OER is about 3.5. Depletion of GSH by pretreatment with DEM or BSO has no, or only a marginal, effect on the induction of SSB both under aerobic and under strongly hypoxic conditions (0.01 μM oxygen initially present in medium). At intermediate concentrations of oxygen (0.5-10 μM), however, the sensitization for the induction of SSB is much larger. With HeLa cells comparable results were obtained, although the oxygen concentration of substantial sensitization was somewhat higher than with CHO cells. The K_m values (see Koch *et al.* 1984) for untreated, BSO- and DEM-

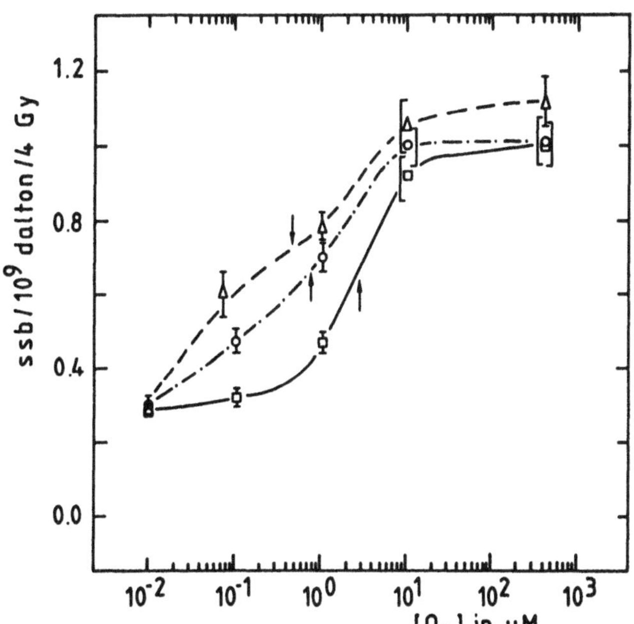

Figure 1. The sensitization of CHO cells by GSH depletion to the induction of SSB in DNA by X-rays as a function of the oxygen concentration: △, GSH depletion with DEM (1 mM, 1h, room temperature); O, GSH depletion with BSO (0.2 mM, 16 h, 37°C); □, control. The arrows indicate the O_2 concentration at which for each curve half of the maximal effect is observed (K_m). Each point represents the averaged slope (+SEM) of the dose-effect curves obtained in two independent alkaline elution experiments. Curves were fitted by eye. (Van der Schans *et al.* 1986).

treated CHO cells are 2.8, 0.75 and 0.45 μM oxygen, respectively, and for untreated, BSO- and DEM-treated HeLa cells 7.5, 2.7 and 3.7 μM oxygen.

In figure 2, the induction of DSB in DNA of CHO cells is shown as a function of the concentration of oxygen in the surrounding medium. Without pretreatment with DEM or BSO the maximum OER is about 4, which is in agreement with our earlier observations. Again, only at intermediate oxygen concentrations does GSH depletion result in a clearly increased induction of DSB. It should be noted that at 0.1 μM oxygen, there is still no significant sensitization of the induction of DSB, whereas at this oxygen concentration the induction of SSB in CHO cells is markedly enhanced by both DEM and BSO (figure 1). This difference can probably be explained by the fact that the high dose rate of the irradiation, which is necessary for DSB analysis, causes an additional reduction in oxygen concentration, due to fast oxygen consumption. The K_m values for untreated, BSO- and DEM-treated CHO-cells are 2.3, 1.2 and 1.0 μM oxygen respectively.

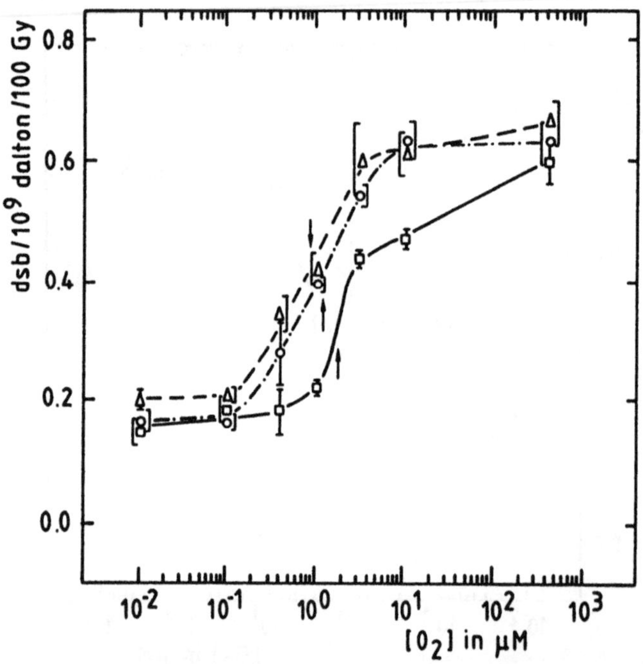

Figure 2. The sensitization of CHO cells by GSH depletion to the induction of DSB in DNA by γ-rays as a function of the oxygen concentration: △, GSH depletion with DEM (1 mM, 1h, room temperature); O, GSH depletion with BSO (0.2 mM, 16 h, 37°C); □, control. The K_m for each curve is indicated by arrows. The points were derived from the averaged slopes of dose effect curves. The vertical bars represent SEM's of four independent experiments (Van der Schans *et al.* 1986).

5.2. *The reduction of cell survival as a function of the concentration of oxygen*

Figure 3 illustrates the reduction of CHO cell-survival by X-rays and the effect of GSH depletion with BSO as a function of the concentration of oxygen in the surrounding medium. The largest sensitization due to the GSH depletion appears to occur in the range between 1 and 10 μM oxygen. The maximum OER (without GSH depletion) in this experiment is about 2.8, which is in reasonable agreement with other values reported in the literature (e.g. Shrieve *et al.* 1985). The K_m values for untreated and BSO-treated CHO cells are 7.0 and 3.5 μM oxygen, respectively, which are higher than the corresponding-values for induction of SSB in the same cell strain (figure 1). These results clearly confirm the importance of intracellular GSH with respect to protection against radiation damage. However, the chemical reconstitution of the target free radical via hydrogen donation by GSH is of importance only when it can *compete* with the reaction of oxygen with the

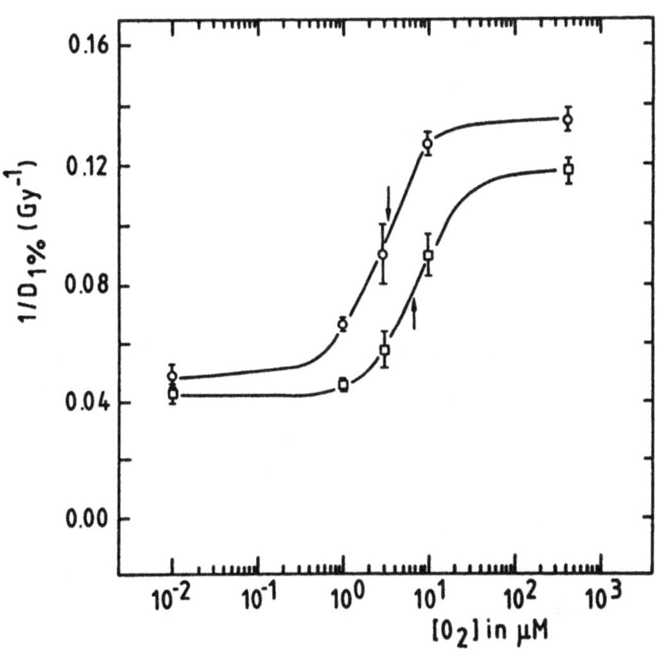

Figure 3. The effect of GSH depletion on the radiosensitivity of CHO cells to X-rays in dependence on the oxygen concentration. The radiosensitivity is expressed as the reciprocal of the one per cent survival dose ($D_{1\ \text{per cent}}$). O, GSH depletion by BSO (0.2 mM, 16 h, 37°C); □ control. The vertical bars represent errors estimated from the range of variation of the corresponding survival curves. (Van der Schans *et al.* 1986).

same free radical in the critical molecules, which is in agreement with the conclusions drawn by other authors from earlier *in vitro* studies.

The sensitizing effect of GSH depletion is more pronounced for the induction of SSB and DSB than for cell death, particularly in CHO cells with enhancement ratios for break induction of up to 1.9. Under the same conditions for cell death it was 1.6. As described in section 3, the oxygen-thiol competition model may not be valid for e.g. base damage. This could explain the fact that cysteamine protects less efficiently against cell killing than against the induction of SSB and DSB (Koch and Painter, 1975, Koch and Howell 1981, Van der Schans *et al.* 1982). Similar observations were made by Koch and Painter (1975) who found an OER of about 4 for the induction of SSB detected by alkaline sucrose-gradient sedimentation, which is higher than the OER for cell killing. Taken together, all these data justify the conclusion that SSB and DSB may not be the only causes of cell death, although it can not be excluded that a small fraction of unrepaired SSB (or DSB) is still responsible for the reduction in cell survival.

Edgren *et al.* (1985) and Révész and Edgren (1984) reported a sensitization under anoxic conditions for the induction of SSB in human cells in which GSH was depleted. In addition, they found a delay of repair of SSB, but only after irradiation on glass Petri dishes on which the cells had been seeded shortly before irradiation; the dosages applied in the repair experiments were rather high. Both factors may have contributed to the rather slow repair of SSB observed by these authors, even in untreated human fibroblasts ($t_{1/2}$ of about 13 min, whereas a $t_{1/2}$ of about 2 min usually found; Van der Schans *et al.* 1982). Evans *et al.* (1984), who used CHO cells growing in suspension, did not find any change in the rate of repair after pretreatment with DEM.

Held (1983) reported that the extent of radiation protection, both with respect to survival of Chinese hamster cells and to induction of SSB, was smaller for aerobically than for hypoxically irradiated cells in the presence of low concentrations of dithiothreitol (DTT). With high DTT concentrations, the protection factors for the hypoxically irradiated cells were smaller. Edgren (1989) and Sölen (1989) confirmed these observations. The experimental data could be explained by assuming a competition between the damage-restituting DTT and the damage-fixing reactants, and the authors postulate that only a certain proportion of the damaged targets may be repairable.

The clonogenic survival of $GSH^{-/-}$ (genetically GSH-depleted cells) and $GSH^{+/+}$ cells was studied after treatment with 1 mM DTT (Edgren, 1987). This concentration caused an increase in the NPSH and GSH content in the nuclei of $GSH^{-/-}$ cells to a similar level as that present in untreated $GSH^{+/+}$ cells. The dose modifying factors (DMF) for anoxically irradiated cells were larger than those for oxically irradiated cells. This implies increased OER for both the proficient and deficient cells. The OER for $GSH^{-/-}$ cells in the presence of DTT reached a value similar to untreated $GSH^{+/+}$ cells, both in the survival experiments and in DNA break-induction assays. It can be concluded that DTT may substitute either directly for the missing GSH in the deficient cells or indirectly by releasing GSH in the nucleus. In contrast, when cysteamine (MEA) was used instead of DTT, the DMF values were larger for oxically than anoxically irradiated cells. The results obtained with 1 mM DTT and 20mM MEA are shown in table 1. The difference between the results obtained with DTT and MEA may be explained by the enhanced enzymatic repair of damage induced under oxic conditions by MEA (Hülsewede and Schulte-Frohlinde, 1986). Edgren (1987) also found that there is a correlation between the *nuclear* GSH concentration and the OER, calculated for the yield of breaks or survival.

Table 1. The DMF and OER values calculated for $GSH^{+/+}$ and $GSH^{-/-}$ cells after treatment with two different radioprotectors (Edgren,1987).

	$GSH^{+/+}$ cells			$GSH^{-/-}$ cells		
Treatment	DMF(O_2)	DMF(Ar)	OER	DMF(O_2)	DMF(Ar)	OER
			DNA breaks			
no DTT	-	-	3.5	-	-	1.3
1 mM DTT	1.8	3.7	8.3	1.5	3.9	3.5
20 mM MEA	2.1	3.5	3.8	1.7	3.8	2.8
			Survival			
no DTT	-	-	2.5	-	-	1.9
1 mm DTT	1.1	1.3	2.9	1.1	1.4	2.5
20 mM MEA	1.6	1.5	2.5	1.7	1.4	1.1

For the anoxic conditions argon (Ar) was applied.

6. PROTECTIVE EFFECT OF DMSO

In an attempt to determine the contribution of OH·-radicals to the induction of damage in DNA, Skov (1984) used DMSO as a scavenger which is believed to be specific for these radicals. She found that DMSO protects mainly under aerobic conditions, with respect to the induction of both SSB and ESS. This is in contrast to the protective effect of SH-compounds which is more pronounced under hypoxic conditions. Moreover, it has been found (Van der Schans *et al.* 1979) that cysteamine does not protect against the induction of ESS. Because both compounds scavenge OH·-radicals, this indicates that the predominant action in the cell of SH-compounds, particularly cysteamine, is not radical scavenging but reaction with a DNA-damage which is modified but not restored.

7. CHEMOPROTECTION AGAINST DNA DAMAGE IN HUMAN BLOOD AND IN THE MOUSE

From the data mentioned above it is clear that oxygen plays an important role in both radiosensitization and radioprotection by many compounds. However, because the effective concentration of oxygen in most *in vivo* situations is not exactly known, it is difficult to predict the sensitizing or protecting efficiency of these compounds under these conditions. Therefore, methods are needed to study the effects of some compounds on radiosensitivity of the cell *in vivo* and to obtain more insight into the molecular processes involved in radioprotection under these conditions. This section deals with the protective action of cysteamine against damage induced by ionizing radiation in the DNA of nucleated cells in human blood irradiated *in vitro*, and also in the DNA of white blood cells (WBC) and bone marrow cells of irradiated mice.

The effect of cysteamine on DNA strand-break induction was studied with alkaline elution (detection of single-strand breaks and alkali-labile lesions; Schutte *et al.* 1988) and by means of an immunochemical method for the detection of radiation-induced single-strandedness which can be amplified with alkali (Van der Schans *et al.* 1989).

7.1. *Application of alkaline elution and ELISA*

We applied these techniques to study the protective action of cysteamine in human blood and *in vivo* in the mouse (blood and bone marrow cells).
Human blood was irradiated in the presence of 10 mM cysteamine. From this the white blood cells were isolated by lysis of the erythrocytes and washing several times.
In table 2 the results are summarized. With both assays, a pronounced protection by cysteamine is observed. However, the protection detected with ELISA is stronger than with alkaline elution. A possible explanation might be that some base damage is

Table 2. Irradiation of human blood *in vitro*

	[cysteamine] (mM)	period before irradiation (min)	DRF
alkaline elution	10	2	2.0 - 3.6
ELISA	10	2	3.5 - 8

converted into breaks during the long exposure at high pH in the case of alkaline elution. The immunochemical method, which involves short term exposure at lower pH, may lead to much less conversion of damage into breaks. It is known that cysteamine protects less efficiently against induction of base-damage (recognized by repair-enzymes of a *M. luteus*-extract) than against induction of sugar-damage.
As stated before, the concentration of oxygen plays an important role in the effects of protective substances. Therefore, it is preferable to study the protective action *in vivo*, because of the unknown, but probably low intracellular concentration of oxygen. Therefore, both alkaline elution and ELISA were applied to investigate the protective action of cysteamine during irradiation *in vivo*.
To this purpose, mice were injected *i.p.* with a sublethal dose of cysteamine (2 mg/20 g body weight) 10 min before irradiation with 300-kV X-rays. Immediately after exposure

the mice were killed and the radiation damage was quantified in the DNA of white blood cells and in that of the bone marrow cells. No significant protection in the DNA of the white blood cells was detected. In the bone marrow, the most critical organ for exposure to ionizing radiation, a small but significant protection with cysteamine was observed. Why there is protection in human blood *in vitro*, and not in the blood of the mouse, irradiated *in vivo*, is not yet clear. It may be due to the continuous supply of oxygen *in vivo*, which results in a (slow) oxidation of cysteamine and loss of its protective capabilities. In the bone marrow this process might be less important due to the low local concentration of oxygen. Further studies are in progress in which the concentration and the time of application is varied, both in human blood as well as *in vivo*.

The protective action of other agents, such as WR2721 and WR1065, will also be studied. Some preliminary results indicate that WR2721 protects both the WBC as well as the bone marrow cells against induction of breaks, provided that the amount injected is in the range of 500 mg/kg. This protection was optimal when the compound was injected 30 min before irradiation.

8. NEW DEVELOPMENTS

A very promising recent development in our laboratory is the detection of radiation-induced base damage after irradiation *in vivo*. To this purpose, some of these base damages are converted into breaks by means of enzymes that specifically recognize these lesions. Subsequently, these breaks are detected as described before, thus representing a measure of these types of lesions. In this way base damages can already be detected in WBC of human blood after exposure to a radiation dose of 3 Gy.

In addition, after irradiation of the mouse with a dose of 5 or 10 Gy, base damages can still be detected at 24 h after exposure.

In the WBC of a leukemia-patient, irradiated with a single dose of 8.6 Gy (over a period of about 1 h), single-strand breaks and base damage could be detected immediately after irradiation. Base damage was still detectable 1.5 h after exposure.

These developments facilitate future studies on the effects of protective agents on the induction of these types of lesions, both after *in vitro* and *in vivo* irradiation.

ACKNOWLEDGEMENTS

Part of the work was sponsored by EURATOM grant B16-E-148-NL. I thank Dr R.A. Baan for critical comments on the manuscript.

REFERENCES

Adams GE, Cooke MS (1969) Electron-affinic sensitization. I. A structural basis for chemical radiosensitizers in bacteria. Int J Radiat Biol 15:457-471

Alper T, Howard-Flanders P (1956) Role of oxygen in modifying the radiosensitivity of E.coli B. Nature 178:978-979

Bump EA, Yu NY, Brown JM (1982) The use of drugs which deplete intracellular glutathione in hypoxic cell radiosensitization. Int J Radiat Oncol Biol Phys 8:439-442

Chapman JD, Sturrock J, Boag JW, Crookal JO (1970) Factors affecting the oxygen tension around cells growing in plastic Petri dishes. Int J Radiat Biol 17:305-328

Edgren MR (1987) Radiation response of human cells genetically deficient in glutathione. Proceedings of the 7th Int. congress Radiation Research, July 1987, Edinburgh, p. 683-688

Edgren MR (1989) Combine radiosensitization by oxygen and radioprotection by thiols studied with clonogenic survival as end point and analysed according to a modifed competition model. Eur Soc Radiation Biology meeting, september 1989, Brussels Book of abstracts p 120

Edgren MR, Nishida T, Scott OCA, Révész L (1985) Combined effect of misonidazole and glutathione depletion by buthionine sulphoximine on cellular radiation response. Int J Radiat Biol 47:463-474

Evans JW, Taylor YC, Brown JM (1984) The role of glutathione and DNA strand break repair in determining the shoulder of the radiation survival curve. British J of Cancer 49, suppl VI:49-53

Griffiths OW, Meister A (1979) Potent and specific inhibition of glutathione synthesis by butionine sulfoximine (S-n-Butyl homocysteine sulfoximine). J Biol Chem 254:7558-7560

Held K (1983) Interactions of oxygen and sulfhydryl-containing radioprotectors in determining radiation-induced mammalian cell killing and DNA damage. Proceedings of the 7th Int. congress of Radiation Research , July 1983, Amsterdam, p. B6-11.

Hülsewede JW, Schulte-Frohlinde D (1986) Radioprotection of E.coli strains by cysteamine in the presence of oxygen. Int J Radiat Biol 50:861-869

Koch CJ, Howell RL (1981) Combined radioprotective and radiation sensitizing agents. II. Radiosensitivity of hypoxic or aerobic Chinese hamster fibroblasts in the presence of cysteamine and misonidazole: Implications for the "Oxygen Effect" (with appendix on calculation of dose-modifying factors). Radiation Res 87:265-283

Koch CJ, Painter RB (1975) The effect of extreme hypoxia on the repair of DNA single-strand breaks in mammalian cells. Radiation Res 63:256-269

Koch CJ, Stobbe CC, Bump EA (1984) The effect on the K_m for radiosensitization at 0°C of thiol depletion by diethylmaleate pretreatment: Quantitative differences found using the radiation sensitizing agent misonidazole or oxygen. Radiation Res 98:141-153

Kohn KW, Erickson LC, Ewig RAG, Friedman CA (1976) Fractionation of DNA from mammalian cells by alkaline elution. Biochemistry 15:4629-4637

O'Neill P (1983) Pulse radiolytic study of the interaction of thiols and ascorbate with OH adducts of dGMP and dG: Implications for DNA repair processes. Radiation Res 96:198-210.

Révész L, Edgren MR (1984) Glutathione-dependent yield and repair of single-strand DNA breaks in irradiated cells. British J of Cancer 49, Suppl. VI:55-60

Schutte HH, Van der Schans GP, Lohman PHM (1988) Comparison of induction and repair of adducts and of alkali-labile sites in human lymphocytes and granulocytes after exposure to ethylating agents. Mutation Res 194:23-37

Shrieve DC, Denekamp J, Minchinton AI (1985) Effects of glutathione depletion by buthionine sulphoximine on radiosensitization by oxygen and misonidazole *in vitro*. Radiation Res 102:283-294

Skov KA, Palcic B, Skarsgard LD (1982) DNA damage and radiosensi-tizers: *M.luteus* sensitive sites for misonidazole-TAN combination. Radiation Res 92:113-119

Skov KA (1984) The contribution of hydroxyl radical to radiosensitization: A study of DNA damage. Radiation Res 99:502-510

Sölen G (1989) Combined radiosensitization by oxygen and radioprotection by thiols studied with DNA breaks as end-point and analysed according to a modified competition model. Eur Soc Radiation Biology meeting September 1989, Brussels p 121

Taylor YC, Evans JW, Brown JM (1983) Radiosensitization by pretreatment with misonidazole: an interaction of damage at the DNA level. Proceedings of the 7th International congress on Radiation Research, Amsterdam, july 1983, p D5-40

Van der Schans GP, Centen HB, Lohman PHM (1979) The induction of gamma-endonucluase-susceptible sites by γ-rays in CHO-cells is not affected by the presence of thiol compounds during irradiation. Mutation Res 59:119-122

Van der Schans GP, Centen HB, Lohman PHM (1982) DNA lesions induced by ionizing radiation. In: A.T. Natarayan *et al.* (eds) Progress in Mutation Res 4:285-299

Van der Schans GP, Van Loon AAWM, Groenendijk RH, Baan RA (1989) Detection of DNA damage in cells exposed to ionizing radiation by use of anti-single-stranded-DNA monoclonal antibody. Int J Radiat Biol 55:747-760

Van der Schans GP, Vos O, Roos-Verheij WSD, Lohman PHM (1986) The influence of oxygen on the induction of radiation damage in DNA in mammalian cells after sensitization by intracellular glutathione depletion. Int J Radiat Biol 50:453-465Ward JF (1987) Radiation chemical mechanisms of cell death. Proceedings of the 8th Int. Congress of Radiation Research, Edingburgh, July 1987 Fielden et al. (eds) Taylor and Francis London p 162-168

Willson RL, Emmerson PT (1970) Reaction of triacetoneamine-N-oxyl with radiation-induced radicals from DNA and from deoxyribonucleotides in aqeous solution. In: Radiation Protection and Sensitization, Moroson EL, Quintiliani M (eds) Taylor and Francis London p 73-79

Wong TW, Whitmore GF, Gulgas S (1978) Studies on the toxicity and radiosenstizing ability of misonidazole under conditions of prolonged incubation. Radiation Res 75:541-549

OXYGEN ENHANCEMENT OF RADIOSENSITIVITY AND NUCLEAR GSH CONTENT

Margareta R. Edgren
Department of Tumor Biology II
Karolinska Institute
S-104 01 Stockholm
Sweden

Contradictory obervations have been reported that, on the one hand, the radical competition processes which determine the oxygen effect, are dependent on the cellular level of non-protein-bound sulphydryls (NPSH) and, on the other hand, oxygen enhancement ratios (OER) can be close to unity allthough sulphydryls are present in a considerable amount in the cells (Edgren 1987a). The controversy can be resolved by assuming a compartmentalization of GSH in the cells, and that the effect of buthionine sulphoximine (BSO), used to deplete GSH from the cell is different in the different compartments (Edgren 1987b). The present experiments were conducted to test this possibility.

Chinese hamster cells (V79-379A) were used with or without pretreatment with 0.1 mmol dm^{-3} D,L-BSO. Whole cells and isolated nuclei were irradiated by X-rays under aerobic and hypoxic conditions. The yield of DNA strand breaks was used as endpoint of the radiation effect determined with the alkaline unwinding method. GSH was measured in acid extracts from cells or nuclei by the DTNB/glutathione reductase assay. After rinsing with saline, nuclei were isolated from monolayer cell cultures by treatment for 4-5 min with 1.5 ml of a hypotonic solution (585 mg dm^{-3} NaCl and 1000 mg dm^{-3} Na-citrate) containing either 0.03 % or 1 % Nonidet P-40. Thereafter 1.5 ml of 0.25 mol dm^{-3} sucrose was added. After a couple of min the nuclei were rinsed once with the hypotonic solution without detergent and twice with saline. In complementary experiments medium containing 2 mmol dm^{-3} GSH or 1 mmol dm^{-3} dithiothreitol (DTT) was added to the isolated nuclei 10 min before irradiation.

NATO ASI Series, Vol. H 54
The Early Effects of Radiation on DNA
Edited by E. M. Fielden and P. O'Neill

The data in tables 1 and 2 indicate that the OER values correlate to the nuclear but not to the cellular GSH content. The OER for isolated nuclei is dependent also on the concentration of detergent used for the isolation. The high detergent concentration leads to a lower OER value conceivably due to leakage of thiols from the nuclei. Thiols added to the medium before irradiation may compensate for the leakage and restore OER.

Table 1. The relationship between OER and the relative GSH content in whole cells and isolated nuclei with and without BSO pretreatment.

Irradiated material	OER		Relative GSH content
cells	3.24 ± 0.13[a]	(14)[b]	100
cells + BSO[c]	2.73	(3)	26
nuclei[d]	3.19 ± 0.23	(10)	100
nuclei + BSO	3.04 ± 0.27	(7)	87

a) Mean ± Standard error
b) Number of replicate experiments
c) Preincubation with 0.1 mmol dm^{-3} D,L-BSO for 3 h
d) Isolation in hypotonic solution with 0.03 % Nonidet P-40

Table 2. OER for nuclei isolated with 1 % Nonidet P-40 in hypotonic solution, untreated or treated with thiols.

Treatment	OER	
none	1.82 ± 0.11[a]	(15)[b]
2 mmol dm^{-3} GSH	3.02	(2)
1 mmol dm^{-3} DTT	3.66	(2)

a) and b) as in Table 1

REFERENCES

Edgren MR (1987a) Radiation response of human cells genetically deficient in glutathione. In: Fielden EM, Fowler JF, Hendry JH, Scott D (eds) Progr 8th Int Congr Radiat Res, Edinburgh July 1987. Francis & Taylor, London, p. 683

Edgren MR (1987b) Nuclear glutathione and oxygen enhancement of radiosensitivity. Int J Radiat Biol 51:3-6

TIMESCALE FOR DEVELOPMENT OF DNA DAMAGE

John M. Warman
IRI, Delft University of Technology
Mekelweg 15
2629 JB DELFT
The Netherlands

During this thursday afternoon session the problems dealt with extended from the primary ionization events within the stacked bases of the DNA molecule itself, studied by multiphoton laser flash photolysis (Schulte-Frohlinde), through the very elegant measurements of the timescale for chemical repair using the gas explosion technique (Michael), to the effects of extracellular chemical agents on the eventual biological end-points (van der Schans, Held and Edgren). In an attempt to provide some cohesion between the talks and to be able to place the various aspects discussed in the general overal framework of cellular radiation damage a physical chemist's schematic view of the biological machinery involved was provided by the chairman. This is reproduced without apology below.

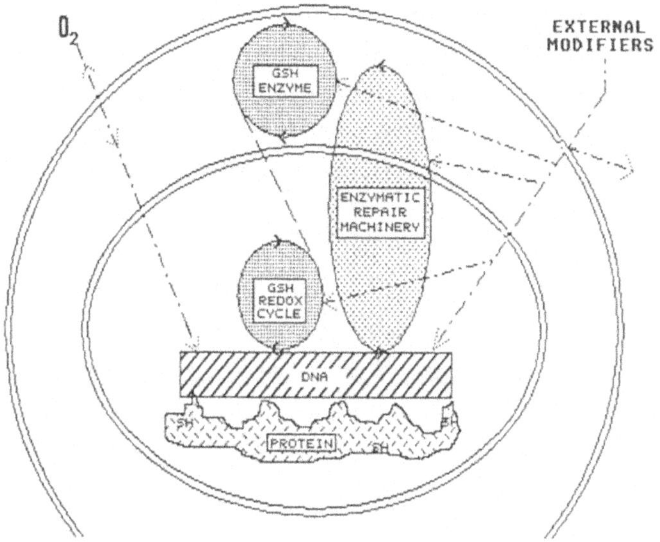

NATO ASI Series, Vol. H 54
The Early Effects of Radiation on DNA
Edited by E. M. Fielden and P. O'Neill

The title of the session is perhaps misleading. If a common thread had to be found between the various presentations then it would be the mechanistic role of natural and artificial thiols in DNA damage and repair rather than the dynamics of these processes per se. However, in keeping with the title (and the interests of the chairman) this summary will concentrate on the kinetic aspects of the data presented and on potential interrelationships within this area between the results obtained using the extremely diverse experimental approaches and systems of the main speakers.

In addition to providing a wealth of information on primary base damage at a molecular level using laser flash photolysis and pulse radiolysis techniques, Schulte-Frohlinde presented in vitro kinetic measurements on the reaction of single-strand and double strand DNA radical sites with the dithiol dithiothreitol (DTT). These showed radical repair by DTT to occur with a rate constant of *ca* $1x10^6$ $dm^3mol^{-1}s^{-1}$ which is many orders of magnitude lower than the diffusion controlled limit. From a study of the effect of DTT on the survival of E. coli cells a rate constant for repair of lethal damage very close to that found in the in vitro experiments and even possibly slightly larger was derived.

Probably due to its larger size and net negative charge the in vitro rate constant for DNA radical repair by the enzymatically produced, endogenous cellular thiol glutathione (a glutamic acid-cysteine-glycine tripeptide, abbreviated GSH) is an order of magnitude less than DTT at *ca* $1x10^5$ $dm^3mol^{-1}s^{-1}$. On the basis of estimates of several millimolar for the cellular concentration of GSH one would therefore expect natural thiol repair processes to occur on a timescale of milliseconds if the in vitro rate constant was applicable.

A timescale of milliseconds for natural repair under hypoxic conditions is in fact found in the experiments reported by Michael who derives values of 1.9, 2.6 and 4.8 milliseconds for repair in cellular systems with as end-points cell survival, double-strand breaks and single-strand breaks respectively. This result would therefore indicate, perhaps surprisingly, that the diffusional mobility, and hence medium viscosity, in the DNA environment in the V79-379A fibroblasts studied is very similar to that in the in vitro, i.e. dilute aqueous solution, case.

A particularly intriguing aspect of the results of the oxygen explosion experiments is that the thiol initiated repair of double-strand breaks occurs with a rate which is twice that for single-strand breaks. This leads Michael to the important conclusion that a double-strand break must have as a precursor initially paired radical sites. On the basis of this a revised view of LET effects on the oxygen enhancement ratio (OER) is given together with a novel symbolic representation of multiple radical sites and a modified formulation of the Alper and Howard-Flanders equation for oxygen enhancement taking into account multiply damaged sites. All very interesting stuff!

On the basis of the close to diffusion controlled rate constant of $1x10^9$ $dm^3mol^{-1}s^{-1}$ for radical reactions with O_2 found in aqueous solution and the millisecond timescale estimated and found for natural thiol repair, one would predict that equal competition for radical sites between O_2 and GSH would occur for O_2 concentrations on the order of micromolar, i.e. two orders of magnitude lower than the fully-air-equilibrated (FAE) concentration. This is in fact found to be the case as is illustrated by the results of van der Schans who has studied oxygen enhancement for O_2 concentration from 10^{-8} to $4x10^{-4}$ molar. For unperturbed (by BSO or other moderators) CHO and HeLa cells the half-enhancement O_2 concentrations (K values) are found to be 2.8 and 7.5 micromolar respectively. Again therefore predictions based on in vitro kinetic data appear to be in reasonable agreement with results obtained for cellular systems.

For all systems studied by van der Schans an O_2 concentration less than one tenth of the FAE value was found to be sufficient to result in saturation of the O_2 enhancement effect. At the other end of the scale it is apparent that an O_2 concentration less than 10^{-4}FAE is necessary in order to ensure that residual O_2 sensitization is absent. This represents a timescale on the order of a tenth of a second which indicates the extreme stability of the (multiple)radical sites involved in O_2 sensitization.

A potentially important, kinetically related effect was a difference found between X- and gamma-irradiated samples. This is tentatively attributed by van der Schans to an oxygen radiation-depletion effect. Since a dose of 10 Gy could result in the complete depletion of several micromolar of O_2 this is something which should be kept in mind particularly for experiments under hypoxic conditions and high dose rates. Unless chemical or high viscosity barriers are present, O_2 replenishment within the nucleus should however be very rapid, with diffusion calculated to take place on a time scale of milliseconds over the micron dimensions involved, as is in fact experimentally verified by the results of Michael.

Many additional interesting experimental results in the area of the effects of external moderating agents on the repair and sensitization processes at a cellular level were presented by van der Schans and by the additional speakers Held and Edgren. The results demonstrated the great difficulty that exists in defining and characterising cellular systems and differentiating the various possible enzymatic and physico-chemical processes influenced by the foreign additive. Clearly much further work will be required before the effects of external modifiers on the complex symbiosis of the intracellular repair and sensitization machinery are fully understood.

PRIMARY DAMAGE AND FIXATION OF CHROMOSOMAL DNA AS PROBED BY MONOCHROMATIC SOFT X-RAYS AND LOW-ENERGY NEUTRONS

Masao S. Sasaki
Radiation Biology Center
Kyoto University
Kyoto 606
Japan

INTRODUCTION

Chromosome structural rearrangements constitute an important genetic alteration in cells exposed to ionizing radiations. Moreover, in some instances, the chromosome rearrangement itself or mechanism involved has been unequivocally demonstrated to be a critical event leading to cell death, induction of mutation, and malignant transformation (e.g., Joshi et al. 1982, Cox and Masson 1978, Kodama and Sasaki 1987). As is the case in the lower eukaryotic cell system, there is also evidence suggesting that DNA double strand breaks are the integral component of the primary lesions responsible for the induction of chromosome aberrations and cell death in mammalian cells (Obe et al. 1982, Natarajan and Obe 1984, Bryant 1984, 1985). However, the primary process of damage fixation as a pathway to the expression of radiation injury is still a matter of considerable debate.

On the basis of the microdosimetric concepts, the "theory of dual radiation action" was developed by Kellerer and Rossi (1972) for the interpretation of biological radiation effects, in which the elementary lesions are fixed as a result of a combination of pairs of primary lesions. The concept met the classical linear-quadratic model ($\alpha D+\beta D^2$) of Lea and Catcheside (1942) and sublesion interaction model of Neary (1965) for the induction of chromosome aberrations. Assuming the pairwise interaction of primary damage in its fixation process, the microdosimetric parameters such as track-based energy transfer, range of charged particle and intranuclear geometry of chromosomal DNA can be correlated with the efficiency of damage induction and fixation leading to the formation of chromosome aberrations and thus relative biological effectiveness (RBE).

NATO ASI Series, Vol. H 54
The Early Effects of Radiation on DNA
Edited by E. M. Fielden and P. O'Neill

Alternatively, there are also some suggestions that the pairwise interaction of sublesions may not be needed for damage fixation. Chadwick and Leenhouts (1978) postulated the chromosome rearrangement as a consequence of incorrect reciprocal recombinational repair of a DNA double strand breakage. The non-linear dose-response relationship was interpreted as a reflection of the two-hit nature of the production of DNA double strand breaks. More recently, in a series of experiments with ultrasoft X-rays, Goodhead (1982) developed the "threshold-energy repair-saturation model", where the relevant lesions could be produced as a one-hit event and the pairwise interaction need not be invoked. The curvilinearity of dose-response relationship is explained by the saturation of repair mechanisms and the presence of classes of lesions having different energy requirement.

The use of ultrasoft X-rays in radiobiology has been regained by Goodhead and colleagues (Goodhead and Thacker 1977, Goodhead et al. 1979) after the pioneering work of Lea (Lea and Catcheside 1942). Monochromatic soft X-rays with sufficiently long wavelength provide useful probes for the submicroscopic dissection of the primary processes of damage induction and fixation, because they generate photoelectrons with discrete initial energy and range.

In this paper, the dose-response relationships of chromosome aberrations have been compared in the cells irradiated with the monochromatic soft X-rays of varying energy. The observations lead to several conclusions relevant to the primary damage and fixation in the chromosomal DNA which are involved in the formation of chromosome aberrations.

MONOCHROMATIC SOFT X-RAYS AND EXPERIMENTAL PROCEDURES

The X-rays used in this experiment were monochromatic soft X-rays obtained from synchrotron radiation produced by the electron storage ring at the Photon Factory, National Laboratory for High Energy Physics, Tsukuba. The irradiation systems and details of the dosimetry have been published by Kobayashi et al. (1987). The intensity of X-rays at the sample position was determined by the ionization chamber and then converted to the absorbed dose in Gy based on the mass energy absorption coefficients estimated by the elemental composition of spleen for the irradiation of human lymphocytes and soft tissue for the irradiation of cultured mouse cells as described previously (Sasaki et al. 1989).

Peripheral blood lymphocytes isolated by the Ficoll-paque gradient sedimentation were suspended in culture medium RPMI-1640 containing 20% fetal bovine serum, and sandwiched as a monolayer between 5 µm thick polycarbonate films or 1.5 µm thick Mylar films before irradiation with X-rays. The transmission of the covering film was measured for the X-rays with specific energy and included in the calculation of dose to the cells. In some experiments, lymphocytes in plasma were used instead of lymphocytes isolated and suspended in culture medium. Cells irradiated in plasma tended to give rather lower aberration yield, but the data were combined with those irradiated in culture medium since the difference was small. Immediately after irradiation, the cells were cultured in the presence of phytohaemagglutinin-M for 50 hours and processed for the chromosome preparations as described previously (Sasaki et al. 1989).

The cultured mouse cell line used in this experiment was the established cell line, m5S, which was characterized by a near-diploid chromosome number and flat cell morphology at confluence (Sasaki and Kodama 1987). The cells were grown to confluence on plastic plates (20mm diameter), covered with 1.5 µm thick Mylar film and exposed to X-rays. The irradiated cells were dispersed by trypsinization, cultured, and processed to harvest the 1st postirradiation mitoses.

Chromosome aberrations were scored and their frequencies were fitted to a linear-quadratic model, $Y=C+\alpha D+\beta D^2$, in which C was the spontaneous frequency and D was dose in Gy. Because of relatively small number of cells available in each experiment, C was adopted from the pooled data in our laboratory, i.e., $C=6.473\times10^{-4}$ for dicentrics and $C=8.496\times10^{-4}$ for dicentrics plus rings in the experiments with human lymphocytes. The dose-response curves were constrained to pass through the origin, i.e., C=0, in the experiments with mouse m5S cells.

CHANGES IN THE INITIAL SLOPE WITH X-RAY ENERGY

The biological effects of soft X-rays can be ascribed to the ionization by photoelectrons and their associated Auger electrons. The initial energy of the photoelectrons is equal to the X-ray energy minus electron binding energy of the target atoms. Therefore, the energy of photoelectrons can be controlled by changing the energy of the X-rays. In the biological material, oxygen constitutes about 75% of the elemental composition and therefore its photoabsorption plays a major role in the biological effects in the cells exposed to soft X-rays.

Here, the induction kinetics of chromosome aberrations were compared in human G_0 lymphocytes irradiated with soft X-rays with varying energy ranging from 1.9 keV to 14.6 keV. The dose-response relationships of exchange aberrations such as dicentrics and rings were better expressed by a linear-quadratic model rather than a linear model for all X-ray energies herein dealt. Table 1 shows the regression coefficients, α and β , of the least square fit for the yield of dicentrics to the linear-quadratic function, $Y=C+\alpha D+\beta D^2$ with $C=6.473\times10^{-4}$.

Table 1. Some physical characteristics and regression coefficients for dicentrics in human Go lymphocytes irradiated with different X-rays.

X-rays* (keV)	Photoelectrons from oxygen Energy(keV)	Range(um)	α $(10^{-2}Gy^{-1})$	β $(10^{-2}Gy^{-2})$
14.59	14.06	4.51	4.32±1.19	8.72±1.22
13.79	13.24	4.06	7.66±3.90	11.73±2.59
12.40	11.87	3.35	5.44±2.98	14.04±2.54
6.89	6.36	1.15	20.95±4.22	15.89±3.94
4.77	4.24	0.58	15.48±4.74	16.55±5.75
3.10	2.55	0.25	5.05±0.45	11.92±0.23
2.14	1.60	0.11	0.78±0.50	12.53±0.23
1.91	1.36	0.087	7.12±0.69	16.71±0.69
^{60}Co γ**	589.05	2171.5	0.86±0.54	6.42±1.19

*)Data on X-rays with energy between 4.77 keV and 14.59 keV are from Sasaki et al. (1989)

**)For ^{60}Co γ-rays, energy and range represent the average kinetic energy of Compton electrons and their corresponding range.

The quadratic components were similar among all soft X-rays while a slight systematic change wa suggested. However, the linear components varied considerably with X-ray energy: they increased with the decreasing energy down to 6.89 keV and then dropped with further decrease of X-ray energy. Since the linear component is the initial slope of the dose-response curve, its energy dependence is important with respect to the intrinsic biological efficiency of low-energy electrons.

The RBE has been correlated with the linear energy transfers (LET). For the various particle radiations the efficiency as expressed by the initial slope increases with LET, reaching a maximum at around LET=100 keV/μm and then falls sharply with further increase in LET (e.g., Cox et al. 1977, Goodhead et al. 1980, Lloyd and Edwards 1983). Recently, in the study on inactivation of yeast cells by low-energy X-rays, Frankenberg (1984) reported the presence of another peak RBE at LET of 2-5 keV/μm. In

Fig. 1, the linear and quadratic coefficients of the dose-response relationships of dicentrics are plotted against the track-average LET of photoelectrons. While the quadratic component stays rather constant for a relatively wide range of LET, the change of the linear component is evident: it increases with the increase of LET, reaches a maximum at the LET of about 5 keV/μm, then falls with further increase of LET. This LET dependence is very comparable to that found in the yeast cells, and indicates that the electrons have their own peak RBE which is different from that of other charged particles such as protons and α particles. The

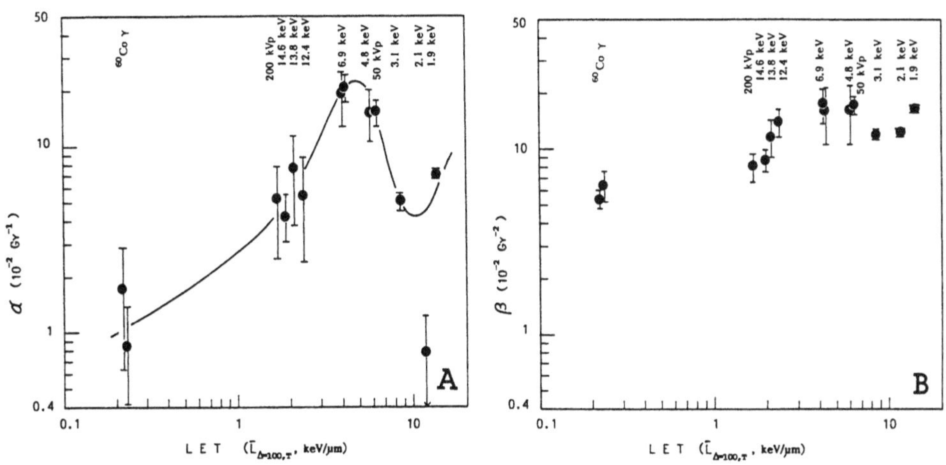

Fig.1. Regression coefficients for dicentrics plotted against the LET. (A) Linear component (α). (B) Quadratic component (β).

2.1 keV X-rays showed lowest value of the linear component, while the 1.9 keV X-rays showed the linear component which was comparable to that of the 3.1 keV X-rays. Although it is premature to conclude such a drastic variation within a small range of energy, yet it may be worth noting that the pattern of cellular distribution of chromosome aberrations was quite dissimilar among these X-rays: the distribution of chromosome aberrations was highly overdispersed in the cells irradiated with 1.9 keV X-rays whereas those in cells irradiated with X-rays with 2.1 or higher energy showed a good fit to the Poisson distribution. The overdispersion may be a consequence of unequal distribution of dose either to the cell populations or within a cell. The overdispersion of chromosome aberration distribution has been also noted by Virsik et al. (1980) in human lymphocytes similarly irradiated with aluminium K- and carbon K-characteristic X-rays.

ENHANCEMENT BY X-RAYS WITH RESONANCE ENERGY

The monochromatic soft X-rays can be used to study the biological consequences of the energy absorption of a specific elemental component in the cells. Here, free bromine in the form of 5-bromodeoxyuridine (BrdU) and phosphorous atoms in the cells were targeted with X-rays with the resonance energy. The energy of the K-shell absorption edge of bromine is 13.47 keV. The K-shell resonance peak of phosphorous was determined by Dr. Kobayashi (National Laboratory for High Energy Physics) to be at 2.151 keV by measuring the absorbance of a thin film of potassium phosphate.

Fig. 2 shows the results of experiment with cells in BrdU. Because of the complexity of energy transfer associated with photoabsorption of the incorporated bromouracil (Mönkehaus and Köhnlein 1973), the BrdU was not

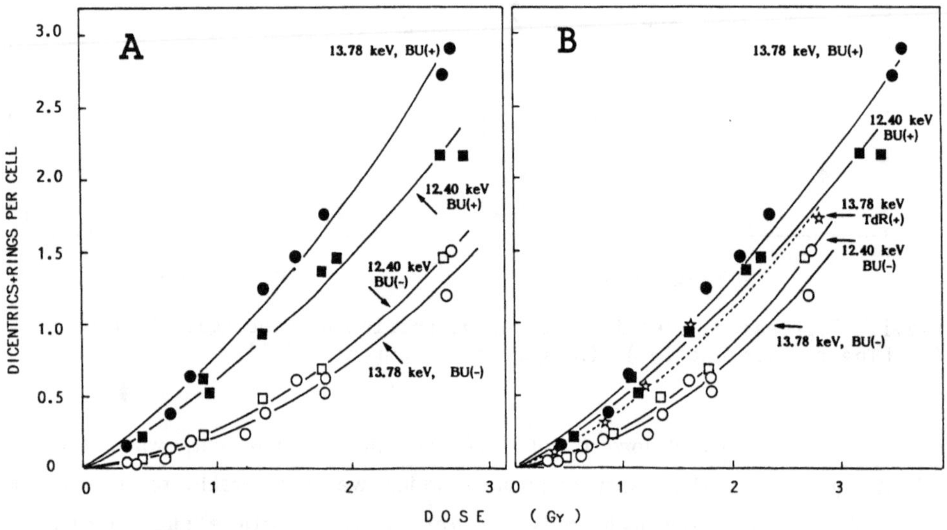

Fig. 2. The frequencies of dicentrics and rings per cell plotted against dose by 13.78 keV and 12.40 keV X-rays in the presence or absence of BrdU. A: The absorbed dose calculated according to the elemental composition of spleen. B: The absorbed dose calculated by including the relative abundance of bromine in the components.

incorporated into DNA but the human peripheral blood lymphocytes in G_0 stage were simply soaked in 0.1 M BrdU before irradiation. X-rays with 13.78 keV were used for resonance absorption, and those with 12.40 keV were used as control energy below the edge. As seen in Fig. 2A, the presence of BrdU sensitized the cells in inducing chromosome aberrations. The sensitization by BrdU was more evident in cells irradiated with 13.78 keV X-rays than those irradiated with 12.40 keV X-rays, clearly showing the edge effect. However, when the absorbed dose was recalculated by including bromine atoms as an elemental component, the difference between the two energies disappeared (Fig. 2B). The aberration yields in the presence of BrdU were not distinguishable from those in cells suspended in the same concentration of thymidine. This indicates that the sensitization by BrdU alone may be associated with the nucleotide pool imbalance (Meuth 1984), and that the enhancement of chromosomal effects with X-rays with resonance energy is mainly due to photoelectrons and the contribution of localized energy deposition by Auger emission may be small, if any.

Fig. 3 shows the induction of dicentrics in the mouse m5S cells irradiated with the X-rays of energy corresponding to the resonance peak of phosphorous (Saigusa, unpubl.). The X-rays with the resonance peak energy (2.151 keV) are highly effective as compared with those with energies below (2.145 keV) and above (2.157 keV) the peak energy. The magnitude of the enhancement can not be explained by the difference in the absorbed dose calculated based on the atomic composition of DNA. Because the enhancement was largely due to the elevated value of the quadratic component, we assume that the dissociation of the sugar-phosphate moiety of DNA due to photo-absorption by phosphorous atoms may develop into double strand breaks when

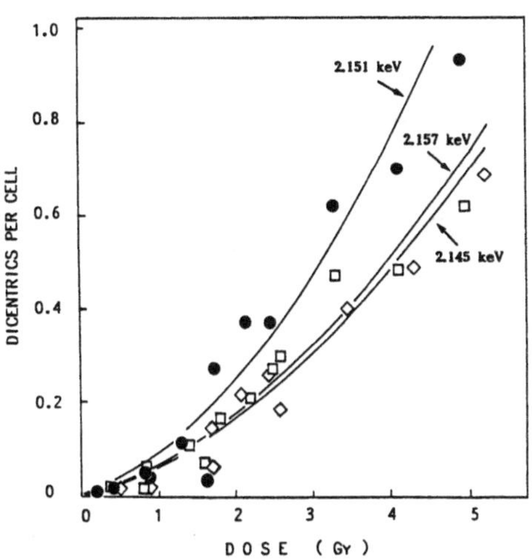

Fig. 3. Frequencies of dicentrics in the mouse m5S cells irradiated with X-rays with resonance peak energy of 2.151 keV and those with below (2.145 keV) and above (2.157 keV) the peak energy.

the photoabsorptions have taken place on opposite complementary strands with close proximity. Such double strand breaks may also be responsible for the formation of chromosome aberrations. The probability of such events should be priportional to the square of the dose, and hence results in the enhancement of the non-linear portion of the dose-response curve.

DISTANCE MODEL FOR THE LIMITATION OF LET-RBE RELATIONSHIP

The change in the linear component (α-term) presented in Fig. 1A is consistent with the pairwise interaction of primary damage. The probability for two damages to be formed by a single electron increases with the ionization density along the electron track. In the heavier charged particles, the rapid fall of RBE after the peak has been interpreted by the over-killing or the unequal distribution of hits among cells. However, the same mechanism may not be likely for the electrons. The efficiency could be due to the consequence of the short-segment track compared with the distance between interactable damages produced by a single pass of an electron. The combined CSDA range of a photoelectron and its associated Auger electron is about 0.8 µm for X-rays corresponding to the peak linear component (5 keV) and about 0.1 µm for softer X-rays showing lowest linear component (2.1 keV). Although the practical range is much shorter (Cole 1969), these ranges are still larger than the dimension of the target DNA duplex. However, assuming that the DNA is packed into a thick elemental fiber of about 34 nm (Griffith 1975, Finch and Klug 1976), it is highly likely that such elemental fiber constitutes a unit of damage interaction in forming chromosome exchange aberrations while DNA is the prime target. If the DNA duplex is packed into 34 nm fiber with a packing ratio of 1:50

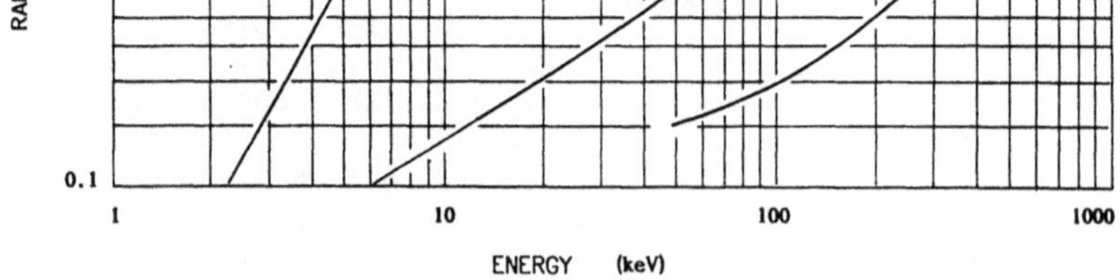

Fig. 4. CSDA range of electrons, protons and α-particles against particle energy. The range calculated according to the equation, $R(\beta^2)=M\cdot F(\beta^2)/z^2$, in which $F(\beta^2)$ is the range of proton with corresponding β^2, M is particle's rest mass, β is relative velocity and z is effective charge.

in a lymphocyte nucleus with a diameter of 6 µm, the mean distance between fibers is about 80 nm. Moreover, the elemental fibers are not homogeneously distributed within a cell nucleus (e.g., Commings 1980), and recent findings using chromosome specific DNA probe indicate that DNA from each chromosome has a rather localized compartment within an interphase nuclei (Pinkel et al. 1986). Such intranuclear geometry of the elemental fibers may play a significant role in the limitation of damage interaction and hence LET-RBE relationship determined by the initial slope of dose-response curves.

The calculated CSDA range in water for electrons, protons and α-particles are shown in Fig. 4. The critical range of about 0.8 µm for electrons corresponds to 60 keV protons and 300 keV α-particles; the LET of these radiations being about 92 keV/µm and 220 keV/µm, respectively. These are also the LET ranges know to show peak RBE (Cox et al. 1977, Thacker et al. 1979, Goodhead et al. 1980, Lloyd and Edwards 1983). These observations pose the limitation of the use of LET as an index of radiation quality in the evaluation of biological effectiveness, and rather suggest that the path length of charged particles is of biological significance. The quality parameter, z^2/β^2, has been suggested to be a better parameter to reconcile the difference in the RBE of different types of radiations (Thacker et al. 1979, Folkard et al. 1989). This is consistent with the distance model since the parameter is inversely proportional to the range of particle.

A single pass of a charged particle with a range below the diameter of elemental fibers (34 nm) may be unable to produce microscopically recognizable chromosome exchange aberrations. However, they may still produce two damages in very close proximity within a fiber, and retain a capacity to induce submicroscopic minute interstitial deletions or mutations (Fig. 5). Therefore, the LET-RBE relationship may not be the same for chromosome aberrations and submicroscopic mutations despite the underlying mechanism being essentially the same.

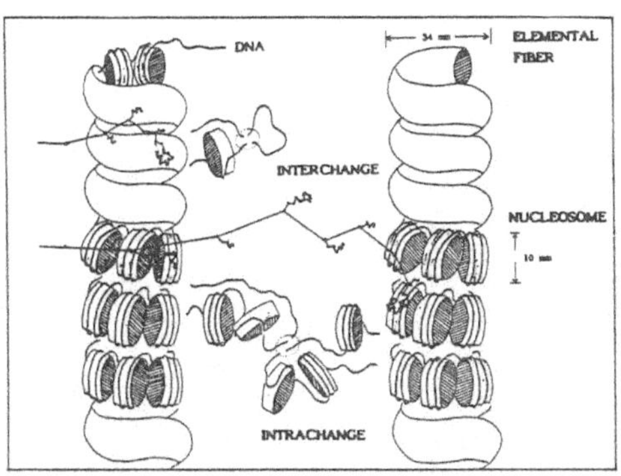

Fig. 5. Schematic representation of damage in chromosomal DNA and their interaction to form exchange aberrations in relation to the elemental fibers.

Carbon K-characteristic X-rays generate L-shell electrons with a very short range of about 7 nm, which is far below the dimension of the elemental fiber. With the present distance model, they are expected to be unable to induce chromosome exchange aberrations as one-track events. On the contrary, the carbon X-rays have been demonstrated to be very efficient in inducing chromosome aberrations (Virsik et al. 1980, Thacker et al. 1986). It is noteworthy, however, that such low-energy X-rays are strongly attenuated during their passage through the cells. Therefore, relatively large fractions of the photons are absorbed in the periphery of the cells giving large amount of dose per unit volume in those regions. Under such conditions, the cellular response may resemble those associated with the heavy charged particles of high LET (Tobleman and Cole 1974). The regain of the apparently high efficiency of X-rays with energy below 2.1 keV may be associated with this localized energy deposition in the cell nucleus. The overdispersion of chromosome aberration distributions is a characteristic feature of high LET radiations (Edwards et al. 1979, 1980, Virsik and Harder 1981, Takatsuji and Sasaki 1984). Therefore, the underlying mechanisms may be different between the primary increase in the initial slope in response to LET and the regain of the efficiency for the ultrasoft X-rays after the breakdown at the energy around 2 keV.

EFFECTS OF LOW-ENERGY NEUTRONS

The dose given to the cells by fast neutrons is due to recoil protons from elastic scattering mainly on hydrogen, while oxygen, carbon and nitrogen are also target atoms. The maximum initial energy of recoil protons is determined by the energy of the neutrons. Thus the neutrons, particularly low-energy neutrons, also provide useful probes for the submicroscopic dissection of damage induction and fixation processes.

To see any effects of neutron energy on the induction of chromosome aberrations, dose-response relationships have been studied in human peripheral blood lymphocytes irradiated in G_0 with fission neutrons having different energy spectra. The fission neutrons used were a) undegraded fission spectrum neutrons with the mean energy of E=1.98 MeV obtained by bombarding the thermal neutrons to ^{235}U energy converter (KUR-Conv), b) uranium fission neutrons degraded to the mean energy of E=0.1 MeV by passing throug graphite reflectors (UTR-Air), c) the latter further

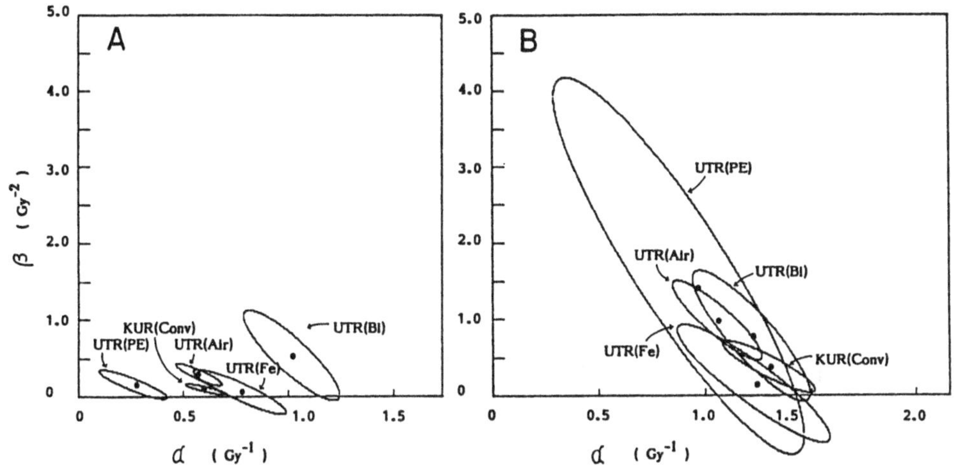

Fig. 6. Regression coefficients, α and β, for dose-response curves of dicentrics and rings in human lymphocytes irradiated with fission neutrons having different energy spectra. Ellipses are 95% confidence limits calculated based on F-distribution, and are shown in half size. A: dose-responses based on the combined neutron and γ-ray dose. B: dose-responses based on neutron dose alone after subtracting the estimated γ-contribution from the total aberration yield.

modified by 5 cm thick bismuth (UTR-Bi), d) iron (UTR-Fe), and e) polyethylene filter (UTR-PE).

The frequencies of dicentrics and rings were fitted to the linear-quadratic model of dose-response relationships. Fig. 6 plots the values of linear (α) and quadratic (β) components together with 95% joint confidence regions. The dose-response relationships varied considerably among types of radiations when the combined neutron and γ-ray doses were used (Fig. 6A). However, when the contribution of γ-rays were subtracted and the remaining aberrations were expressed as a function of neutron dose alone, the dose-response relationship was found to be refractory to the neutron spectra and the dose-response relationship was refractory to the neutron spectrum and the efficiencies were very similar to each other in spite of the large differences in their energy distribution (Fig. 6B).

Fig. 7 compares the energy spectra and kerma rate in KUR-Conv and UTR-Air neutrons. It is evident that in both cases most of the dose is given by high energy neutrons (>100 keV). In a separate experiment, we observed that the thermal neutrons were no more efficient than fission spectrum neutrons in inducing chromosome aberrations. Therefore, the observation that the chromosomal effectiveness is refractory to the changing energy spectrum indicates that the high energy flux is the dominant contributor to the biological effectiveness. In agreement with the distant model for the damage interaction, this indicates that nuclear interactions which produce short-ranged protons are less efficient in inducing one-track aberrations. The RBE of monoenergetic neutrons have been reported by Hall et al. (1975) and Miller et al. (1989) for the survival of cultured Chinese hamster V79 cells and malignant transformation of mouse C3H10T½ cells, respectively. The energy range was between 0.22 MeV and 15 MeV. In those experiments, the RBE was highest for 0.35 MeV and tended to fall at 0.22 or 0.23 MeV, which produced recoil protons with mean energy of about 100 keV. Recently, an iron-filtered 24 keV neutron beam was reported to be highly efficient in

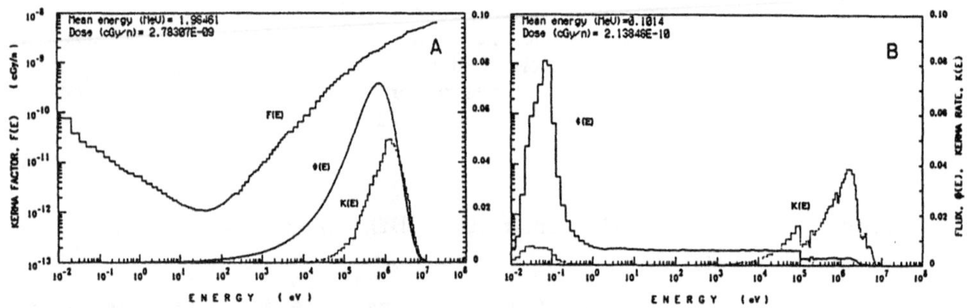

Fig. 7. The neutron energy spectra and kerma rate of fission neutrons. A: Fission spectrum neutron (KUR-Conv) obtained by bombarding the ^{235}U with thermal neutrons. B: Fission neutrons with degraded energy spectrum (UTR-Air). (E): Neutron flux. K(E): Kerma rate. F(E): Kerma factors calculated by the estimates of Howerton (1986) and elemental composition of spleen.

inducing chromosome aberrations; the efficiency being comparable to that of fission neutrons (Lloyd et al. 1988, Aghamohammadi et al. 1989). These observations do not meet with the prediction of the present distance model. The recoil protons have the energy far below the proton edge of LET and can travel not further than 0.5 um. However, these neutrons have very low kerma factors, and therefore require about $4x10^3$ neutrons per μm^2 area to deposit a dose of 1 Gy. The density of damage may be high enough to interact and be fixed into chromosome exchange aberrations even if each damage is formed by a single short track. Aghamohammadi et al. (1989) noted the presence of a quadratic component in their dose-response curve.

POSSIBLE BIOLOGICAL MECHANISMS OF DAMAGE INTERACTION

The question is then raised as to why and how the "breakage-and-rejoining" or "damage interaction" holds for the fixation of primary damage into exchange aberrations. DNA double-strand breaks are generally assumed to be repaired by recombination process (Resnik 1976, Krasin and Hutchinson 1977). In bacteria, recA protein has been suggested to play a role in the repair of double-strand breaks (Sargentini and Smith 1986, Takahashi and Kobayashi 1990). Triggered by damage in DNA, recA protein promotes a search for homology, stimulates unwinding and initiates heteroduplex formation by strand uptake (Cunningham et al. 1979, DasGupta et al. 1981, Howard-Flanders et al. 1984). Assuming a similar repair process in mammalian cells, the chromosome exchange aberration could be the result of illegitimate recombination even when it involves only one lesion (Fig. 8, pathway A). However, the dual nature of aberration formation observed in the present experiments rather suggests that such a repair process is essentially error-free, and illegitimate recombination could be promoted when another DNA strand taken up by the recA-like protein also contains a lesion (Fig. 8, pathway B). While it is highly speculative at this stage, such mechanisms of genetic exchange constitute a base-line concept for the damage fixation leading to chromosome exchange aberrations and probably also for the dual nature of radiation action in general.

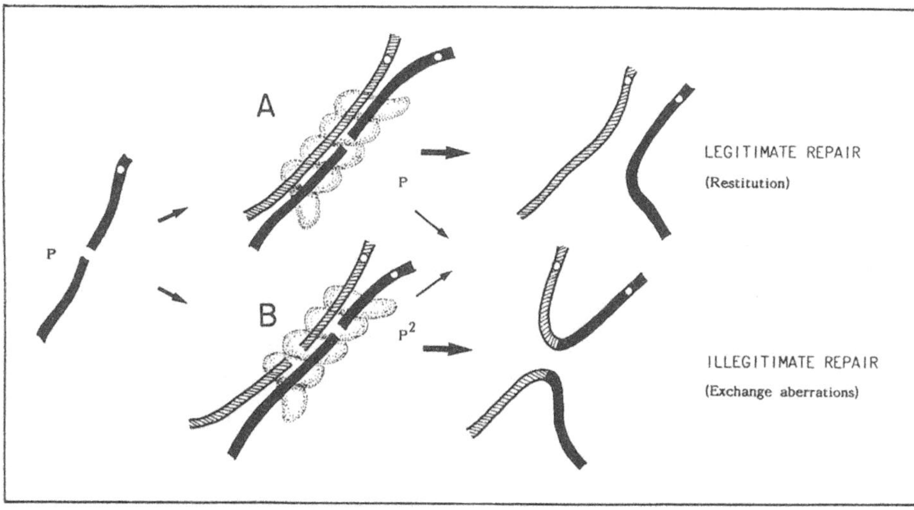

Fig. 8. Recombinational model for the damage interaction in the formation of chromosome exchange aberrations.

ACKNOWLEDGMENTS

This work was supported in part by the Grant-in-Aid for Scientific Research from the Ministry of Education, Science and Culture, Japan. The studies on the monochromatic soft X-rays has been performed under the approval of the Photon Factory Program Advisory Committee (Proposal No. 88-036). The experiments with fission neutrons has been performed within the framework of the Visiting Researchers Program at the Research Reactor Institute, Kyoto University, and Visiting Researcher System of the Institute of Nuclear Energy, Kinki University.

REFERENCES

Aghamohammadi SZ, Goodhead DT, Savage JRK (1989) Production of chromosome aberrations, micronuclei, and sister-chromatid exchanges by 24-keV epithermal neutrons in human G_0 lymphocytes. Mutation Res 211:225-230.

Bryant PE (1984) Enzymatic restriction of mammalian cell DNA using PvuII and BamH1: Evidence for the double strand break origin of chromosomal aberrations. Int J Radiat Biol 46:57-65.

Bryant PE (1985) Enzymatic restriction of mammalian cell DNA: Evidence for the double strand breaks as potentially lethal lesions. Int J Radiat Biol 48:55-60.

Chadwick KH, Leehouts HP (1978) The rejoining of DNA double-strand breaks and a model for the formation of chromosomal rearrangements. Int J Radiat Biol 33:517-529.

Cole A (1969) Absorption of 20-eV to 50,000-eV electron beams in air and plastic. Radiation Res 38:7-33.

Comming DE (1980) Arrangement of chromatin in the nucleus. Hum Genet 53: 131-143.

Cox R, Masson WK (1978) Do radiation-induced thioguanine-resistant mutants of cultured mammalian cells arise by HGPRT gene mutation or X-chromosome rearrangement? Nature 276:629-630.

Cox R, Thacker J, Goodhead DT (1977) Inactivation and mutation of cultured mammalian cells by aluminium characteristic ultrasoft X-rays and radiations of different LET. Int J Radiat Biol 31:561-576.

Cunningham RP, Shibata T, DasGupta C, Radding CM (1979) Single strands induce recA protein to unwind duplex DNA for homologous pairing. Nature 281:191-195.

DasGupta C, Wu AM, Kahn R, Cunningham RP, Radding CM (1981) Concerted strand exchange and formation of Holliday structure by _E. coli_ RecA protein. Cell 25:507-516.

Edwards AA, Lloyd DC, Purrott RJ (1979) Radiation induced chromosome aberrations and the Poisson distribution. Rad Environ Biophys 16:89-100.

Edwards AA, Purrott RJ, Prosser JS, Lloyd DC (1980) The induction of chromosome aberrations in human lymphocytes by alpha-radiation. Int J Radiat Biol 38:83-91.

Finch JT, Klug A (1976) Solenoid model for superstructure in chromatin. Proc Nat Acad Sci USA 73:1897-1901.

Folkard M, Prise KM, Vojnovic B, Davies S, Roper MJ, Michael BD (1989) The irradiation of V79 mammalian cells by protons with energies below 2 MeV. Part I: Experimental arrangement and measurements of cell survival. Int J Radiat Biol 56:221-237.

Frankenberg D (1984) Interpretation of the dose and LET dependence of RBE values for lethal lesions in yeast cells. Radiation Res 97:329-340.

Goodhead DT (1982) An assessment of the role of microdosimetry in radiobiology. Radiation Res 91:45-76.

Goodhead DT, Thacker J (1977) Inactivation and mutation of cultured mammalian cells by aluminium characteristic ultrasoft X-rays. I. Properties of aluminium X-rays and preliminary experiments with Chinese hamster cells. Int J Radiat Biol 31:541-559.

Goodhead DT, Thacker J, Cox R (1979) Effectiveness of 0.3 keV carbon ultrasoft X-rays for the inactivation and mutation of cultured mammalian cells. Int J Radiat Biol 36:101-114.

Griffith JD (1975) Chromatin structure: deduced from a microchromosome. Science 187:1202-1203.

Hall EJ, Novak JK, Kellerer AM, Rossi HH, Marino S, Goodman LJ (1975) RBE as a function of neutron energy. I. Experimental observations. Radiation Res 64:245-255.

Howard-Flanders P, West SC, Stasiak A (1984) Role of RecA protein spiral filaments in genetic recombination. Nature 309:215-220.

Howerton RJ (1986) Calculated neutron KERMA factors based on the LLNL ENDL data file. UCRL-50400, Vol. 27.

Joshi GP, Nelson WJ, Revell SH, Shaw CA (1982) X-ray-induced chromosome damage in live mammalian cells, and improved measurements of its effects on their colony-forming ability. Int J Radiat Biol 41:161-181.

Kellerer AM, Rossi HH (1972) The theory of dual radiation action. Current Topics in Radiation Research Quarterly 8:85-158.

Kobayashi K, Hieda K, Maezawa H, Ando M, Ito T (1987) Monochromatic X-ray irradiation system (0.08-0.4nm) for radiation biology studies using synchrotron radiation at the photon factory. J Rad Res 28:243-253.

Kodama S, Sasaki MS (1987) The involvement of chromosome 13 in the X-ray-induced _in vitro_ transformation of mouse m5S cells. Jpn J Cancer Res 78:372-381.

Krasin F, Hutchinson F (1977) Repair of DNA double-strand breaks in _Escherichia coli_, which requires _rec_A function and the presence of duplicate genome. J Mol Biol 116:81-98.

Lea DE, Catcheside DG (1942) Induction by radiation of chromosome aberrations in Tradescantia. J Genet 44:216-245.

Lloyd DC, Edward AA (1983) Chromosome aberrations in human lymphocytes: effect of radiation quality, dose, and dose rate. In: Ishihara T, Sasaki MS (eds) Radiation-induced Chromosome Damage in Man. Alan R Liss, New York, pp23-49.

Lloyd DC, Edwards AA, Prosser JS, Finnon P, Moquet JE (1988) In vitro induction of chromosomal aberrations in human lymphocytes, with and without boron 10, by radiations concerned in boron neutron capture therapy. Brit J Radiol 61:1136-1141.

Meuth M (1984) The genetic consequences of nucleotide precursor pool imbalance in mammalian cells. Mutation Res 126:107-112.

Miller RC, Geard CR, Brenner DJ, Komatsu K, Marino SA, Hall EJ (1989) Neutron-energy-dependent oncogenic transformation of C3H10T½ mouse cells. Radiation Res 117:114-127.

Mönkehaus F, Köhnlein W (1973) Single- and double-strand breaks in 5-bromouracil-substituted DNA of _B. subtilis_ and phage PBSH after irradiation with long-wave length UV and their correlation to intramolecular energy transfer. Biopolers 12:329-340.

Natarajan AT, Obe G (1984) Molecular mechanisms involved in the production of chromosomal aberrations: III. Restriction endonucleases. Chromosoma 90:120-127.

Neary GJ (1965) Chromosome aberrations and the theory of RBE. I. General considerations. Int J Rad Biol 9:477-502.

Obe G, Natarajan AT, Palitti F (1982) Role of DNA double-strand breaks in the formation of radiation-induced chromosome aberrations. Prog Mut Res 4:1-9.

Pinkel D, Straume T, Gray JW (1986) Cytogenetic analysis using quantitative, high-sensitivity, fluorescence hybridization. Proc Nat Acad Sci USA 83:2934-2938.

Resnick MA (1976) The repair of double-strand breaks in DNA: a model involving recombination. J. Theor Biol 59:97-106.

Sargentini NJ, Smith KC (1986) Involvement of genes uvrD and redB in separate mutagenic deoxyribonucleic acid repair pathways in Escherichia coli K-125 uvrb5 and B/r uvrA155. J Bacteriol 143:212-220.

Sasaki MS, Kobayashi K, Hieda K, Yamada T, Ejima Y, Maezawa H, Furusawa Y, Ito T, Okada S (1989) Induction of chromosome aberrations in human lymphocytes by monochromatic X-rays of quantum energy between 4.8 and 14.6 keV. Int J Radiat Biol 56:975-988.

Sasaki MS, Kodama S (1987) Establishment and some mutational characteristics of 3T3-like near-diploid mouse cell line. J Cell Physiol 131:114-122.

Takahashi N, Kobayashi I (1990) Evidence for the double-strand break repair model of bacteriophage λ recombination. Proc Nat Acad Sci USA 87:2790-2794.

Takatsuji T, Sasaki MS (1984) Dose-response relationship of chromosome aberrations induced by 23 MeV alpha particles in human lymphocytes. Int J Radiat Biol 45:237-243.

Thacker J, Cox R, Goodhead DT (1980) Do carbon ultrasoft X-rays induce exchange aberrations in cultured mammalian cells? Int J Radiat Biol 38: 469-472.

Thacker J, Stretch A, Stephens MA (1979) Mutation and inactivation of cultured mammalian cells exposed to beams of accelerated heavy ions. II. Chinese hamster V79 cells. Int J Radiat Biol 36:137-148.

Thacker J, Wilkinson RE, Goodhead DT (1986) The induction of chromosome exchange aberrations by carbon ultrasoft X-rays in V79 hamster cells. Int J Radiat Biol 49:645-656.

Tobleman WT, Cole A (1974) Repair of sublethal damage and oxygen enhancement ratio for low-voltage electron beam irradiation. Radiation Res 60: 355-360.

Virsik RP, Harder D (1981) Statistical interpretation of the overdispersed distribution of radiation-induced dicentric chromosome aberrations at high LET. Radiation Res 85:13-23.

Virsik RP, Schafer CH, Harder D, Goodhead DT, Cox R, Thacker J (1980) Chromosome aberrations induced in human lymphocytes by ultrasoft Al_K and C_K X-rays. Int J Radiat Biol 38:545-557.

CHROMOSOMAL DAMAGE IN HUMAN LYMPHOCYTES: EFFECT OF RADIATION QUALITY

A.A. Edwards and D.C. Lloyd
National Radiological Protection Board
Chilton
Didcot
Oxon, OX11 0RQ
United Kingdom

INTRODUCTION

For many years the cytogenetics group at NRPB has studied the induction of chromosomal aberrations in human lymphocytes by ionising radiations. The effects of radiation quality and dose rate have been studied extensively and a summary of results will be presented in this paper.

THE BIOLOGICAL SYSTEM

Whole venous blood is irradiated, sometimes in a tube, sometimes in a thin flat layer depending on the absorption characteristics of the radiation. Full details of the culture method are described in Lloyd et al (1982). Briefly, the lymphocytes are stimulated to start their division cycle by adding phytohaemagglutinin. They are cultured in medium containing bromodeoxyuridine which allows a check on cell cycling and ensures that cells only in the first in-vitro metaphase are analysed. After 45 hours at 37°C colcemid is added to block the cell cycle at metaphase. At 48 hours the cells are treated with a hypotonic salt solution and metaphase spreads are prepared and stained with fluorescence plus Giemsa. The duplicated chromosomes are easily seen and any aberrations are noted. Only complete cells are scored and dicentrics, centric rings and excess acentrics are recorded.

Because of their greater ease of scoring we normally specify dose effect relationships in terms of the dicentric yield. Centric rings occur with a frequency between 5 and 10% of the dicentrics and so it makes little difference whether or not these are included. Excess acentrics are more difficult to score, have a higher background and are more variable among the cells. Thus not so much emphasis is placed on results from these.

NATO ASI Series, Vol. H 54
The Early Effects of Radiation on DNA
Edited by E. M. Fielden and P. O'Neill

ANALYSIS OF DICENTRIC YIELDS

The yield of dicentrics, Y, is related to absorbed dose, D, by equation (1) where C, α and β are fitted coefficients.

$$Y = C + \alpha D + \beta D^2 \qquad (1)$$

The coefficient C represents the control level of aberrations and lies in the region of 0.0005 to 0.001 dicentrics per cell. The coefficient α dominates the relationship at low doses and β that at high doses. The coefficients α and β are both dependent on radiation quality. Fig. 1 shows some representative yield curves while Table 1 gives the coefficients α and β for all radiations measured at NRPB. Lloyd et al (1976, 1978, 1984a, 1986, 1988), Prosser et al (1983), Edwards et al (1980, 1986), Purrott et al (1980).

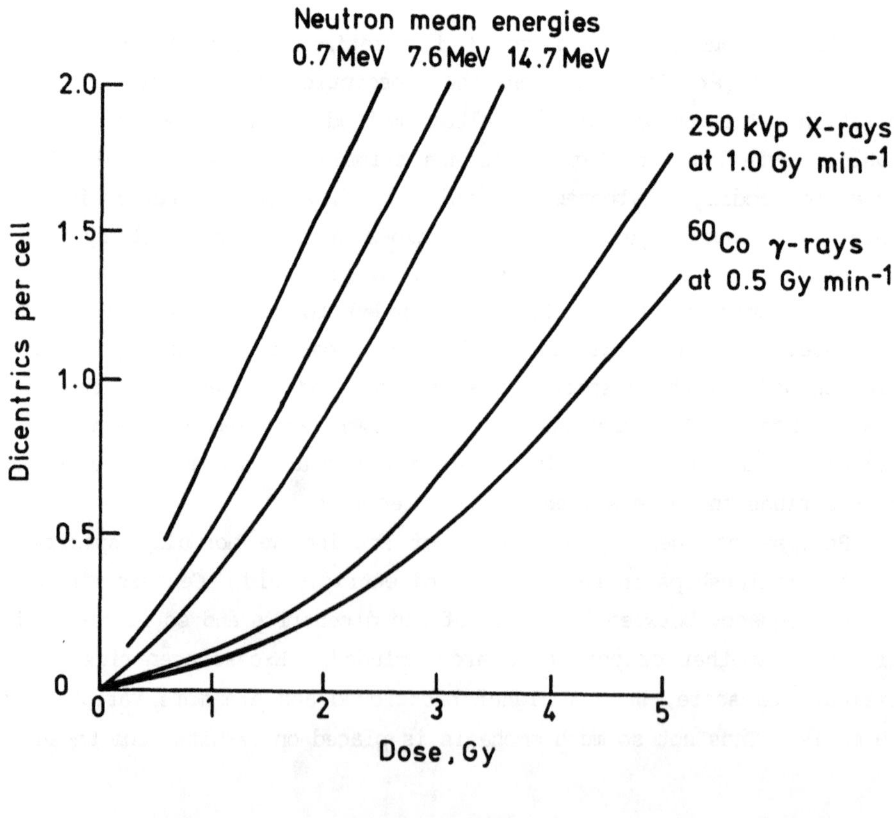

Fig 1. Dose response curves for some high and low LET radiations.

General features can be seen from the Table. For low LET radiations the coefficient α is dependent upon the energy of the electrons which deposit the dose. Low energy electrons from tritium produce a higher value of α than the high energy electrons released by cobalt-60 γ-rays. The value for intermediate energy x-rays lies in between. The value for β, however, is little changed. For neutrons the value for α increases as neutron energy decreases but the coefficient β drops to zero at lower neutron energies. The variations of both α and β for neutron radiations can be better understood by presenting an analysis in terms of LET.

TABLE 1. Observed values of α and β for all radiations

Radiation	$\alpha \pm$ SE (10^{-2} Gy^{-1})	$\beta \pm$ SE (10^{-2} Gy^{-2})
^{60}Co	1.42 ± .44	7.59 ± 0.27
250 kVp x-rays	3.64 ± 0.53	6.67 ± 0.22
^{3}H β-particles	5.37 ± 0.58	6.60 ± 0.41
14.9 MeV neutrons	19.5 ± 1.8	11.9 ± 2.0
MRC cyclotron generated neutrons ($\bar{E}$ = 7.6 MeV)	48.2 ± 4.0	6.35 ± 2.2
^{252}Cf fission neutrons	60.0 ± 1.9	-
AWRE fission neutrons ($\bar{E}$ = 0.9 MeV)	73.3 ± 2.4	-
BEPO fission neutrons ($\bar{E}$ = 0.7 MeV)	83.2 ± 3.2	-
24 keV neutrons	82.1 ± 3.1	-
^{242}Cm alpha particles	28.6 ± 1.5	-
^{239}Pu alpha particles	37.5 ± 2.4	-
8.7 MeV protons (L = 5.1 keV.μm^{-1})	4.36 ± 0.75	5.81 ± 0.57
23.5 MeV ^{3}He ions (L = 22 keV.μm^{-1})	39.4 ± 1.8	-

The α coefficient

Figure 2 presents graphs of the distributions of dose as a function of LET for two of the neutron energies, a fission spectrum ($\bar{E}$ = 0.7 MeV) and 14.7 MeV derived using the program of Edwards and Dennis (1975). They are plotted in such a way that the dose deposited in any LET interval is proportional to the area under the graph. Thus for the fission spectrum neutrons (Fig. 2a) about 50% of the dose is deposited at LET values between 50 and 100 keV/μ. For 14.7 MeV neutrons the corresponding value is only 6%. Because the value of α for fission neutrons is about a factor 4 higher

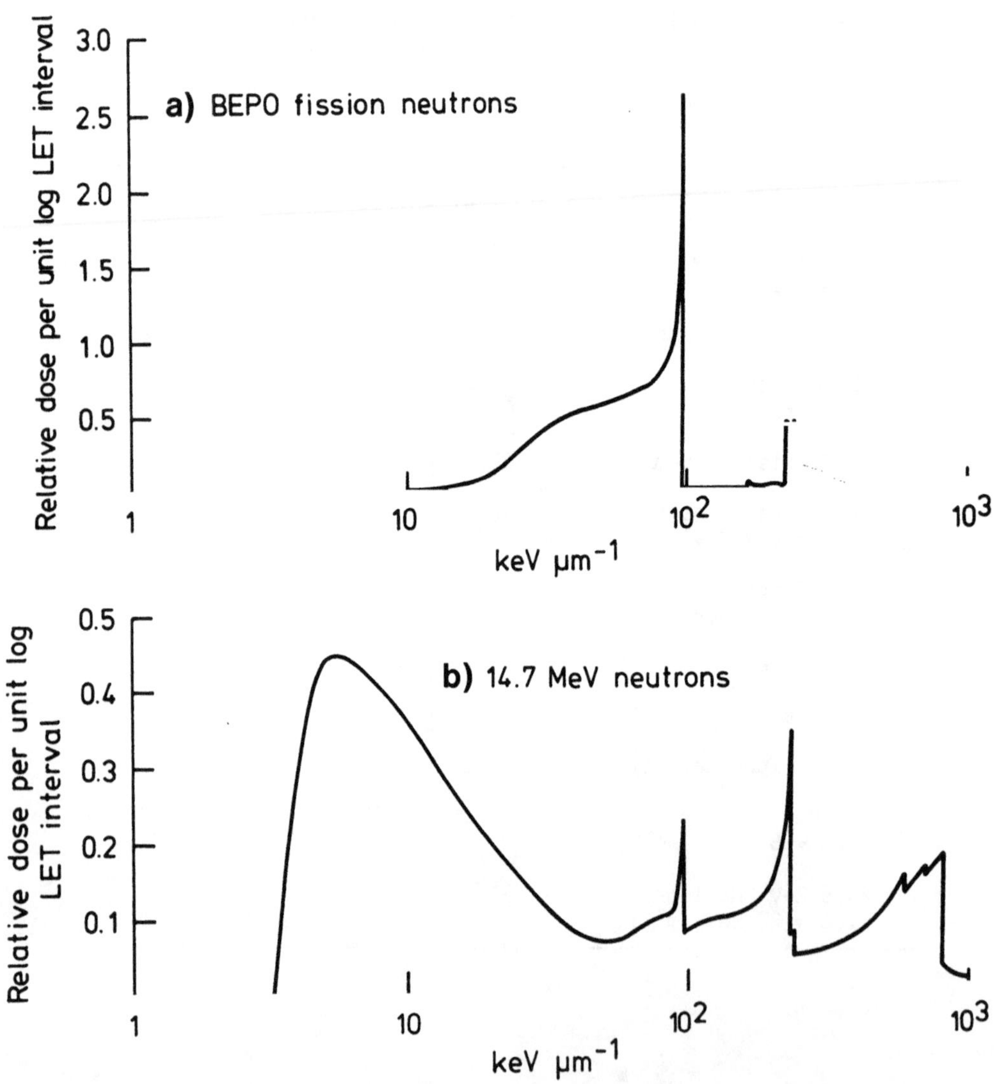

Fig 2. Dose distributions in LET for two neutron spectra

than for 14.7 MeV neutrons, the implication is that particles in the LET range 50 to 100 keV/μm are particularly effective for producing dicentrics at low doses compared with other particles. Protons of energy 10 to 500 keV and alpha particles of 4 to 10 MeV fall into this category. Further analysis of neutron values of α produced a proposed relationship between α and LET which when combined with the appropriate dose distribution in LET predicted the observed values of α for each of the neutron spectra reasonably well, Edwards et al (1985). This relationship is shown in Fig. 3. The most effective particles are those with an LET in the range 60-70 keV/μm. Note the rapid decrease of α as LET increases above 100 keV/μm. This occurs because of the effect of interphase death and mitotic delay at high LET. A high LET particle (for example an alpha particle) when crossing a lymphocyte nucleus has a high chance of killing the cell or delaying its progress through the cell cycle thereby preventing it from reaching metaphase at 48 hours. If a cell does not reach metaphase at 48 hours then aberrations formed in that cell will not be seen. The only cells seen are those that have not been irradiated (and therefore will have no aberrations), plus those which have been irradiated and survive to the first metaphase. Thus the yield of aberrations observed is critically affected by the chance that a single particle will prevent the cell reaching metaphase. As LET increases the energy deposited in the cell per particle increases and the probability of death or delay increases. This causes the rapid drop of α with LET shown in Fig. 3. The other feature that α rises proportionally to LET is in agreement with the theory of aberration production given by Neary (1965).

We have attempted to verify the relationship in Fig. 3 by performing track segment experiments with accelerated particles. Results for two such experiments using 8.7 MeV protons and 23.5 MeV helium-3 ions are given in Table 1. The value of α for the helium-3 ions (LET = 22 keV/μm) agrees very well with that shown in Fig. 3. For the protons (LET = 5.1 keV/μm) the observed yield is about a factor 2 lower than that suggested by Fig. 3. Further experiments with LET values in this region are necessary.

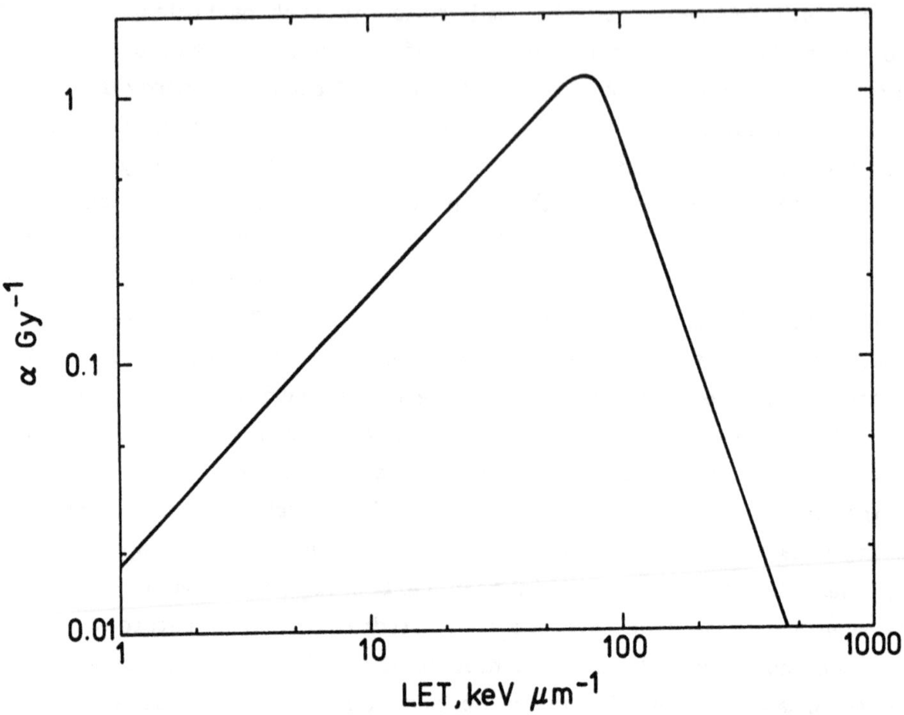

Fig 3. A derived relationship between the initial slope coefficient, α, and LET_∞ in blood

The β coefficient

The variation of the coefficient β with radiation quality can be seen from Table 1. For the low LET radiations (photons, β particles and the high energy protons) the coefficient β is constant at about 6×10^{-2} Gy^{-2}. For high LET radiations (alpha particles and fission neutrons) β is zero. For these radiations more than 80% of the dose is deposited at LET values greater than 20 keV/μm, Edwards et al (1985). It therefore appears that β is constant at low LET, zero at high LET and falls sharply at about 20 keV/μm. Such a relationship was predicted by Neary (1965). There are some anomalies however. For 14.7 MeV neutrons where only 60% of the dose is deposited at 20 keV/μm or less, the value for β is high. For the cyclotron generated neutrons ($\bar{E}$ = 7.6 MeV) where only 50% of the dose is deposited at less than 20 keV/μm the value for β is perhaps a factor 2 too high. For the helium-3 ions at 23 keV μm^{-1} the observed value of zero seems a little low. However, for all of these radiations the uncertainties on the measurements of β are fairly large, Table 1.

THE EFFECT OF DOSE RATE

In radiation protection we are preoccupied with risks at low doses and low dose rates. This end-point of chromosome damage allows quite accurate measurements of the effects at low doses and a significant increase in dicentric yield has been observed at doses as low as 30 mGy compared with zero dose. At low dose rates the practical limit for irradiating whole blood held at 37°C is about 2 days. However, Doloy et al (1983) have succeeded in irradiating lymphocytes with fission neutrons for up to 7 days by maintaining the cells in culture medium. Her experiment compared dicentric yields for the 7 day irradiations with acute irradiations and found no difference in the linear yield. It appears that the coefficient α is not sensitive to changes in dose rate which is consistent with the notion of linearity at low doses. Lloyd et al (1984b) drew a similar conclusion for samples irradiated with cobalt-60 γ-rays but for irradiation times up to 12 h. The coefficient β decreased by a factor of 4 when irradiation times increased from acute (<15 minutes) to 12 hours. Thus measurements of α obtained from low dose and high dose rate data appear to be applicable at low dose rates. However, for acute irradiation with x-rays the lowest dose which was used was 50 mGy. In view of recent work at very low doses Wolff et al (1989), Pohl-Rüling et al (1983), Moquet et al (1989) which indicates that repair phenomena may be stimulated, the statement above should be viewed with caution if extrapolation to <20 mGy is contemplated.

Despite this proviso the measured alpha to beta ratio has been used to derive a dose and dose rate effectiveness factor as defined by ICRP (1990). Using a former γ-ray curve from this laboratory Rossi (1990) has calculated a factor between 4-17 as dose increases from 1.0-5.0 Gy. Reworking his calculation using the new γ-ray data shown in Table 1 we derive factors of 6-20 in the same dose range. These values are somewhat higher than the value of 2 proposed by ICRP (1990).

VARIATIONS WITH PHOTON ENERGY

Table 1 shows that the coefficient α is higher for 250 kVp x-rays than for cobalt-60 γ-rays by a factor of 2-3. This effect has been seen for other end-points, for example, stamen hair mutation in Tradescantia and cell killing in mouse oocytes as summarised by ICRU (1986). Although with respect to γ-rays the maximum RBE (ratio of the α coefficients) for fission neutrons is about 50, ICRP (1990) are proposing to disregard the difference between X and γ-rays and select a standard low LET radiation somewhere between the two. If applied here maximum RBE's for the present neutron data would then reduce to about 30.

Attempts to calculate dose distributions in LET for photons have been made despite the acknowledged limitations of the continuous slowing down concept when applied to electrons. In ICRU (1970), a method of calculating distributions in LET for photons is given as well as a table of average values. Because x-rays produce lower energy electrons than cobalt-60 γ-rays, average LET values are higher. The ratio of the dose averaged LET restricted to ignore single energy transfers greater than 100 eV for x-rays and γ-rays is given as 1.36 which does not seem large enough to explain a maximum RBE value in the range 2-3.

Probably a better representation of radiation quality is to make measurements with a low pressure proportional counter which essentially produces distributions in lineal energy, y, ICRU (1983). Fig. 4 shows the distribution of dose in y for cobalt-60 γ-rays and 36 keV monoenergetic photons taken from Kliagua and Dvorak (1978) where y is determined in a sphere of 1 μm diameter. These authors show that the distributions for monoenergetic photons for 36, 60 and 140 keV do not vary greatly so that the spectrum for 36 keV photons is representative of that for 250 kVp x-rays which have a mean photon energy of 90 keV. Both of the curves shown in Fig. 4 are drawn so that the area under the curve is proportional to the dose in any interval of y. The different patterns of energy deposition of cobalt-60 γ-rays and 250 kVp x-rays are clearly seen. The ratio of the dose mean value of y for these two radiations is given as 2.2 by Kliagua and Dvorak (1978) which is in closer agreement with the ratio of α coefficients given in Table 1 than the 1.36 based on LET.

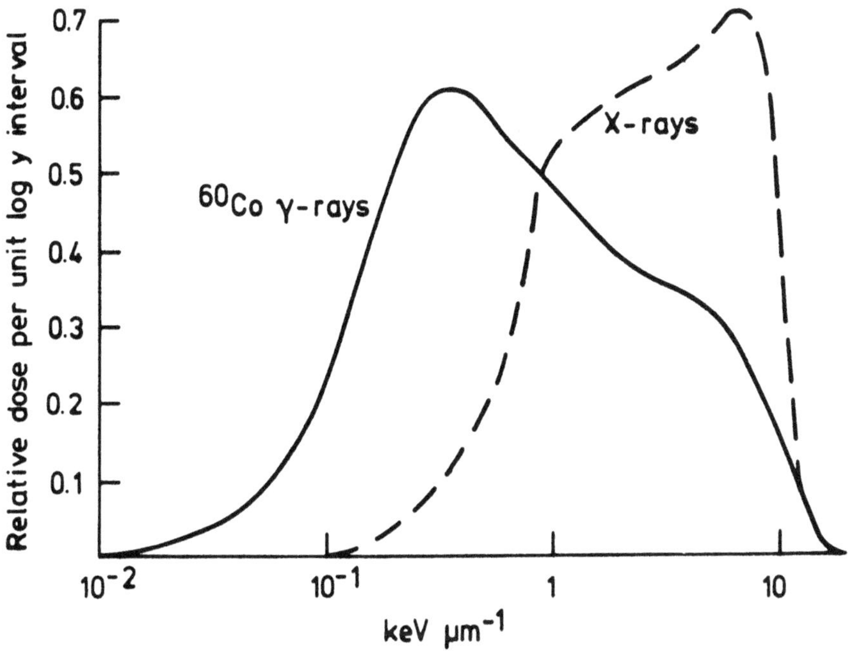

Fig 4. The dose distribution in lineal energy in a sphere of 1 µm diameter for two photon radiations

For very low energy photons (<2 keV) such as the characteristic x-rays obtained from aluminium or carbon, Virsik-Peuckert (1983) has shown that the linear coefficient α increases further as photon energy decreases. Aluminium x-rays of energy 1.5 keV produce electrons of about 1.5 keV which have a range much less than 1 µm and are therefore mainly "insiders". The value of y is therefore about 2.2 keV/µm. This value lies between the dose average y values for cobalt-60 γ-rays and x-rays (1.6 and 3.5 respectively). Thus the choice of a diameter of 1 µm as the volume in which y is determined fails to predict the experimental values for α for very low energy x-rays.

It is possible to measure or calculate a dose distribution in y for neutron radiations as demonstrated in ICRU (1983). In fact there are similarities to the plots in terms of LET given in Figs. 2a and 2b. Using the same ideas which produced Fig. 3, a relationship between α and y (in

1 μm diameter) can be derived as shown by Edwards et al (1985). When this is done the neutron and photon values of α given in Table 1 are predicted reasonably well and are therefore seen to be consistent. However, recently 24 keV neutrons have been used to irradiate blood and these neutrons deposit energy by recoil protons of range much less than 1 μm which was used to determine y. Under these circumstances we find that predictions of α in terms of y do not agree with the measurement shown in Table 1 but predictions in terms of LET from Fig. 3 do agree. The implication is that in order to have a universal relationship between the coefficients α and the linear energy y, the y has to be determined in volumes of diameter much less than 1 μm.

CONCLUSIONS

There is a good body of chromosomal aberration data available now for radiations of different quality which can be used to test theories of biological effect. We have attemped to use LET_∞ and lineal energy y as a basis for modelling variations with radiation quality. There appears to be no single model which explains all the observations. LET gives a satisfactory description for charged particles such as protons, alpha particles and the lighter ions, although there could be doubts at high energies when δ-rays have ranges greater than the radius of the lymphocyte nucleus (say 3 μm). LET is a poor descriptor of electron transport and hence does not predict photon responses very well. Conversely lineal energy y is the better descriptor of electron energy deposition but its ability to explain biological effect breaks down when the electron track is small compared with the 1 μm site size. This implies that it is necessary to consider energy deposition in much smaller volumes.

Experimentally, it is intended in the near future to obtain data for more energetic accelerated ions of helium-4 and carbon-12. For these particles the track width will be of the order of the diameter of the lymphocyte nucleus. The yield curves for these particles may then contribute further information in the modelling of radiation effects.

REFERENCES

Doloy MT, Ducatez G, Lepetit J, Bourguignon M, Reillaudou M, Chaput, B (1983) Effets de faibles debits de dose de neutrons de fission sur l'induction d'anomalies chromosomiques dans les lymphocytes humains. In: Booz J, Ebert HG (eds) Microdosimetry. Proceedings of the Eighth Symposium. EUR 8395 EN, CEC Luxembourg, p 835

Edwards AA, Dennis JA (1975) The calculation of charged particle fluence and LET spectra for the irradiation of biologically significant materials by neutrons. Phys Med Biol 20: 395-409

Edwards AA, Lloyd DC, Prosser JS (1985) The induction of chromosome aberrations in human lymphocytes by accelerated charged particles. Radiat Prot Dosim 13: 205-209

Edwards AA, Lloyd DC, Prosser JS, Finnon P, Moquet JE (1986) Chromosome aberrations induced in human lymphocytes by 8.7 MeV protons and 23.5 MeV helium-3 ions. Int J Radiat Biol 50: 137-145

Edwards AA, Purrott RJ, Prosser JS, Lloyd DC (1980) The induction of chromosome aberrations in human lymphocytes by alpha-radiation. Int J Radiat Biol 38: 83-91

ICRP (1990) Recommendations of the Commission. Draft report of the International Commission on Radiological Protection February 1990

ICRU (1970) Linear energy transfer. Report No 16, International Commission on Radiation Units and Measurements, Washington

ICRU (1983) Microdosimetry. Report No 36, International Commission on Radiological Protection, Bethesda

ICRU (1986) The quality factor in radiation protection. Report No 40, International Commission on Radiation Units and Measurements, Bethesda

Kliagua P, Dvorak R (1978) Microdosimetric measurement of ionisation by monoenergetic photons. Radiat Res 73: 1-20

Lloyd DC, Edwards AA, Prosser JS (1986) Chromosome aberrations induced in human lymphocytes by in vitro acute x and gamma radiation. Radiat Prot Dosim 15: 83-88

Lloyd DC, Edwards AA, Prosser JS, Bolton D, Sherwin AG (1984a) Chromosome aberrations induced in human lymphocytes by D-T neutrons. Radiat Res 98: 561-573

Lloyd DC, Edwards AA, Prosser JS, Corp MJ (1984b) The dose response relationship obtained at constant irradiation times for the induction of chromosome aberrations in human lymphocytes by cobalt-60 gamma rays. Radiat Environ Biophys 23: 179-189

Lloyd DC, Prosser JS, Purrott RJ (1982) The study of chromosome aberration yield in human lymphocytes as an indicator of radiation dose: revised techniques. NRPB M-70, National Radiological Protection Board, Chilton.

Lloyd DC, Purrott RJ, Dolphin GW, Edwards AA (1976) Chromosome aberrations induced in human lymphocytes by neutron irradiation. Int J Radiat Biol 29: 169-182

Lloyd DC, Purrott RJ, Reeder, EJ, Edwards AA, Dolphin GW (1978) Chromosome aberrations induced in human lymphocytes by radiation from californium-252. Int J Radiat Biol 34: 177-186

Lloyd DC, Edwards AA, Prosser JS, Finnon P, Moquet JE (1988) In vitro induction of chromosomal aberrations in human lymphocytes, with and without boron 10, by radiations concerned in boron capture therapy. Br J Radiol 61: 1136-1141

Moquet JE, Prosser JS, Edwards AA, Lloyd DC (1989) Sister-chromatid exchanges induced by mitomycin C after acute or chronic exposure of human lymphocytes to a low dose of x-rays. Mutation Res 227: 207-213

Neary GJ (1965) Chromosome aberrations and the theory of RBE. 1. General considerations. Int J Radiat Biol 9: 477-502

Pohl-Rüling J, Fischer P, and 16 others (1983) Effect of low-dose acute x-irradiation on the frequencies of chromosomal aberrations in human peripheral lymphocytes in vitro. Mutation Res 110: 71-82

Prosser JS, Lloyd DC, Edwards AA, Stather JW (1983) The induction of chromosome aberrations in human lymphocytes by exposure to tritiated water in vitro. Radiat Prot Dosim 4: 21-26

Purrott RJ, Edwards AA, Lloyd DC, Stather JW (1980) The induction of chromosome aberrations in human lymphocytes by in vitro irradiation with α-particles from plutonium-239. Int J Radiat Biol 38: 277-284

Rossi HH (1990) The dose rate effectiveness factor. Health Physics 58: 359-361

Virsik-Peuckert RP (1983) Inducibility of chromosome aberrations by ultrasoft x-rays. In: Ishihara T, Sasaki MS (eds) Radiation induced chromosome damage in man. Liss, New York, p 51

Wolff S, Wiencke JK, Afzal V, Youngblom J, Cortés F (1989) The adaptive response of human lymphocytes to very low doses of ionising radiation: a case of induced chromosomal repair with the induction of specific proteins. In: Baverstock KF, Stather JW (eds) Low dose radiation: biological bases of risk assessment. Taylor and Francis, London New York Philadelphia, p 446

HEAVY-ION INDUCED CHROMATIN DAMAGE AND REPAIR: PCC AND CHROMOSOME PAINTING [1]

E. H. Goodwin[2] and E. A. Blakely
Cell and Molecular Biology Division
Lawrence Berkeley Laboratory
Berkeley, CA 94720
USA

The experimental technique known as "premature chromosome condensation (PCC)" allows detection of interphase chromosome breakage or "chromatin breaks" at a time when repair processes are still at work. Using the PCC technique it has been shown that x-rays induce breakage linearly with dose and that these breaks rejoin with kinetics similar to repair of potentially lethal damage. Because they lack a clearly defined centromere, it has been difficult to detect aberrations in G_1-phase PCC chromosomes; an important exception being the ring chromosomal aberration. Certain misrepair events can also be detected by *in situ* hybridization of labeled DNA probes and visualization by fluorescent staining. Our goal was to examine chromatin breakage, rejoining, and misrepair in cells exposed to radiations of differing quality.

We used a human/hamster hybrid cell line (UV24C2-3), with a karyotype composed of 21 hamster chromosomes plus a single copy of human chromosome #2 that accounts for 4% of the total DNA. Confluent serum-deprived cultures were exposed to either 150 kVp x-rays or to 425 MeV/u Ne ions near the Bragg peak where the LET was 183 keV/µm. Either immediately or after a repair interval, irradiated cells were fused to mitotic cells to induce PCC. Microscope slides were prepared and stained with Giemsa. In addition, some slides were processed by a species-specific "chromosome painting" technique using a biotinylated human genomic DNA probe, plus fluorescent staining.

Scoring Giemsa-stained slides, it was found that x-rays induced breakage at a rate of 0.0754 chromatin breaks/cell/cGy in the total genome. Ne ions were 1.5

1 Supported by NIH-RR-05918, NIH CA15184 and DOE-DE-AC03-76SF00098.
2 Present address: Los Alamos National Laboratory, Life Science Division, Los Alamos, NM 87545 USA.

NATO ASI Series, Vol. H 54
The Early Effects of Radiation on DNA
Edited by E. M. Fielden and P. O'Neill

times more effective. Choosing doses yielding the same initial number of breaks (121/cell), rejoining kinetics were examined in cells exposed to each radiation. Results (Fig. 1) indicated a time-dependent decrease in the number of breaks that could be fitted adequately by first order kinetics. The time constant for the rejoining process was similar in cells exposed to either x-rays or Ne ions. However, 4.4 times as many Ne-induced breaks failed to rejoin after long repair periods compared to x-rays. If these unrejoined breaks are truly unrepairable and not just slowly repairing, then they are likely to result in chromosomal deletions which are lethal lesions. Misrepair kinetics were also examined by counting ring PCC chromosome fragments (Fig. 2). The yield of rings per cell increased steeply during the first hour and more slowly afterwards until an apparent plateau was reached beyond 6 hours. Far fewer rings formed in the Ne-irradiated cells which suggests that chromosomal exchange aberrations form by a misrepair of chromatin breaks. The chromosome painting technique confirmed a linear dose response for breakage in the single human chromosome. Fifty percent of Ne-induced breaks in the human chromosome did not rejoin, a value similar to that for breaks in the entire genome. Preliminary evidence indicates that misrepair events, detectable as red-yellow fusions between hamster and human chromatin, increase in a time-dependent fashion (Goodwin et al 1989).

Results of studies presented here demonstrate that cells exposed to Ne ions suffer both increased chromosomal damage per unit dose and a shift towards a greater proportion of the more severe, unrejoinable lesions. The high cell-killing effectiveness of Ne-ions is undoubtedly related to these dual burdens placed on the cellular repair system.

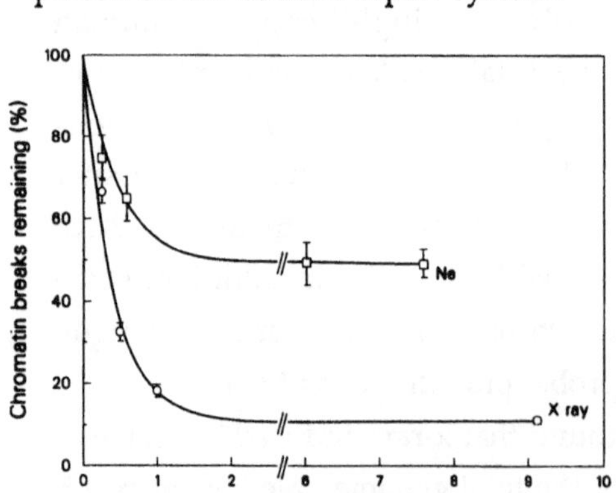

Fig.1 Rejoining kinetics of chromatin breaks in G_1 phase UV24C2-3 cells exposed to either 1600cGy Xrays or 1062cGy Neon ions.

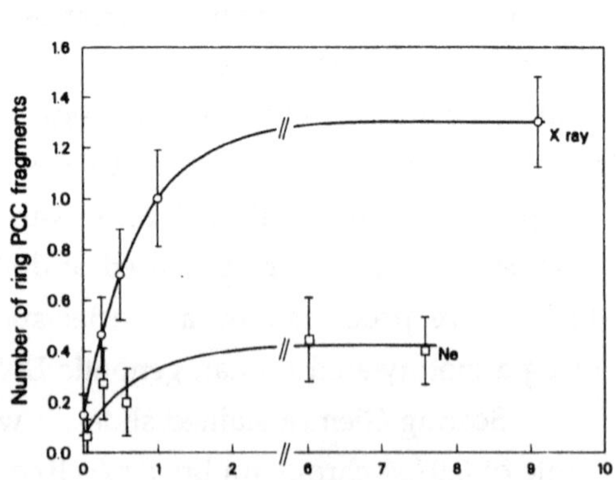

Fig. 2 Kinetics of the formation of PCC chromosomal ring aberrations in G_1 phase UV24C2-3 cells

Reference

Goodwin E, Blakely E, Ivery G and Tobias C (1989) Repair and misrepair of heavy-ion-induced chromosomal damage. Adv. Space Res., 9: (10)83-(10)89.

DISCUSSION SESSION ON LET AND CHROMOSOME DAMAGE

John H Peacock
Radiotherapy Research Unit
Institute of Cancer Research
15 Cotswold Road
Belmont, Sutton
Surrey SM2 5NG

Chromosome damage may be said to represent the final manifestation of the early effects of ionizing radiation in that chromosomal events can probably be linked more closely to cell death than any of the other measures of radiation effect so far discussed. Furthermore, until the advent of clonogenic survival assay in the 1950's, the study of chromosome damage formed the basis for most of our knowledge of the cellular effects of irradiation and even today conventional chromosome studies still provide much information with regard to the underlying mechanisms of radiation action.

Studies of chromosome damage are usually performed on metaphase chromosomes prepared at the first cell division following irradiation and this means that damage being measured is that which remains after normal cellular processing such as repair and fixation has occurred. Metaphase chromosome aberrations can be divided into two classes depending on the cells position in the cell cycle at the time of irradiation. Chromosomal aberrations such as dicentric and ring chromosomes result when cells are irradiated in the G1 phase of the cell cycle before replication has taken place. Chromatid aberrations occur following irradiation in the S and G2 phases of the cycle.

One of the most common observations following G1 irradiation is the formation of acentric fragments. Such fragments may result either directly from a single break in a chromosome arm and the subsequent separation of the two pieces of chromosome or more commonly from an interaction between two damaged chromosomes which mis-rejoin, resulting in the loss of part of one or both chromosomes.Acentric fragments are almost always lost from the daughter nucleii during mitosis and this results in the formation of micronuclei at the completion of mitosis. This property has been exploited in a very careful complete study of the

NATO ASI Series, Vol. H 54
The Early Effects of Radiation on DNA
Edited by E. M. Fielden and P. O'Neill

relationship between chromosome damage and cell kill in mammalian cells. Revell and coworkers (1983) studied synchronized BHK cells irradiated in G1 phase. They followed individual cells from the time of irradiation right through to colony formation. Their results showed that in diploid cells the loss of a chromosome fragment assessed by the occurrence of a micronucleus was closely related to inhibited colony formation. A tetraploid variant showed similar relative fragment loss per unit radiation dose but was much less sensitive to this loss when cell survival was measured. In addition a hypotetraploid line also showed an aberration frequency proportional to chromosome number but was even more resistant to fragment loss than the tetraploid line. They concluded that in the different cells the initial damage inflicted by any given dose was similar but that the cells differed in their ability to tolerate this damage. This study alone provides considerable incentive to continue the study of chromosome damage and the interphase events which lead to its expression.

As stated above, the majority of chromosome fragments arise from exchange events following radiation induced breakage of two chromosomes, rather than the straightforward loss of part of a single chromosome. Coupled with the conclusions from the work of Revell this could be interpreted as showing that cell death results from the interaction of two separate events (see Brenner, this volume). In addition it is not yet clear whether the appearance of exchange aberrations result from straightforward ligation of two damaged chromosomes or a more complex recombination repair process. Although the use of plasmid DNA artificially damaged by endonucleases has shown that simple ligation of DNA does occur in cells, it is not yet clear whether this occurs following the more complex damage induced by radiation. Certainly this is an area that requires elucidation.

Until recently it has not been possible to directly relate metaphase chromosome damage to earlier biochemical events such as DNA damage. In 1970, however, Johnson and Rao described the technique of premature chromosome condensation. This phenomenom results from the fusion of the interphase cell of interest with a mitotic cell arrested in metaphase. After fusion, soluble factors in the mitotic cell result in the condensation of the interphase chromatin of the cell under study. The resulting prematurely condensed chromosomes (PCCs) differ in their appearance depending on the cells position in the cell cycle. PCCs from cells in G1 or G2 form as recognisable structures in which breaks can be measured whereas S-phase PCCs tend to condense as small fragments of chromatin which are unsuitable for analysis.

A number of workers have used this technique in an attempt to study early effects of radiation on interphase chromosomes. Cornforth and Bedford (1985) have studied differences in chromosome damage between normal human fibroblast cells and radiosensitive Ataxia Telangectasia fibroblast cell lines. At the earliest time at which PCC damage could be measured initial chromosome damage was similar in the two lines. Incubation of the cells at 37°C for various times before analysis showed that in both cell types the majority of PCC breaks were rejoined at a similar rate. When repair was complete however there was a difference between the residual number of breaks in the two lines. The frequency of residual PCC fragments correlated well with the number of metaphase chromosome breaks which in turn agreed with the predicted number of lethal lesions calculated from the survival data. In this case sensitivity appeared to be determined by the cellular capacity to repair damage.

With this background the three papers documenting the effect of variation in LET on chromosome damage were presented. The discussion of these papers concentrated on the suitability of LET as a measure of radiation quality when studying mechanisms of radiation-induced cell kill. The use of heavy particles as a source of high LET radiation (Blakely) results in a relatively uniform track structure, LET (an average value) therefore provides a relevant measure of energy deposition. However, many "low" LET radiations (such as ^{60}Co α-radiation), are heterogeneous in terms of track structure and much of the energy is deposited in track ends or clustrers of ionisation. These high LET events (up to 30% of total energy deposition for ^{60}Co α-radiation) may provide a disproportionate contribution to cell kill and so provide complications when relating LET to biological effect.

References

Cornforth MN and Bedford JS (1985) On the nature of the defect in cells from individuals with ataxia-telangiectasia. Science 227:1589-1591

Johnson RT and Rao RN (1970) Mammalian cell fusion: Induction of premature chromosome condensation in interphase nuclei. Nature (London) 226:717-722

Revell SH (1983) Relationship between chromosome damage and cell death. Radiation-induced Chromosome Damage in Man. Liss AR, New York (ed), pp 215-233

RADICAL OXIDATION OF THE PURINE MOITIES OF DNA AND RELATED NUCLEOSIDES

J. Cadet, M. Berger, J.F. Mouret, F. Odin, M. Polverelli, J.L. Ravanat.
Laboratoires de Chimie
Département de Recherche Fondamentale
Centre d'Etudes Nucléaires - 85X
F-38041 Grenoble Cedex, France.

Over the past five years relevant information on the chemical reactions of OH adducts to purine components and related radical cations has become available by using the powerful redox titration technique [O'Neill & Chapman, 1985; Jovanovic & Simic, 1989, Steenken, 1989]. More recently the main decomposition products arising from the fate of the corresponding 2'-deoxyribonucleoside radicals were isolated and characterized. It was shown that the reaction of hydroxyl radicals with the base moiety of 2'-deoxyguanosine in aerated aqueous solution gave rise predominantly to an oxazolone type rearrangement product via an unstable precursor [Cadet et al, 1990]. It is noteworthy that these modified nucleosides may also be produced by photooxidation processes [Cadet & Vigny, 1990] involving the type I photodynamic mechanism through the transient formation of a guanyl radical cation. In addition, the same oxidation products were found to be generated by far-UV irradiation which likely leads to photoionisation of the purine ring. These observations may be rationalyzed in terms of the initial formation of either an OH adduct or a radical cation which subsequently undergoes fast dehydration or deprotonation. The resulting neutral radical which may be produced by hydrogen abstraction of the guanine moiety by triplet excited ketone photosensitisers, may exist in various tautomeric forms as recently suggested by Steenken [1989]. It is worth noting that molecular oxygen was found to react with one of these radical intermediates as inferred from $^{18}O_2$ labelling experiments which clearly show the incorporation of one atom of oxygen in the oxazolole and in its precursor.

Hydration of the 2'-deoxyadenosine radical cation generated by either direct photoionisation or triplet photosensitisation may explain the substantial formation of the 8-hydroxylated derivative (which is also generated by OH radicals [Cadet & Berger, 1985]). On the other hand, this process was not detected under conditions of radical oxidation (OH radical, type I photosensitisation) when 2'-deoxyguanosine was used as the substrate.

NATO ASI Series, Vol. H 54
The Early Effects of Radiation on DNA
Edited by E. M. Fielden and P. O'Neill

However 8-hydroxy-2'-deoxyguanosine was found to be the main oxidation product of the Fenton reaction which may involve perferryl ions or ferroxo complexes as the reactive species. It is interesting to note that the presence of a reducing agent such as ascorbate in the gamma irradiated solution of 2'-deoxyguanosine also led to an enhancement of the formation of 8-hydroxy-2'-deoxyguanosine together with a complete disappearance of the oxazolone decomposition product. The latter observation may be rationalyzed in terms of reduction of the neutral guanine oxidizing radical which leads to its repair. This may indirectly enhance the hydroxylation reaction at carbon C(8) due the higher availability of the reducing 8-hydroxy-7,8-dihydro-2'-deoxyguanosinyl radical for further oxidizing reactions.

The reactivity of the two purine bases within DNA, with respect to the formation of the corresponding 8-hydroxylated derivatives upon exposure to either OH radicals or the reactive species of the Fenton reaction, was investigated by using two analytical methods involving an HPLC-electrochemical assay [Berger et al, 1990] and a sensitive HPLC-^{32}P-postlabeling technique [Mouret et al, 1990]. It is interesting to note that 8-hydroxyguanine was found to be predominant over 8-hydroxyadenine irrespective of the conditions used (OH radicals or Fenton reaction).

Berger M, Anselmino C, Mouret J-F, Cadet J (1990) High performance liquid chromatography-electrochemical assay for monitoring the formation of 8-oxo-7,8-dihydroadenine and its related 2'-deoxyribonucleoside. J Liquid Chromatog (in press)

Cadet J and Berger M (1985) Radiation-induced decomposition of the purine bases within DNA and related model compounds. Int J Radiat Biol 47:127-143

Cadet J, Vigny P (1990) The Photochemistry of Nucleic Acids. In: Morrison H (ed) Photobiochemistry of Nucleic Acids. Wiley and Sons, New York (in press).

Cadet J, Berger M, Joshi PC, Shaw A, Voituriez L, Kan L-S, Wagner R (1990) Comparative effects of ultraviolet and ionizing radiations on nucleic acids. In: Riklis E (ed) Photobiology: The Science and its Application. Plenum Press, New York (in press).

Jovanovic SV, Simic MG (1989) The DNA guanyl radical: kinetics and mechanisms of generation and repair. Biochim Biophys Acta 1008: 39-44

Mouret J-F, Odin F, Polverelli M, Cadet J (1990) ^{32}P-Postlabeling measurement of adenine N-oxide in cellular DNA exposed to hydrogen peroxide. Chem Res Toxicol (in press)

O'Neill P, Chapman PW (1985) Potential repair of free radical adducts of dGMP and dG by a series of reductants. A pulse radiolysis study 47: 71-80

Steenken S (1989) Purine bases, nucleosides and nucleotides: Aqueous solution redox chemistry and transformation reactions of their radical cations and e^- and OH adducts. Chem Rev 89:503-520

RADICAL-INDUCED BASE DAMAGE IN AQUEOUS DNA SOLUTIONS

A. F. Fuciarelli
Division of Radiation Medicine
Massachusetts General Hospital
Boston MA USA 02114

W. F. Blakely and B. J. Wegher
Department of Radiation Biochemistry
Armed Forces Radiobiology Research Institute
Bethesda MD USA 20814-5145

M. Dizdaroglu
Center for Chemical Technology
National Institute of Standards and Technology
Gaithersburg MD USA 20899

Several techniques are currently available to assess the yield of free radical-induced damage to nucleic acid constituents. Few techniques thus far developed, however, have achieved the exquisite specificity and high sensitivity of methodology incorporating gas chromatography-mass spectrometry with selected ion monitoring (GC/MS-SIM).

GC/MS-SIM methodology has been described for the analysis of purine and pyrimidine products and pyrimidine-amino acid crosslinks resulting from radical-induced damage to DNA constituents (Dizdaroglu and Gajewski 1990). These advances, in addition to the identification of internal standards and measurement of molar response factors (Fuciarelli et al. 1989), permit quantitative analysis. The inherent specificity of GC/MS-SIM is achieved through chromatographic separation, and mass spectral data provide detailed information regarding chemical structure. The GC/MS-SIM technique exhibits a high degree of sensitivity with respect to product detection for DNA (Fuciarelli et al. 1989; Fuciarelli et al. 1990) and chromatin (Gajewski et al. 1988; Dizdaroglu et al. 1989; Gajewski et al. 1990) irradiated in solution.

For the above mentioned reasons, GC/MS-SIM methodology permits unprecedented opportunities for the analysis of microenvironmental factors on the yield of radiation-induced DNA damage. Irradiation of DNA in aqueous solutions saturated with various gases permits an analysis of the effects of the radical environment on product yields. Physical factors, such as DNA conformation, have also been shown to affect product yields. The influence of these factors on the yields of radiation-induced DNA base

NATO ASI Series, Vol. H 54
The Early Effects of Radiation on DNA
Edited by E. M. Fielden and P. O'Neill

products was examined in aqueous solutions containing native or heat-denatured DNA saturated with N_2O, N_2, air, or N_2O/O_2 (4:1) in a dose range from 20 Gy to 100 Gy. The products measured were: 5,6-dihydrothymine, 5-hydroxy-6-hydrothymine, thymine glycol, 5-hydroxy-6-hydrocytosine, cytosine glycol, 4,6-diamino-5-formamidopyrimidine, 8-hydroxyadenine, 2,6-diamino-4-hydroxy-5-formamidopyrimidine, and 8-hydroxyguanine. The yields of DNA base products were markedly influenced by DNA conformation. The yields of the pyrimidine products measured in samples of DNA irradiated in the heat-denatured conformation were greater than those measured in native DNA using any of the aforementioned gases to saturate the DNA solutions. However, this observation was not consistent for the purine products, which were measured simultaneously in each DNA sample during the analysis, and varied somewhat with respect to the type of gas used to saturate the solutions. For example, the yields of the formamidopyrimidines were generally lower in samples of DNA irradiated in the native as compared to the heat-denatured conformation. On the other hand, formation of 8-hydroxypurines was favored in the native conformation in deoxygenated solutions, yet in aerated solutions, their yields were either not greatly influenced by DNA conformation (air) or were greater in the heat-denatured conformation (N_2O/O_2). The type of gas used to saturate the aqueous DNA samples also influenced the yields of DNA base products. In oxygenated solutions, 5,6-dihydrothymine and the 5-hydroxy-6-hydropyrimidines were not detected. The 8-hydroxypurines and pyrimidine glycols dominated the spectrum of products observed. However, under deoxygenated conditions, the yields of 8-hydroxypurines and pyrimidine glycols were reduced and 8-hydroxyguanine was not detected in N_2-saturated solutions. The yields of all products, except 5,6-dihydrothymine, were greater in irradiated DNA samples from N_2O-saturated solutions than from N_2-saturated solutions. These data indicate that DNA conformation and type of gas used to saturate the irradiated solutions profoundly influenced yields of base products in DNA (Fuciarelli et al. 1990).

Analysis of radical-induced DNA base damage by GC/MS-SIM has also been reported in DNA treated with hydrogen peroxide (Aruoma et al. 1989a; Blakely et al. 1990), hypoxanthine/xanthine oxidase (Aruoma et al. 1989b) and stimulated neutrophils (Jackson et al. 1989). With respect to radical induced damage by H_2O_2, the yields of the products depend upon the H_2O_2 concentration and increased in the following order: 8-hydroxyadenine, cytosine glycol, 2,6-diamino-4-hydroxy-5-formamidopyrimidine, 8-hydroxyguanine, thymine glycol, and 4,6-diamino-5-formamidopyrimidine (Blakely et al 1990). Hydrogen peroxide may participate in a Fenton-like reaction with transition metal ions (e.g., iron and copper), which are readily bound to DNA in trace quantities, in a process that may result in the production of hydroxyl radicals in close proximity to the DNA. Alternatively, the DNA-metal ion complexes can interact with H_2O_2 to form a DNA-metal ion-peroxo species, which has properties reportedly similar to hydroxyl radicals. The role of radicals in H_2O_2-induced base damage was examined using a radical scavenger, dimethyl sulfoxide (DMSO). The yields of the DNA base products were reduced by 40% to 60% with 10 mmol dm^{-3} DMSO. Consistent with the expected properties of a site-specific reaction, a 100-fold increase in the concentration of DMSO did not further reduce H_2O_2-induced base damage. The influence of metals on this reaction was examined by treating DNA with metal chelators (e.g., EDTA, DETAPAC, and desferal). However, to avoid potential radical scavenging by chelators, the metal-chelate complexes were removed by exhaustive dialysis before H_2O_2 treatment. The yields of DNA base products measured in DNA freed from bound metal ions were not significantly different from the levels measured in untreated DNA. These results illustrate that metal ions bound to DNA can act as reactive centers for free radical-induced damage to the bases in DNA.

Detectable quantities of DNA base products were found in untreated DNA samples. These products represent only those that are typically formed following irradiation under oxygenated conditions. Products formed exclusively following irradiation under deoxygenated conditions, such as 5,6-dihydrothymine and the 5-hydroxy-6-hydropyrimidines, are not detectable in untreated DNA or chromatin samples. These observations indicate that a significant amount of autooxidation occurs in DNA samples prior to analysis which is unrelated to the conditions of sample preparation (Dizdaroglu and Bergtold 1986; Fuciarelli et al. 1989; Gajewski et al. 1990).

Therefore, the GC/MS-SIM methodology is useful for the analysis of radical-induced DNA damage, including base damage measured following acid or enzymatic hydrolysis and DNA base-amino acid crosslinks (Dizdaroglu and Gajewski 1990). This technique provides an exquisite opportunity to study the influence of microenvironmental factors that might include chemical constituents present within close proximity to radical targets, such as oxygen, organic molecules, and metal ions (Aruoma et al. 1989a; Aruoma et al. 1989b; Blakely et al. 1990; Fuciarelli et al. 1990; Gajewski et al. 1990) and physical factors such as DNA conformation (Dirksen et al. 1988; Fuciarelli et al. 1990) and chromatin structure (Dizdaroglu et al. 1989; Gajewski et al. 1990) on the types and yields of free radical-induced products.

ACKNOWLEDGMENTS

Financial support to attend this meeting is gratefully acknowledged from the National Institutes of Health (AFF), U. S. Department of Energy (WFB), and NATO. Research support was provided by the U. S. Defense Nuclear Agency (00150, 00105) and National Institute of Standards and Technology.

REFERENCES

Aruoma OL, Halliwell B, Gajewski E, Dizdaroglu M (1989a) Damage to the bases in DNA induced by hydrogen peroxide and ferric ion chelates. J Biol Chem 264:20509-20512

Aruoma OL, Halliwell B, Dizdaroglu M (1989b) Iron ion-dependent modification of bases in DNA by the superoxide radical-generating system hypoxanthine/xanthine oxidase. J Biol Chem 264:13024-13028

Blakely WF, Fuciarelli AF, Wegher BJ, Dizdaroglu M (1990) Hydrogen peroxide-induced base damage in deoxyribonucleic acid. Radiat Res 121:338-343

Dirksen ML, Blakely WF, Holwitt E, Dizdaroglu M (1988) Effect of DNA conformation on the hydroxyl radical-induced formation of 8,5'-cyclopurine 2'-deoxyribonucleoside residues in DNA. Int J Radiat Biol 54:195-204

Dizdaroglu M, Bergtold DS (1986) Characterization of free radical-induced base damage in DNA at biologically relevant levels. Anal Biochem 156:182-188

Dizdaroglu M, Gajewski E (1990) Selected ion mass spectrometry: Assays of oxidative DNA damage. Methods in Enzymology 186:530-544

Dizdaroglu M, Gajewski E, Reddy P, Margolis, SA (1989) Structure of hydroxyl radical-induced protein crosslinks involving thymine and tyrosine in nucleohistone. Biochemistry 28:3625-3628

Fuciarelli AF, Wegher BJ, Gajewski E, Dizdaroglu M, Blakely WF (1989) Quantitative measurement of DNA base products using gas chromatography-mass spectrometry. Radiat Res 119:219-231

Fuciarelli AF, Wegher BJ, Blakely WF, Dizdaroglu M (1990) Yields of radiation-induced base products in DNA: Effects of DNA conformation and gassing conditions. Int J Radiat Biol (In press)

Gajewski E, Fuciarelli AF, Dizdaroglu M (1988) Structure of hydroxyl radical-induced DNA-protein crosslinks in calf thymus nucleohistone *in vitro*. Int J Radiat Biol 54:445-459

Gajewski E, Rao G, Nackerdien Z, Dizdaroglu M (1990) Modification of DNA bases in mammalian chromatin by radiation generated free radicals. Biochemistry (In press)

Jackson JH, Gajewski E, Schraufstatter IU, Hyslop P, Fuciarelli AF, Cochrane CG, Dizdaroglu M (1989) Damage to the bases in DNA induced by stimulated human neutrophils. J Clin Invest 84:1644-1649

PRIMARY AND SECONDARY RADICALS IN THYMINE DERIVATIVES: SOLID STATE ESR/ENDOR STUDY OF 1-ME-THYMINE AND THYMIDINE

E.O. Hole and E Sagstuen, University of Oslo, Norway
W.H. Nelson, Georgia State University, Atlanta, GA
D.M. Close, East Tennessee State University, Johnson City, TN

Singlé crystals of **thymidine (Td)** and **1-methylthymine (1MT)** have been studied using ESR, ENDOR and FSE techniques subsequent to x-irradition at 8 K. Four base-centered radicals were identified in both systems. In addition, an alkoxy radical and a C1' H-abstraction radical were observed in **Td** (Sagstuen <u>et</u> <u>al</u>. 1989).

<u>O4 protonated anion</u>: In both **Td** and **1MT** the anionic product was a dominant radical at 8 K. It decayed with no apparent successor upon warming to 40 K. Four proton couplings (C6-H, N3-H, C4-O4-H, C5-CH_3) were associated with the **Td**-anion. Due to a non-planear geometry (a C4-O4---H dihedral angle of 156°) the H(O4) proton coupling was easily observed, thus proving the protonation state. Furthermore, one of the couplings was assigned to a C5-CH_3 β-methyl interaction from the excellent agreement between calculated and experimental dipolar tensors. The coupling was reconstructed assuming 6% <u>negative</u> spin density at C5 and ca 45% spin density at both C4 and C6. This spin density distribution agrees with the INDO RHF/CI MO-calculations for the O4-protonated anion. For the negatively charged anion, however, the spin density at C5 is predicted to be <u>positive</u> (.09), with only 15% at C4. The features of the C5-CH_3 coupling may thus be used as an indicator of the protonation state of the thymine anion radical.

In **1MT** the protonation state of the anion is not unambigously determined like in **Td**. This follows since the C4-O4---H dihedral angle is 90°, thus restraining the observation of a possible H(O4) proton. Nevertheless, there is a very close correspondence between the other three proton couplings in the two systems. In particular, the couplings to the methyl group have almost identical dipolar tensors. This strongly suggests that similar radicals (O4 protonated anions) are formed in both **Td** and **1MT**.

<u>5-yl (C6 H-addition) radical</u>: At 8 K, the 5-yl radical was

NATO ASI Series, Vol. H 54
The Early Effects of Radiation on DNA
Edited by E. M. Fielden and P. O'Neill

present in only **Td**. In **1MT** it was formed upon warming to 250 K.

6-yl (C5 H-addition) radical: The 6-yl radical was observed in both systems at 8 K. As a minor product in **Td** it decayed about 60-90 K. In **1MT** it reportedly transforms into the 5-yl radical upon warming or white light bleaching at 77 K (Flossmann 1980).

7-yl (C7 H-abstraction) radical: The well known 7-yl radical was observed to be a major product in both **Td** and **1MT** at 8 K. It remained stable until room temperature.

Conclusions: Relating the present data with previous studies of thymine and oriented DNA, two factors are important:

1) The thymine anion protonates at O4 below 8 K in **Td**, and there is strong evidence that this occurs in **1MT** as well.
2) There is no distinct connection between the decay of the anionic product and the formation of the 5-yl radical.

In all but one of the pyrimidines/purines studied protonation of the anion below 8 K is the reaction by which the system re-establishes charge balance subsequent to reduction. In aqueous solution protonation at O4 in pyrimidines is fast and reversible, whereas protonation at C6 is slow and irreversible (Deeble et al. 1985). In solids, however, only O4 has a ready supply of nearby protons. This may lead to a low probability for C6-protonation of the anion. In partially deuterated crystals the 5-yl radical was mainly formed by D-addition in **Td**, and mainly by H-addition in **1MT**. This indicates different mechanisms where protons from both exchangeable and non-exchangeable sites are involved in the 5-yl radical formation in **Td** and **1MT**. Both 5-yl and 6-yl radicals may be formed either by ·H addition, or by H^+ addition followed by electron capture during radiation.

The 7-yl radicals may be formed by deprotonation of the cation, although other mechanisms are possible.

References

Deeble DJ, Das S, von Sonntag C (1985) Uracil derivatives: Sites and kinetics of protonation of the radical anions and the UV spectra of the (C5) and (C6) H-atom adducts. J Phys Chem 89:5784-5788

Flossmann H, Zehner H, Müller A (1980) Radicals produced in single crystals of thymine derivatives by uv and x-rays. Z Naturforsch C35:20-29

Sagstuen E, Hole EO, Nelson WH, Close DM (1989) Structure of the primary reduction product of thymidine after x-irradiation at 10 K. J Phys Chem 93:5974-5977

RADIATION EFFECTS ON TRANSCRIPTION IN CELLS AND ISOLATED CELL NUCLEI

K.J. Weber
Strahlenzentrum
Justus-Liebig-Universität Gießen
Leihgesterner Weg 217
6300 Gießen
Germany

Radiation effects on gene transcription represent an early cellular response to DNA damage. Considering the basic processes of RNA synthesis, polymerase binding/initiation and elongation may be affected by distinct mechanisms. Inhibition of the elongation reaction is generally assumed to occur at the site of critical DNA lesions - operating as termination signals - and the measurement of RNA synthesis (ribosomal RNA in particular) was used to quantitate the induction and eventually repair of transcription terminating pyrimidine dimers from UV-light exposure. With ionizing radiation a corresponding mechanism of elongation inhibition had been concluded for in vitro systems and some microorganisms - presumably due to the induction of single-strand breaks and/or base damage. More complex phenomena - however - are observed for the inhibition of template activity with mammalian cells that cannot readily be explained by the above mechanism. Evidence has accumulated that the expressable state of eucaryotic genes correlates to the topology of larger functional chromatin domains (matrix-attached loops)(e.g. Weisbrod, 1982; Goldman, 1988) and radiation induced structural transitions of the chromatin have been discussed with respect to loss of DNA functionality (Villeponteau and Martinson,1987; Luchnik et al., 1988).

The problem of transcription inhibition in mammalian cells was readressed in the present investigation using cultured V79 cells and nuclei isolated from cells following exposure to UV-light (254 nm) and X rays (80 kV). Whereas in washed nuclei (in the presence of alpha-Amanitin) RNA synthesis is restricted to polymerase I elongation (rRNA) the assay conditions for the integral cells were adjusted (by variable labelling intervals) that allowed detection of 28S and 18S rRNA synthesis originating mainly from newly initiated templates.

A drastically higher sensitivity of transcription inhibition in cells when compared to the nuclei assay could be demonstrated and was identical for the 28S and 18S ribosomal species. This finding indicates that the observed loss of template activity may not be attributed to a termination mechanism. At the respective D_0-value about one SSB is

NATO ASI Series, Vol. H 54
The Early Effects of Radiation on DNA
Edited by E. M. Fielden and P. O'Neill

induced per repeated ribosomal DNA unit of 44 Kbp that contains a 13 Kbp rRNA gene. It is inferred that this unit can be identified with a chromatin loop whose structural configuration (nucleosome opening due to torsional stress) had been proposed to reflect the expressable state of gene(s) residing in this loop (see above). SSBs are likely candidates to cause relaxation of this chromatin structure and are therefore assumed to act over a distance interferring with the process of promotor recognition or polymerase binding/initiation. The time course of recovery from transcription inhibition observed during post-irradiation incubation of cells indicates that single-strand break repair must be followed by a slower process of chromatin reformation before transcription can be resumed. Such a mechanism would be reminiscent of a reported restoration of DNase I sensitivity in actively expressed genes that had been lost due to treatment with single-strand nicking agents (Villeponteau and Martinson, 1987).

Remarkebly, post-irradiation incubation of UV exposed cells lead to a further time-dependent reduction of rRNA synthetic activity rather than expression of a recovery phenomenon - despite the well documented repair of UV lesions in actively expressed genes in rodents (Smith, 1987). According to the above hypothesis it is assumed that endonucleolytic incisions - occuring when lesion repair is attempted - prevent reinitiation via relaxation of active chromatin domains. The lack of subsequent recovery may be due to the high frequency of initial damage at which this phenomenon was detected and a limited cellular repair capacity.

References

Goldman MA (1988) The chromatin domain as a unit of gene regulation. BioEssays 9:50-55

Luchnik AN, Hisamutdinov TA, Georgiev GP (1988) Inhibition of transcription in eucaryotic cells by X-irradiation: Relation to loss of topological constraint in closed DNA loops. Nucl Acids Res 16:5175-5184

Smith CA (1987) DNA repair in specific sequences in mammalian cells. J Cell Sci Suppl 6:225-241

Villeponteau B, Martinson HG (1987) Gamma rays and bleomycin nick DNA and reverse the DNaseI sensitivity of beta-globin gene chromatin in vivo. Mol Cell Biol 7:1917-1924

Weisbrod S (1982) Active chromatin. Nature 297:289-295

SUMMARY COMMENTS FROM A PHYSICIST

DUDLEY T. GOODHEAD
Medical Research Council Radiobiology Unit
Chilton
Didcot OX11 ORD
United Kingdom

In considering the mechanisms of radiation damage to DNA and final cellular effect, it should be the primary task of the physicists to describe the initial physical insult to a sufficient degree of accuracy so that all relevant chemical and later processes can be followed through to their cellular consequences. Formally, the physicists´ task should then be done and the problem can be handed over to the chemists, biochemists and biologists. But, of course, the physicists need knowledge of these later effects so that they can provide, and refine, their description in the best ways.

Physicists also seem, however, to have largely taken on the roll of quantitative modellers of the overall processes linking the initial physics to the final cellular effects. This is probably because they should best know the initial starting conditions (physical). There are far too many gaps in current scientific knowledge for detailed modelling of all the intermediate stages; here we should look to close co-operation with chemists, biochemists and biologists to close the gaps. But despite these gaps clearly-defined models have been presented, based on specified simplified assumptions, and these serve well to clarify particular steps and options along the way.

If chemists were given the atomic position and orbital of each initial ionisation, they should in principle be able to describe the reactions and the types and probabilities of the final products of DNA in cells. I have heard a good deal this week about the reactions and products, but I personally remain unable to judge how far they are from achieving this difficult goal.

NATO ASI Series, Vol. H 54
The Early Effects of Radiation on DNA
Edited by E. M. Fielden and P. O'Neill

We were told that the total variety of base-damage products is quite small, and that these each have specific abilities to cause strand breaks. In this case, the goal may be sufficiently within sight for realistic quantitative modelling of the damage, starting with the full track structures of radiation tracks. The chemists claimed strongly, and without a dissenting voice I believe, that even when the initial physical insult produces highly clustered ionisations (and excitations) in and/or near the DNA, the ensuing chemistry of the DNA is a linear sum of the individual radical reactions. If this is truly the case, it simplifies the chemistry and it implies that the well-established enhanced biological effect of clusters is not due to the chemical stage. I wonder if this assumption of purely linear chemistry, with no non-linear products, can be tested further by focussed studies on radicals and yields with high-LET radiations for which non-linear effects should be greatest. Assuming that the linear assumption remains valid, does this mean that the enhancement due to the physical clusters is visible first at the stage of biochemical damage, due say to spatial association of individual DNA strand breaks, base damages, etc.? Would cross-links, with protein or DNA, also arise solely from this linear chemistry? It may be hoped that in future more detailed molecular characterisation of the complexity of the initial biochemical damage, and the cell´s ability to repair it, will assist in pointing back to particularly relevant chemical products and processes.

But how far are the physicists from being able to deliver to the chemists a fully adequate description of the initial state from which the chemistry must follow? Existing track structure codes provide considerable detail, but they are by no means complete or perfect. Many low-energy processes are not considered, and existing codes are mainly for water, and have restricted limits on particle type and energy. Codes for other materials are being developed, including DNA and protein. I am less unhappy now, than I was prior to the meeting, that our main descriptions to date have been confined to water - it was emphasized throughout the meeting that water molecules are always in great abundance in or near the DNA, in the grooves and as a water shell, and that the water chemistry of these adjacent molecules must be of considerable relevance to the DNA. Nevertheless, an adequately complete description given to the chemists must include the substantial proportion of ionisations (and excitations) which the radiation produces in the DNA itself, so further development of the track codes must continue. Present approaches tend also to neglect fast physico-chemical processes, such as energy transfer, migration and relaxation which may take place between the initial physical interactions and subsequent chemistry. These processes may be of some relevance, but it is clear that they do not so greatly disseminate the energy that the initial track structure differences between radiations are lost; the differences persist right through to the biological consequences.

We learnt well during the week to avoid misleading descriptions such as "direct" or "indirect" effect, unless we clearly specified our context and meaning. One factor which tends to make these terms particularly unhelpful as general descriptions is that, in the cellular environment, it is the ionisations in molecules only very near to, or within, the DNA that cause damage to the DNA. "Scavengable" and "unscavengable" damage may be more explicit terms for particular experimental observations, but even then it should be recognised that a multiple biochemical lesion such as a DNA double-strand break may have been caused by a hybrid of scavengable and unscavengable damage.

So it seemed to be generally agreed that it is the ionisations within a very small volume around, and including, the DNA that are important for DNA damage in cells. And it has been hypothesized that localised clusters of ionisations within such volumes are particularly important in producing the biochemical damage which the cells find most difficult to handle. If this hypothesis turns out to be correct (and whether these events are called ionisation clusters, electron track-end clusters, local multiply damaged sites, or any other name) it should be recognised that the hypothesis was stated and quantified by Howard-Flanders as early as 1958. He arrived at it from physical analysis of radiation tracks and a simple model correlating track parameters with final biological effect in simple eukaryotic systems exposed to different radiations. It may be remarkable that he arrived at quantities such as 6-8 ionisations within distances of 4 nm despite the fact that there were at that time no Monte-Carlo track-structure codes and he made no _a priori_ assumptions as to the nature or dimensions of the target molecules - DNA was not assumed. Howard-Flanders´ simple LET-with-clusters model by-passed virtually all the chemistry and biochemistry to link directly initial physical damage to final cellular effect. Yet it served as a valuable pointer as to what track features may be of most relevance to the chemistry and biochemistry of cellular damage.

Finally, I should recall two other experimental questions discussed at this meeting which may be important to attempts to model the processes from initial physical damage to final cellular effect. It was stated that a dsb can be produced by the reaction of a single OH radical with DNA. What proportional contribution does this mechanism make to total dsb in cells irradiated with low-LET radiation? This question is important because this single-OH mechanism runs counter to the generally observed increased biological effectiveness of ionisation clusters, as in track-end irradiations or high-LET irradiations. The second question is what is the true shape of the yield versus dose curve for dsb induction by low-LET radiation at the biologically-relevant doses where curved dose responses are obtained for cellular effects? This was discussed in considerable detail but a consensus was not achieved, so we can each only make our own best judgement from the evidence presented. Resolution of the question is critical to almost all biophysical models of radiation action on cells because the majority of them are dependent on the assumption that the initial molecular damage is produced linearly with dose, while a few models are totally dependent on non-linearity of initial damage as the prime cause of non-linearity of finally cellular responses.

SUMMARY FROM A CHEMIST'S POINT OF VIEW

JOHN F. WARD
Department of Radiology
School of Medicine M010
UCSD
La Jolla
CA 92093, USA

The Proceedings of this Workshop outline the current understanding of the topic from the viewpoints of the physicists, chemists and biologists who were present. They also delineate the probable fruitful directions for future research. At the Workshop there were frequent energetic interchanges among the disciplines resulting in the development of a good understanding of the early events of radiation - physical, chemical and biochemical - which are the origins of the biological effects of ionizing radiation. The definition of the mechanisms of DNA damage intracellularly appears, as our understanding develops, to be extremely complex:

Radiation energy is deposited in discrete events, amounts of energy deposited and the volume in which the energy is deposited have broad spectra.

Within each event multiples of ions/radicals are formed, the number depending on the energy deposited.

The radicals (which lead to DNA damage) are located on the DNA, on molecules immediately next to the DNA and on molecules which can subsequently diffuse to and react with the DNA.

Damage can be produced with a variety of identities on the deoxyribose moieties and all the DNA bases.

Thus there is a spectrum of possibilities at each of the stages between energy deposition and damage induction. The resolution of all of these steps will require the continued contributions of the disciplines of physics, chemistry and biology to achieve a solution sufficiently detailed to be capable of predicting variations of biological response. Thus it is hoped that in the future, opportunities will continue to be made available for the collegial exchange of ideas, for the definitions of the different jargons and for the forging of intellectual collaborations among the disciplines. This must lead to the

NATO ASI Series, Vol. H 54
The Early Effects of Radiation on DNA
Edited by E. M. Fielden and P. O'Neill

most rapid progress in the most rewarding directions in this field.

It became apparent during the Workshop that as well as areas in which a consensus exists there are several areas in which major questions remain about the physics and chemistry involved in DNA damage induction in the cell by ionizing radiation. The formal presentations cover individual points of view, but the disagreements do not, of course, appear. Some of the disputes lay in the model system mechanisms proposed and some in the validity of the extrapolations from model DNA to cellular DNA. A discussion of the more general points is presented below.

A concern of the chemists is the validity of extrapolation of calculations of energy deposition patterns for unit density gas to liquid water. Will the polarity of the medium, the presence of neighboring molecules of solvent (hydrogen bonded) have effects? What can be projected to energy deposition mechanisms which will occur at the DNA-water interface?

Radiation chemical data from a variety of model systems were reported - single crystals, frozen solutions, dry DNA, dilute solutions and solutions containing scavengers. The inter-relationships among the mechanisms occurring in the various systems are not straightforward; the same mechanisms of radical migration and consequent damage induction do not occur in all of the systems. Solid systems (frozen or crystals) are ideal for the studies of the structures of the radicals induced by radiation, but are limited in considerations of kinetics or of subsequent radical reactions. There is concern about the total yields of radicals measured by ESR systems; are other species present? No sugar radicals are detected in irradiated DNA by ESR, yet energy is deposited in the deoxyribose moieties, ·OH radicals do react with sugars (at room temperature), so how is the absence of sugar radicals explained? If radical-radical reactions have occurred (e.g. when radicals are formed in close proximity in regions of high energy deposition) they cannot be detected. What are the fates of diamagnetic species so produced?

There seems to be general agreement on the mechanisms by which the bases of DNA are damaged and what the damaged products are. There is less consensus on the suggestion that transfer of a radical from base to sugar can occur and lead to a strand break. The total picture of strand break production (single and particularly double) is still elusive.

A focal point of the Workshop was the topic of DNA double strand breaks as a type of damage which has been shown to be biologically significant. Three non-exclusive mechanisms were proposed, by among others, Schulte-Frohlinde, Symons and Ward. The resolution of the contributions of these mechanisms to intracellular DNA is of prime importance if we are to use this lesion as a focus or link between the physics and chemistry and the biological effects of radiation. Physics gives us the energy distribution patterns, chemistry the radicals formed as a result of energy deposition and their reactions - but we need to know how this progression continues to the formation of the dsb if we are to make the link between biological effects and the physics and chemistry.

In extrapolation of model system data to the cell there is a need to know the details of the immediate environment of the DNA in the cell nucleus. Indeed this topic was the most frequently mentioned in the list requested from participants prior to the Workshop. This is understandable since this information is necessary to extrapolate from models to the cells. What is the extent of the oriented water molecules? From what distances can ·OH radicals diffuse? What other molecules are in the vicinity? One consequence of these considerations was a suggestion that damage to DNA in cells should be classified as coming from two sources - scavengable damage and non-scavengable damage (i.e. that which can be removed by ·OH radical scavengers and that which can not) - these terms could replace indirect and direct effect, since there is no way of distinguishing the latter in the cell. Thus scavengable damage results from ·OH radicals and non-scavengable damage is the sum of that from direct ionizations, from reactions of water molecules ionized next to DNA and from radicals formed on other molecules in close proximity to the DNA.

Another topic which raised some controversy, was transfer of excitation/ionization over distances of tens to hundreds of base pairs. Several lines of evidence on each side of this topic were presented. The existence of such transfer over long distances would require long life times for the species involved and would depend on the environment in which the radicals are formed: If there were an alternate route to radical destruction then long distance movement would not be expected to occur. Within the discussion of charge separation and rejoining a mechanism was suggested for the formation of those strand breaks which cannot be scavenged by chemical repair (e.g. thiols): If the hole (located on the deoxyribose) and the electron are separated after the initial energy deposition/ionization event by distances too short for the electron to escape, they may recombine with sufficient energy to initiate the formation of a strand break.

SUMMARY OF THE WORKSHOP FROM THE VIEW OF A BIOLOGIST

D.Frankenberg
Institut für Biophysikalische Strahlenforschung
Gesellschaft für Strahlen-und Umweltforschung mbH München
Paul-Ehrlich-Str. 20
D-6000 Frankfurt 70

Dr.Fielden and Dr. O'Neill invited me to summarize the biological aspects of the contributions discussed at this workshop. Since a summary cannot be exhaustive, I take the freedom to pick up special topics and, therefore, I have to apologize in advance for the personal views in selecting the papers to be mentioned.

Radiation biology starts with the interaction of a radiation field with biological matter. Figure 1 shows a comprehensive hierarchy of reactions starting from the structure of the energy deposition imparted to matter and leading to biological reactions such as cell death, cell transformation or mutation . At the end of the "chemical" phase of radiation action (in the order of seconds after the traversal of a particle) cells respond to the stable, however, chemically altered state of the DNA with enzymatic repair processes to eliminate the radiation induced DNA changes (beginning of the "biological " phase of radiation action). Enzymatic repair processes are not always error-free. The reasons for this might be the structure of the energy deposition of radiation, the ultrastructure of the target (chromosomes), and the enzymatic repair process itself.

A very important parameter which might affect the induction and repair/ misrepair of DNA lesions is the structure of the environment of the DNA. In his contribution, O.Sapora demonstrated that the induction of DNA double-strand breaks (DSB) is by a factor of 4 less in DNA associated with proteins in polynucleosomes compared to naked DNA. The relative contributions of direct and OH-mediated radiation effects on DNA (for example the induction of single-strand breaks (SSB) and DSB) may also be influenced by the nucleosomal structure of the chromatin. However, as pointed out by D.Schulte-Frohlinde both the directly ($DNA^{+\cdot}$) and OH-mediated ($DNA^{\cdot}$) DNA radicals result in the DNA- OH adduct via reaction with surrounding water molecules. Thus, there would be no need to distinguish

NATO ASI Series, Vol. H 54
The Early Effects of Radiation on DNA
Edited by E. M. Fielden and P. O'Neill

between direct and OH-mediated radiation effects. Considering the ionic bonds between the "acid" DNA and the "basic" histones only a fraction of the DNA is in contact with water molecules and would not lead to the DNA-OH adduct. However, J.Ward mentioned that several layers of water molecules might be found between histones and the DNA double helix. Only a few data are known relating chromatin structure to the induction of DNA lesions and their repair/misrepair. So, induction and repair of definite DNA lesions are different in transcriptionally active compared to bulk chromatin. The discussion highlights our ignorance about the influence of the chromatin structure on the induction and repair/ misrepair of DNA lesions.

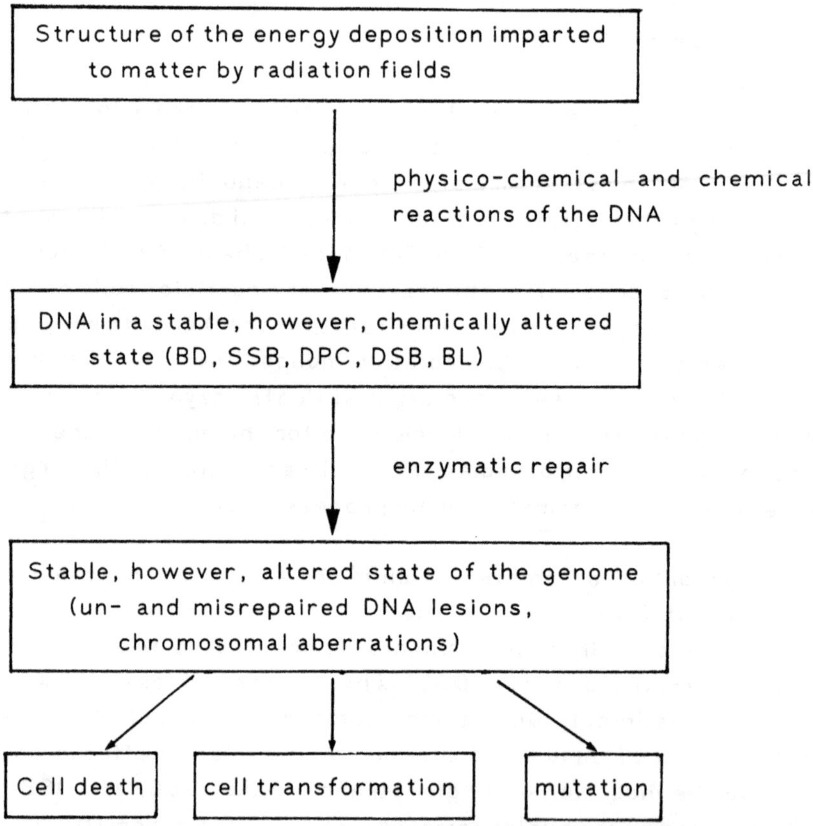

Figure 1 . Hierarchy of reactions starting from radiation physics and leading to biological effects. BD: base damage, SSB: single-strand break, DPC: DNA - protein crosslink , DSB: DNA double -strand break, BL: bulky lesion

The effect of endogenous thiols on molecular and cellular responses and OER was addressed by several papers (M.Edgren, K.Held, B.Michael, D.Schulte-Frohlinde, G. van der Schans). It was demonstrated by M.Edgren that the nuclear GSH content rather than the overall cellular GSH content correlates with OER-values. Considering the different chemical repair kinetics of DSB and SSB, a new mechanism for the dependence of OER on LET was offered by B.Michael.

Experimental data are accumulating which suggest that DSB may be the lesions which lead to chromosomal aberrations. Chromosomal aberrations in turn are often presumed to be the most important cause of radiation induced inactivation and transformation of cells. Therefore, quantitative data about the induction and repair/ misrepair of DSB as a function of radiation quality, oxygen status, content of endogenous thiols, chromatin structure etc. are important prerequisites for our understanding of the mechanism leading to cellular radiation effects. Besides sedimentation of the DNA in a neutral sucrose gradient, neutral elution, pulse field gel electrophoresis and premature chromosome condensation are used to determine DSB. In this context papers were presented by M.Belli (neutral elution and neutral sedimentation), E. Blakely (premature chromosome condensation), K.Held (neutral elution), G.Iliakis (neutral elution and pulse field gel electrophoresis), K.Prise (neutral elution), I.Radford (neutral elution) and D.Frankenberg (neutral sedimentation). Table 1 compares the different methods with respect to dose-dependence and RBE- values for initial and unrejoined DSB. The most outstanding discrepancies bet-

Table 1. Comparison of different methods for the measurement of DSB. PFGE: pulse field gel electrophoresis, PCC: premature chromosome condensation

	Neutral sediment.	Neutral elution	PFGE	PCC
DSB_{ind} vs dose for low LET	linear	non-linear or linear	linear	linear
RBE for DSB_{ind}	> 1	1	----	> 1
RBE for DSB_{unrej}	increase	increase	----	increase*

* approach those for chromosomal aberrations and cell killing

ween neutral elution on one hand and neutral sedimentation, pulse field gel electrophoresis and premature chromosome condensation on the other hand are found for the dose-effect relationships for sparsely ionizing radiations and RBE-values of densely ionizing radiations for initial DSB. On the basis of the following considerations a non-linear relationship between initial DSB and dose after irradiation with low LET radiations as observed with the neutral elution method is unlikely. D.Goodhead pointed out the low probability to find several ionizations within a volume of some nm cord length originating from two *different* electron tracks. The exponential survival curves of yeast mutants totally deficient in the repair of DSB do not allow a D^2 - term for the induction of DSB (D Frankenberg). In order to explain the exponential survival curves of AT cells after irradiation with low LET radiations , despite the non-linear induction of DSB in these cells, an increase in target size and/or change of the regulation of misrepair processes is postulated (I.Radford). However, cytogenetic and molecular data suggest that the exponential survival curves of AT cells are due to a defect in the repair of DSB (less fidelity compared to normal fibroblasts), so that in AT cells single DSB may be lethal which is in accordance with linearly induced DSB.

The RBE-values of densely ionizing radiations compared to low LET radiations are approximately 1 for the induction of DSB when the neutral elution method is used, however, RBE-values significantly higher than 1 are observed with the neutral sedimentation (D.Frankenberg) or the premature chromosome condensation method (E.Blakely). The results of K.Prise, that the RBE-values of DSB induction are independent of LET in the range of 17 keV/µm (1.9 MeV protons) up to 104 keV/µm (α-particles) is in disagreement with radiation physics. With ^{60}Co-γ-rays or hard x-rays about 25% of the total energy is imparted to matter by slow electrons with LET-values in the range from 10 to 20 keV/µm. Approximately 75% of the energy is imparted to matter by clusters of 1 to 3 ionizations with a much lower probability to induce DSB compared to slow electrons. With 1.9 MeV protons, however, 100% of the energy is imparted to matter at 17 keV/µm. Since ^{60}Co-γ-rays and 1.9 MeV protons differ mainly in the fraction of energy which is imparted to matter around LET-values of 17 keV/µm, different RBE-values for DSB induction should be observed.

Pulse field gel electrophoresis for the determination of DSB is still in a rapid development and data about RBE-values are not available yet. As for the neutral elution method the lysis conditions are critical parameters and have to be controlled carefully to achieve relevant data (G.Iliakis). This short discussion of the different methods for the determination of DSB demonstrates that further experimental data are needed to clarify the different results obtained with the neutral elution and the neutral sedimentation method. It has to be elucidated whether the neutral elution method may be developed to a routine method for DSB measurement. As pointed out by D.Charlton there exists a large variation in yields of DSB in mammalian cells for low LET radiations. This is in contrast to DSB determinations in yeast. The yields of DSB as determined with the neutral sedimentation method by four different groups are equal within 20%. Efforts have to be made to find out the reasons for the large variations in DSB yields in mammalian cells. Systematic comparisons of the methods and standardizations should be taken into account.

The last session of the workshop was concerned with chromosomal damage and its dependency on radiation quality (M.Sasaki, D.Lloyd, E.Blakely). Chromosomal damage represents the stable, however irreversibly altered state of the genome which may lead to cell death, cell transformation or mutation (fig. 1). The data on chromosomal aberrations in dependence on LET (soft x-rays, neutrons) as shown by M.Sasaki are in agreement with the distance model of pairwise interaction of primary damage being most probably DSB. The curvilinear relationship between chromosomal aberrations and dose is mirrored by the shoulders of survival curves after low LET exposure assuming chromosomal aberrations as critical lesions for cell death. These experimental data and the observation of exponential kinetics (mono- or biphasic) for DSB repair is in disagreement with the saturable repair model as an alternative explanation of shouldered survival curves. In her presentation E.Blakely demonstrated that premature chromosome condensation and chromosome painting are powerful tools in radiobiological research.

I hope that in a few years there will be another opportunity to meet again and to discuss the progress achieved in the areas mentioned.

SUBJECT INDEX

NATO ASI Series H

Vol. 1: **Biology and Molecular Biology of Plant-Pathogen Interactions.** Edited by J. A. Bailey. 415 pages. 1986.

Vol. 2: **Glial-Neuronal Communication in Development and Regeneration.** Edited by H. H. Althaus and W. Seifert. 865 pages. 1987.

Vol. 3: **Nicotinic Acetylcholine Receptor: Structure and Function.** Edited by A. Maelicke. 489 pages. 1986.

Vol. 4: **Recognition in Microbe-Plant Symbiotic and Pathogenic Interactions.** Edited by B. Lugtenberg. 449 pages. 1986.

Vol. 5: **Mesenchymal-Epithelial Interactions in Neural Development.** Edited by J. R. Wolff, J. Sievers, and M. Berry. 428 pages. 1987.

Vol. 6: **Molecular Mechanisms of Desensitization to Signal Molecules.** Edited by T. M. Konijn, P. J. M. Van Haastert, H. Van der Starre, H. Van der Wel, and M. D. Houslay. 336 pages. 1987.

Vol. 7: **Gangliosides and Modulation of Neuronal Functions.** Edited by H. Rahmann. 647 pages. 1987.

Vol. 8: **Molecular and Cellular Aspects of Erythropoietin and Erythropoiesis.** Edited by I. N. Rich. 460 pages. 1987.

Vol. 9: **Modification of Cell to Cell Signals During Normal and Pathological Aging.** Edited by S. Govoni and F. Battaini. 297 pages. 1987.

Vol. 10: **Plant Hormone Receptors.** Edited by D. Klämbt. 319 pages. 1987.

Vol. 11: **Host-Parasite Cellular and Molecular Interactions in Protozoal Infections.** Edited by K.-P. Chang and D. Snary. 425 pages. 1987.

Vol. 12: **The Cell Surface in Signal Transduction.** Edited by E. Wagner, H. Greppin, and B. Millet. 243 pages. 1987.

Vol. 13: **Toxicology of Pesticides: Experimental, Clinical and Regulatory Perspectives.** Edited by L. G. Costa, C. L. Galli, and S. D. Murphy. 320 pages. 1987.

Vol. 14: **Genetics of Translation. New Approaches.** Edited by M. F. Tuite, M. Picard, and M. Bolotin-Fukuhara. 524 pages. 1988.

Vol. 15: **Photosensitisation. Molecular, Cellular and Medical Aspects.** Edited by G. Moreno, R. H. Pottier, and T. G. Truscott. 521 pages. 1988.

Vol. 16: **Membrane Biogenesis.** Edited by J. A. F. Op den Kamp. 477 pages. 1988.

Vol. 17: **Cell to Cell Signals in Plant, Animal and Microbial Symbiosis.** Edited by S. Scannerini, D. Smith, P. Bonfante-Fasolo, and V. Gianinazzi-Pearson. 414 pages. 1988.

Vol. 18: **Plant Cell Biotechnology.** Edited by M. S. S. Pais, F. Mavituna, and J. M. Novais. 500 pages. 1988.

Vol. 19: **Modulation of Synaptic Transmission and Plasticity in Nervous Systems.** Edited by G. Hertting and H.-C. Spatz. 457 pages. 1988.

Vol. 20: **Amino Acid Availability and Brain Function in Health and Disease.** Edited by G. Huether. 487 pages. 1988.

NATO ASI Series H

NATO ASI Series H

Vol. 42: **Biomechanics of Active Movement and Deformation of Cells.**
Edited by N. Akkaş. 524 pages. 1990.

Vol. 43: **Cellular and Molecular Biology of Myelination.**
Edited by G. Jeserich, H. H. Althaus, and T. V. Waehneldt. 565 pages. 1990.

Vol. 44: **Activation and Desensitization of Transducing Pathways.**
Edited by T. M. Konijn, M. D. Houslay, and P. J. M. Van Haastert. 336 pages. 1990.

Vol. 45: **Mechanism of Fertilization: Plants to Humans.**
Edited by B. Dale. 710 pages. 1990.

Vol. 46: **Parallels in Cell to Cell Junctions in Plants and Animals.**
Edited by A. W. Robards, W. J. Lucas, J. D. Pitts, H. J. Jongsma, and D. C. Spray. 296 pages. 1990.

Vol. 47: **Signal Perception and Transduction in Higher Plants.**
Edited by R. Ranjeva and A. M. Boudet. 357 pages. 1990.

Vol. 48: **Calcium Transport and Intracellular Calcium Homeostasis.**
Edited by D. Pansu and F. Bronner. 456 pages. 1990.

Vol. 49: **Post-Transcriptional Control of Gene Expression.**
Edited by J. E. G. McCarthy and M. F. Tuite. 671 pages. 1990.

Vol. 50: **Phytochrome Properties and Biological Action.**
Edited by B. Thomas and C. B. Johnson. 337 pages. 1991.

Vol. 51: **Cell to Cell Signals in Plants and Animals.**
Edited by V. Neuhoff and J. Friend. 404 pages. 1991.

Vol. 52: **Biological Signal Transduction.**
Edited by E. M. Ross and K. W. A. Wirtz. 560 pages. 1991.

Vol. 53: **Fungal Cell Wall and Immune Response.**
Edited by J. P. Latgé and D. Boucias. 472 pages. 1991.

Vol. 54: **The Early Effects of Radiation on DNA.**
Edited by E. M. Fielden and P. O'Neill. 448 pages. 1991.

Vol. 55: **The Translational Apparatus of Photosynthetic Organelles.**
Edited by R. Mache, E. Stutz, and A. R. Subramanian. 260 pages. 1991.

Vol. 56: **Cellular Regulation by Protein Phosphorylation.**
Edited by L. M. G. Heilmeyer, Jr. 520 pages. 1991.

Vol. 57: **Molecular Techniques in Taxonomy.**
Edited by G. M. Hewitt, A. Johnston, and J. P. W. Young. 420 pages. 1991.

www.ingramcontent.com/pod-product-compliance
Lightning Source LLC
Chambersburg PA
CBHW061153240926
51846CB00006B/1006
* 9 7 8 3 6 4 2 7 5 1 5 0 9 *